Contents

How to use this book
Mike Ashworth 5

Foreword
Hans Rat 5

Introduction
Early railroads 6

Introduction
Urban rail transit 7

Introduction
Early railroad maps 8

Introduction
From maps to diagrams 9

Zone 1
Berlin 012
Chicago 016
London 020
Madrid 024
Moscow 028
New York 032
Paris 036
Tokyo 040

Zone 2
Barcelona 046
Boston 048
Budapest 050
Buenos Aires 052
Hamburg 054
Hong Kong 056
Lisbon 058
Mexico City 060
Montreal 062
Munich 064
Osaka 066
San Francisco 068
Seoul 070
St. Petersburg 072
Washington DC 074

Zone 3
Amsterdam 078
Athens 079
Beijing 080
Bilbao 081
Brussels 082
Bucharest 083
Copenhagen 084
Delhi 085
Glasgow 086
Kiev 087
Kuala Lumpur 088
Los Angeles 089
Lyon 090
Nagoya 091
Newcastle 092
Oslo 093
Philadelphia 094
Prague 095
Rio de Janeiro 096
Rome 097
Rotterdam 098
São Paulo 099
Singapore 100
Stockholm 101
Taipei 102
Toronto 103
Valencia 104
Vienna 105

Zone 4
Atlanta 108
Baltimore 108
Bangkok 109
Cairo 109
Guangzhou 110
Kharkiv 110
Lille 111
Marseille 111
Milan 112
Naples 112
Newark 113
Nizhniy Novgorod 113
Recife 114
Santiago 114
Shanghai 115
Warsaw 115

Zone 5
Cologne & Bonn 118
Cleveland 118
Dublin 118
Frankfurt 119
Hannover 119
Jacksonville 119
Liverpool 120
Manchester 120
Melbourne 121
Miami 121
Nuremburg 121
Pittsburgh 122
Porto 122
Rhine-Ruhr 122
Stuttgart 123
Sydney 124
Tunis 125
Zurich 125

Zone 6
Adana–Belgrade 128
Belo Horizonte–Calgary 129
Caracas–Daejeon 130
Dallas–Fukuoka 131
Genoa–Houston 132
Incheon–Kobe 133
Kolkata–Manila 134
Maracaibo–Nottingham 135
Novosibirsk–Rennes 136
Rouen–San Juan 137
Santo Domingo–Tbilisi 138
Tehran–Yokohama 139

Appendix
Image credits 142
Bibliography 142
Index 143

About the author

Mark Ovenden

Writer and Broadcaster

Mark was born in London in 1963. His cartographic leanings probably began when his mother thrust maps at him in the vain hope that he might fidget less while traveling on the Underground toward downtown, where his father worked in a West End store.

Instead of reading comic books he pored over old cast-off road maps. On one occasion, armed solely with a Tube map and the fearlessness of a seven-year-old, he left his grandparents and made it home alone, ten miles across London. Using a dog-eared collection of transit fliers, he would spend hours doodling fantasy extensions and didn't see how an entire new city could be built without regard to how its 200,000 residents would travel around; hence Milton Keynes gained a fictitious urban transit system at his twelve-year-old hand (1).

While more "outdoor" boys played war games, Mark would buy a single ticket for the Tube and spend hours voyaging to the terminals of each line, scribbling impossible extensions or creating whole new imaginary mass-transit systems like the one whisked up for Liverpool and Manchester one rainy afternoon (2) .

A magazine feature on the Paris Métro map tantalized him to paint a geometric but more topographically accurate version of London's sacred tube map (3). The messy spaghetti taught him why Harry Beck had altered those features but gained Mark a place at art school studying graphic design.

Though he later pursued a career in media and music, he never lost his fascination with maps and urban transit, and he continued collecting treasures from around the world until assembling them into this compendium.

Since its first British publication in 2003, he's been thrilled to find that he wasn't the only geeky kid fascinated by maps or by making fantasy rail plans—designers of which now form an online army of passionate inventors.

Mark is motivated by the idea of good design helping to increase the use of mass transit, both for aesthetic reasons and for the benefits it brings in decreasing pollution to our planet's environment. He intends this book to be a pleasurable reference resource and homage to historic and contemporary excellence in map or diagram design.

1.

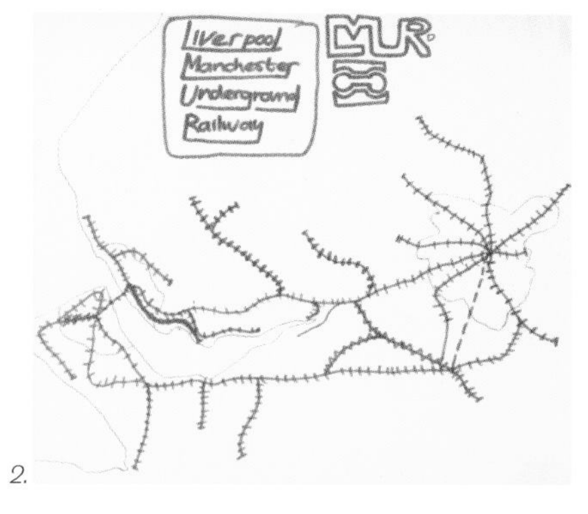

2.

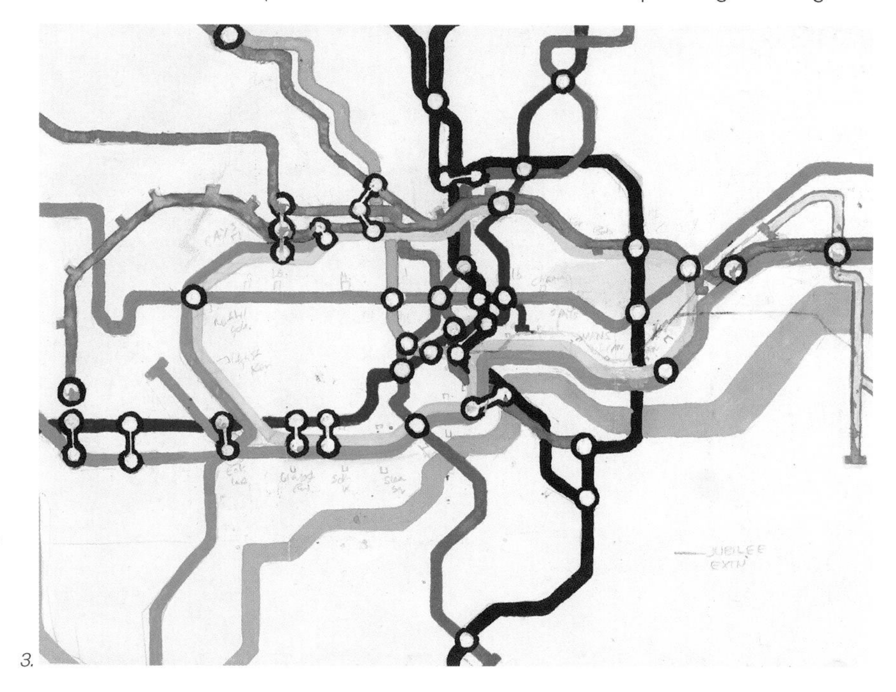

3.

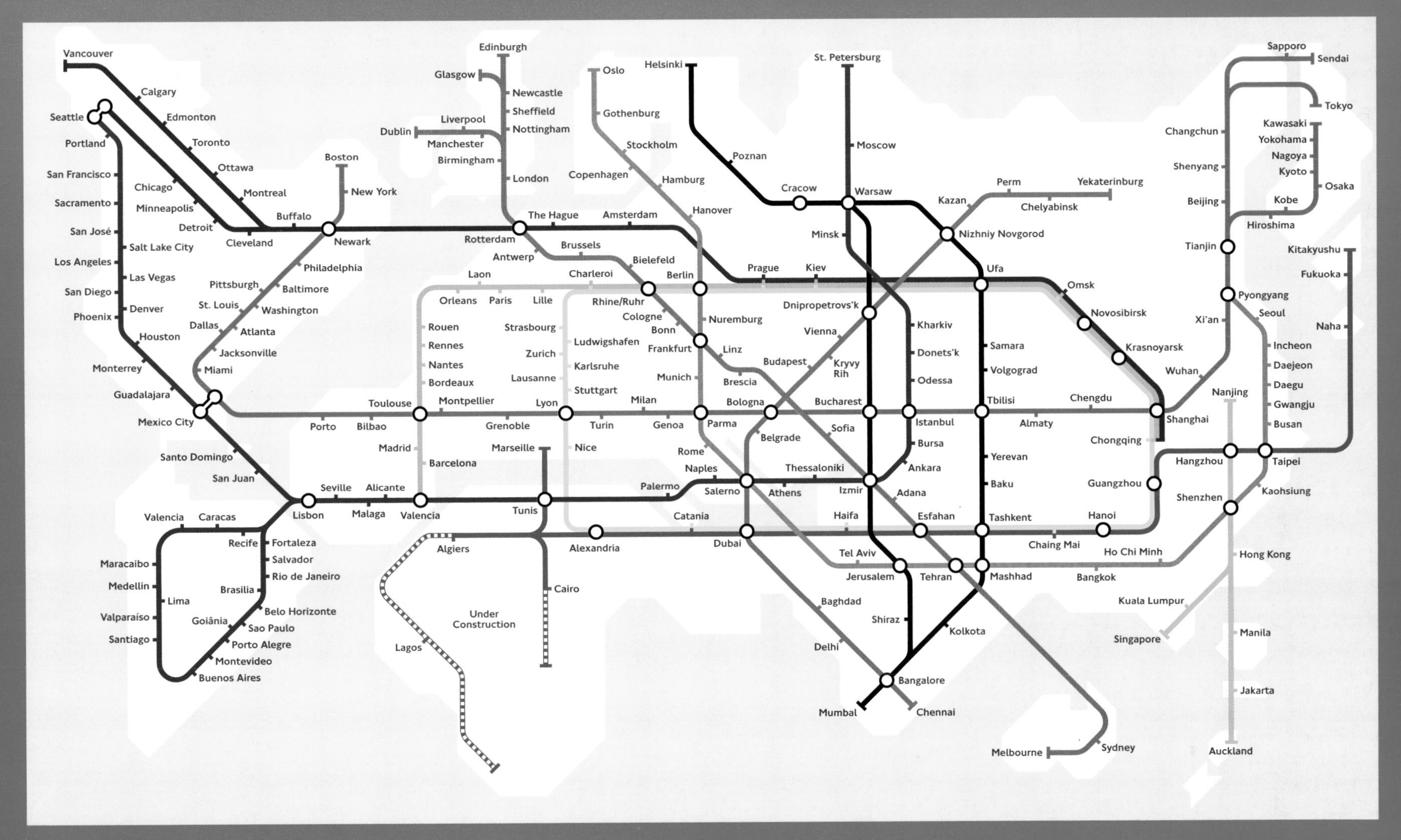

Transit Maps of the World

Mark Ovenden

Edited by Mike Ashworth

Consulting Editors Paul E. Garbutt and Robert Schwandl

This captivating diagrammatic view of the cities included in this book is in the style of Harry Beck's classic London Underground diagram. It was conceived by the author and executed at LS London by Alan Foale, who is responsible for updating the London diagram. It is also available as a full-size wall poster from London's Transport Museum shop or online at www.ltmuseum.co.uk/shop.

** Note: This diagram is not intended as a definitive guide to all operating systems since it includes some proposed and planned networks that may not be constructed.*

PENGUIN BOOKS

PENGUIN BOOKS
Published by the Penguin Group
Penguin Group (USA) Inc., 375 Hudson Street, New York, New York 10014, U.S.A.
Penguin Group (Canada), 90 Eglinton Avenue East, Suite 700, Toronto, Ontario, Canada M4P 2Y3
(a division of Pearson Penguin Canada Inc.)
Penguin Books Ltd, 80 Strand, London WC2R 0RL, England
Penguin Ireland, 25 St Stephen's Green, Dublin 2, Ireland (a division of Penguin Books Ltd)
Penguin Group (Australia), 250 Camberwell Road, Camberwell, Victoria 3124, Australia (a division of Pearson Australia Group Pty Ltd)
Penguin Books India Pvt Ltd, 11 Community Centre, Panchsheel Park, New Delhi – 110 017, India
Penguin Group (NZ), 67 Apollo Drive, Rosedale, North Shore 0745, Auckland, New Zealand
(a division of Pearson New Zealand Ltd.)
Penguin Books (South Africa) (Pty) Ltd, 24 Sturdee Avenue, Rosebank, Johannesburg 2196, South Africa

Penguin Books Ltd, Registered Offices:
80 Strand, London WC2R 0RL, England

First published in Great Britain under the title *Metro Maps of the World* by Capital Transport Publishing 2003
Revised and expanded edition published 2005
This second revised and expanded edition published in Penguin Books 2007

20 19 18 17 16 15 14 13 12 11

Most historic maps are from the Mike Ashworth collection, UITP (International Association of Public Transport) Library, Hans Reidel or the author's collection. Image credits appear on page 142.
While gathering material for this book, every attempt has been made to contact copyright owners; where this has not been possible, we would like to extend our apologies and thank them for their contribution to this history.

CIP data is available

ISBN 978-0-14-311265-5

Printed in Mexico

Acknowledgments

Dr. Alan K. Hogenauer
Alistair Meek
Andre Baradat
Annie Kilvington
Annie Mole
Antonious Kotsonis
Ben Baldry
Benoit Clairoux
Boaz Tal
Brian Smith
Carlo Pensotti
Christel Proust
Christina Ristol
Christopher Saynor
Dan Lou
Darren Tossell
David Petherick
Dmitry Zinoviev
Dr. Guy Slatcher
Flip van Doorn
Erika Wyss
Gay Reineck
Geoffrey Bryson
Graham Garfield
Gregory Stepanek
Hugh Robertson
Ian Farmer
Jennie Doble
Joerg Thomas
Jon Livingston
Josh Lehan
Julian Worricker
Lorenzo Shakespear
Lucy Yates
Marco Danzi
Maria Alderin
Mathieu Posthuma
Michael Kerins
Michael Walton
Nick Agnew
Noah Schuitemaker
Pablo Rosales
Paul Simmons
Pawel Janczyk
Peter Olsen
Peter Woods
Quentin Nield
Robert Reynolds
Roger Torode
Rose Josiane
Serge Savard
Serhii Pakhomov
Thierry Marechal
Vasily Tikhonov
Alexandre Bigle
Anabela Castro
Angela Smith
Anna Rotondaro
Artemy Lebdev
Barbara Moulton
Bettyanne Crawford
Brendan Baker
Christian Kaiser
Colin Kemp
Danielle Wasiewicz
David Dunne
David Ellis
Eddie Jabbour
Eddy Konijnendijk
Evalotta Lamm
Frances Hernandez
Geoff Edwards
Gladys Hansen
Hans-Ulrich Riedel
Jack Reineck
Julian Pepinster
Ken Williams
Lillian Yap
Louise Connell
Marcus Boelt
Mark Kavanagh
Matthew Conil
Dr. Maxwell Roberts
Michael Jurzok
Mike DeToma
Nemone Metaxas
Niels Wellendorf
Oliver Green
Olivier Lacheze-Beer
Pat Chessell
Paul Mijksenaar
Paul Moffat
Peter B Lloyd
Pierre Rijckaert
Reinhard Pankow
Robert Somora
Roman Hackelsberger
Ross Verdich
Sarah Ainley
Sergey Miroshkin
Simon Sadler
Struan Kerr
Susanna Kwan
UITP Library
Vladimir Simko
Yee Shin Tian
Yana Rodye

How to use this book

Mike Ashworth

Design and Heritage Manager,
London Underground

Mass transit helped make possible the expansion of modern cities, many of which are intimately intertwined with their transportation infrastructures.

This book celebrates the diversity of rail-based transit systems in urban environments by collecting together for the first time their cartographic evolution. These include remarkable and innovative design practice conveying complex information in accessible form. Intended for the widest possible audience, this book makes no attempt to resolve any controversy as to what makes a true subway, underground, or metro. Rather than accepting the narrower definitions of the railway enthusiast or industry specialist, the book deliberately includes systems that have chosen to market themselves as urban-transit networks. This book simply indulges in the joy of the graphic design employed to show the traveler how to get from station A to station B.

The zones and the details

Drawing on the symbolism of many networks' ticketing regions the book is laid out in "zones." The placing of each system *does not imply any particular significance as to the quality of the service provided or the importance of the city.* The location has more to do with the availability of historic material. The older, traditional underground railways occupy the first few zones and have more space simply because they have had the time and the need to produce many different maps. Some of the more modern and hybrid systems, including those whose maps have a distinct urban-rail feel, need less space and are in zones toward the back.

Although much technical work went into the reproduction of images, some allowance should be made for the enhancement of antique maps and the multiplicity of sources with varying degrees of print quality.

The figure given as "Urban population" is not simply the number of people living within official city limits but includes the wider metropolitan hinterland, usually that served by the transit system, ticketing structure or operating authority.

"Route length" is the passenger service, not total miles of all track.

Interchanges are counted only once under "Stations." "Open" is when the first section was brought into public service.

Updates

Because new extensions are frequently added, this book should be taken as a snapshot of 2007. For updates see individual operators' Web sites. Robert Schwandl's authorative site (www.urbanrail.net) is highly recommended, as is the Web site of the New York Transit Museum (www.mta.info/mta/museum/) and London's Transport Museum (www.ltmuseum.co.uk). Most operators permit personal use of their diagrams, but reproduction for commercial gain or without permission is forbidden.

Whether you consult this book from a transit perspective, as a graphic design resource, or just to locate a station, it's hoped that you can share the excitement of the beauty, differences and similarities displayed throughout the urban transit maps of the world.

Foreword

Hans Rat

UITP Secretary General

Subway, Métro, Underground, and U-Bahn are designated by UITP (International Union of Public Transport) as "urban rapid-transit" or "metropolitan railway" systems, the city-based counterpart of the mainline and inter-urban railways. Whereas globally around 80 percent of passengers are transported by bus, within the array of existing major urban-transit modes, the subway is the absolute heavyweight champion in terms of capacity, speed, and frequency of daytime service.

At its origin, urban rail was conceived to liberate city centers from congested surface road networks, at that time mostly horses, handcarts and pedestrians. Today it is also a tool for urban reshaping, an instrument to enhance the city's ecological balance sheet, a means to design or reengineer a livable city. Mass transit is the heart of sustainable ecology policies, and, rightly, urban rapid transit lines dominate the journey market.

Urban-transit diagrams are so influential that over time they substitute themselves in the customer's mind as mental maps of a city. UITP is therefore extremely pleased to be associated with the publication of this, the first world atlas of urban rail maps.

Introduction: early railroads

1. Surrey Iron Railway on 1822 map.

Railroads have been evolving since Greek slaves first pushed boats overland along grooves cut into limestone, around 600 BC. The Romans even had horses pull wagons on wheels along stone furrows. Since the 1600s in Europe and the 1700s in North America, basic wooden wagon ways transported heavy loads at mines, quarries, and construction sites. Built in 1604 the two-mile Woolaton Wagonway near Nottingham, England, is credited with being the first surface-level version. A wooden gravity railroad was used at Lewiston, New York, in 1764. As the Industrial Revolution gathered pace, iron tracks appeared in the late 1700s. A horse-drawn freight line, the Surrey Iron Road, which opened in 1803, linking early factories to London's river, is possibly the first railway ever to be depicted on a public map, the British Ordnance Survey of 1822 (1).

The first steam-powered engine to run on rails was at Pen-y-darren ironworks, South Wales, in 1804. In 1807 the horse-drawn Mumbles Railway in Swansea started carrying fare-paying passengers. American engineer Oliver Evans predicted in 1812 that steam railways would link cities via separate parallel tracks with trains running in opposite directions. The Stockton to Darlington became the first to partially do this, in 1825. The first regular passenger steam line was the Liverpool and Manchester Railway, whose thirty-five-mile double tracks linked the industrial city to the Mersey port in 1830. Still in use, it would now be classed an "interurban" line; an 1826 engineer's plan of it (2) also shows George Stephenson's Bolton and Leigh Line, which opened in 1828.

The Baltimore & Ohio, which began in 1830, evolved into a major system (3). Railways were proving for the first time in history that people could travel long distances between city and country expeditiously and in comfort. An intense period of "railwaymania" ensued. In 1832 the world's first street railway, the New York and Harlem Railroad (map on p. 8) encouraged short trips inside the urban area and part of it is still in use as Metro North Railroad. Initially a horse-drawn service, it was the forefather of the streetcar, aka "horsecar," or in Europe, "tram." Belgium, Germany, Austria, Russia, France, and Italy opened lines between 1832 and 1839. Despite a short route from Bermondsey to Deptford, it was not until 1838 that London was connected into a rapidly expanding UK rail network. In 1840 London's rope-hauled Blackwall Railway began frequent passenger service, with closely spaced stations from Tower Gateway to West Ferry. Now used by Docklands Light Railway, this was the world's first "metro-like" service.

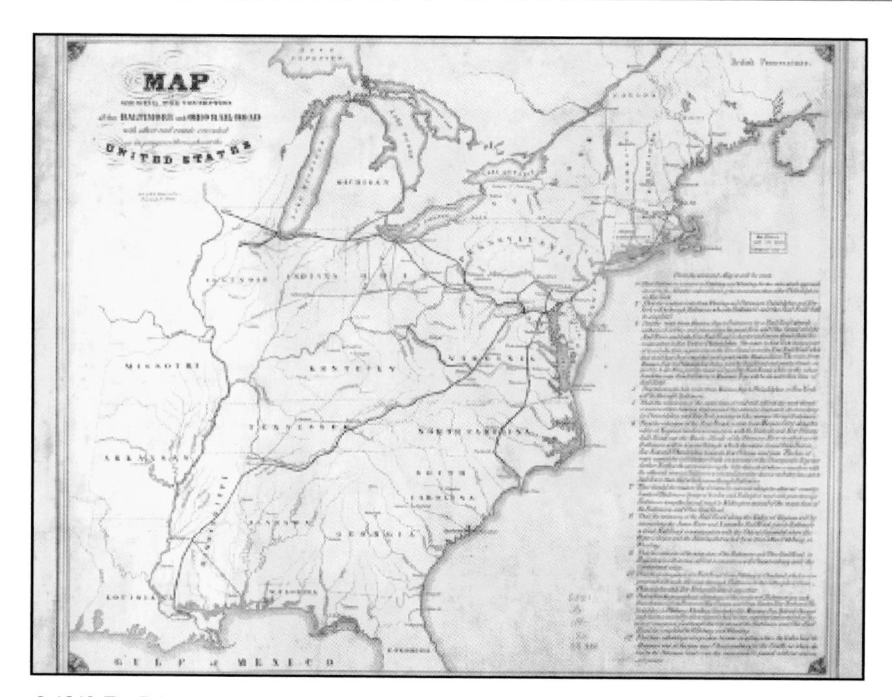

3. 1840: The Baltimore & Ohio evolved into one of the earliest major systems in the world.

By 1850, 9,000 miles of track had been laid in the United States. In comparison, Britain had 6,900 miles by 1852 and Germany 5,000 by 1855. As railwaymania took off during the 1840s and '50s, every private operator wanted their route to converge on London, at that time the world's largest city. However, in order to avoid whole-scale demolition, their termini had to be sited at frustrating distances from the center. For twenty years Paddington, Euston, and King's Cross emptied their passengers at the edge of London to face an arduous walk or a bumpy horse-drawn "Hackney" carriage ride to change trains or get to the center. Proposals to connect them to the City by an entirely new surface railway, or even an elevated one, all failed on grounds of parliamentary disapproval or cost. It was City official Charles Pearson who had the brilliant idea of putting a new railway *under* the ground! The engineer's plan from 1859 (4) shows the full route of the world's first "Metropolitan Railway" in red and how it might have been built with tracks running directly from the termini into the tunnels. The line was duly laid in the bottom of a vast trench largely following the route of London's New Road. The excavation, when re-covered and track positioned below, would bring easy connectivity to all the major mainline rail

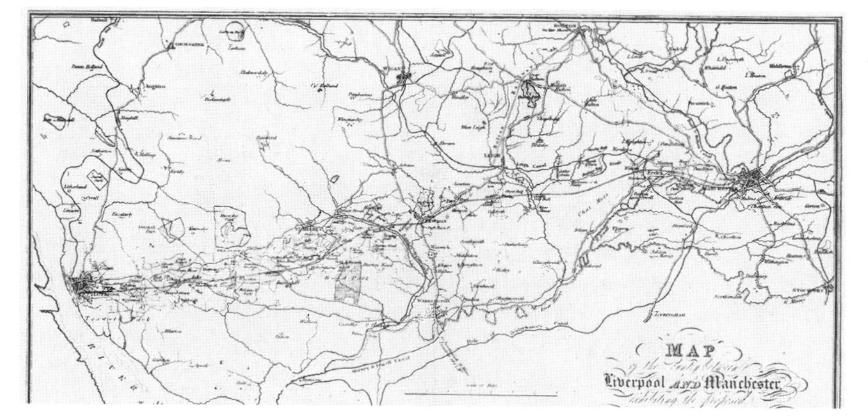

2. 1826: Plan of the Liverpool and Manchester Railway, which required major engineering feats for its time.

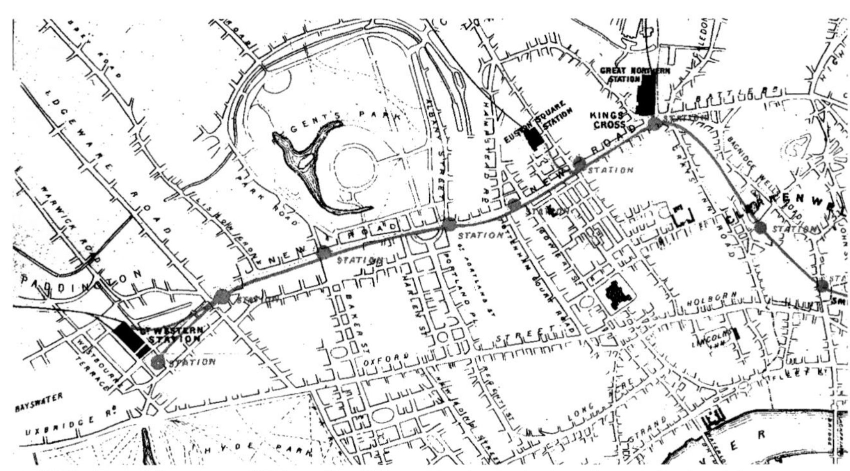

4. 1859: Map of the proposed Metropolitan Railway, showing tracks from the mainline termini.

terminals from Paddington to Farringdon. Despite the "unimaginable chaos of construction" and the lack of a foolproof method of removing exhaust smoke from the steam locomotives, the first underground railway opened on January 10, 1863, and became an instant success! Inadvertently, this also created another significant opportunity, that of allowing passengers to make short-distance rail journeys within the existing built-up area. The Inner Circle loop around London's heart was completed in 1884, by which time the Metropolitan and District railways were throwing out long surface tendrils to distant suburbs.

The far-flung outposts of London's fringes and their Metropolitan railway connections represent early examples of what we now know as the worldwide habit of "commuting." A similar process happened with the Long Island Rail Road and with the El into northern Manhattan, the Bronx, and Queens, where empty land was bought and railways constructed in speculation that the city would spread. As other cities tackled the problem of linking mainline termini or traversing built-up centers without knocking them down, crowded historic centers like Paris, New York, Berlin, and Madrid realized the potential of the Metropolitan-style railway. Initially, elevated systems like New York's Manhattan Railway (5) were favored in North America, but Budapest, Glasgow, and Paris attempted the cut-and-cover method of construction in their own urban rail systems.

Surface disruption was eventually bypassed with the development of the deep tube "tunneling shield." It was first utilized to construct London's short-lived Tower Subway (1870). The first passenger-carrying tube railway to be constructed by the new shield, however, was the City & South London, still operating today as part of the Northern Line. Opening in 1890, it was also the first electrically operated subterranean line, alleviating the problem of noxious exhaust gases. Urban railroads under the ground were about to become big news.

Urban rail in this book

Most early metropolitan urban railways like those in London, New York, Berlin, Chicago, Paris, or Budapest used contemporary rail-industry construction techniques, infrastructure, and operating systems. Their Undergound, Subway, U-Bahn, El, Métro, and Földalatti are categorized as "heavy rail" systems. In their wake came hundreds of other schemes, some wonderfully, farcically untenable, all proposing to move people around, over or under major urban areas. As technology evolved, so too did new methods of construction, types of vehicles, system scale, and cost.

In the last fifty years especially, this has led to a significant blurring of what constitutes an "urban railway." In the twenty-first century there is now a dazzling array of diverse system types. Many have at their heart traditional subway stations in dense downtown areas, but some surface in the suburbs to behave more like street running tram systems. San Francisco's Muni, many German cities' Stadtbahns, and Brussels' Prémétro are good examples. Newark recently converted its old subway into part of a new light-rail system. Liverpool and Sydney chose to bury some of their heavy-rail lines in city-center tunnels, while heavy-rail commuter systems in Paris (RER), major Spanish cities (Cercanías), Dublin (DART), Budapest (HEV), Hong Kong (EastRail), some Japanese commuter lines, and many German cities (S-Bahn) have a metro-like ambience. Others, such as St. Louis, Newcastle, or Manchester, have rebuilt abandoned or underutilized suburban heavy-rail trackbed into metro-like systems using modern light rail.

Purists may argue some of these have no place in a book of urban-rail transit maps. However, all three latter systems embody the word "Metro" in their name and strive through their publicity to convey an impression they are indeed metro-esque, as they operate a "turn-up-and-go" service of frequent trains that do not require consultation of a timetable. Perversely, traditional subway systems can have tracks on the surface and/or elevated high above streets. Even in cities such as New York, London, and Paris, sections of the systems take to the air.

Almost all urban-rail trains run steel wheels on steel rails, but Mexico City, Paris, and Montreal, among others, have pneumatic-tire stock too. Urban-rail tracks are generally exclusive to urban-rail trains, but occasionally other vehicles, such as trams, utilize the same track. The latest development is the "tram-train," popular in Holland and Germany. One common attribute that distinguishes urban rail from, say, a street trolley is that it is segregated from other traffic, though some systems break even this rigid definition.

Worldwide, each system utilizes graphic and cartographic design skills, with varying degrees of effectiveness, to help passengers through what might otherwise be confusing labyrinths with the aid of clear and often beautiful or innovative maps and diagrams. It is the use of these rather than a dogmatic classification of system types that has been employed to select entries for this book.

We live in a perplexing and diverse world, full of dilemmas and differences; try as we might, some things just refuse to fit neatly into categories. So rather than accepting any technical definitions, just think of this book as a pleasurable graphic guide to the wonders of urban-rail map and diagram design.

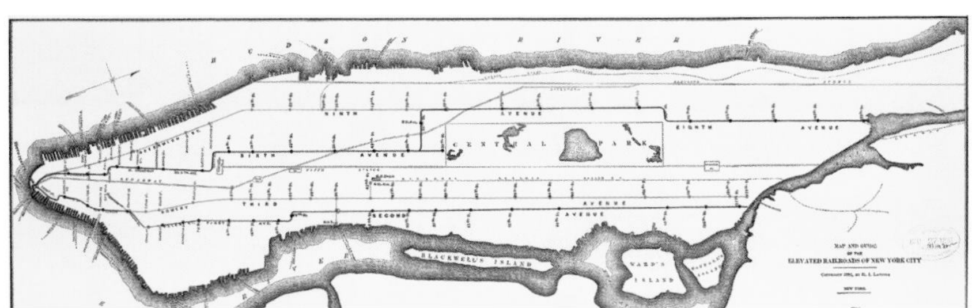

5. 1881: New York's first mass-transit network was elevated, thanks to the Manhattan Railway.

Introduction: early railroad maps

1. 1830: The Liverpool & Manchester overprinted.

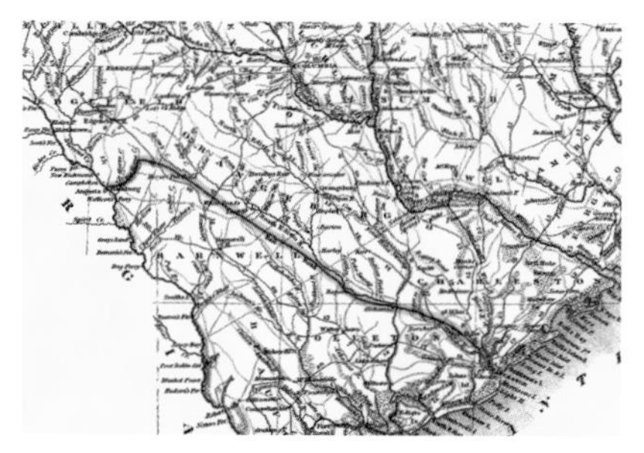

2. 1833: South Carolina Canal & Railroad Co.

3. 1847: New York and Harlem ran on the street.

4. 1869: Railroads helped open up the West.

Maps are simply tools for helping us to navigate spaces. In their crudest form they probably existed before the written word. Beautiful hand-drawn maps of great kingdoms date back to the dawn of modern civilization, when detailed maps of land and coast became crucial for seafarers.

The European colonial powers took maps very seriously. As they voraciously laid claim to the Americas, Africa, and most of Asia, nothing said "we own this territory" better than a detailed map of it. Faced with a war against France and civil unrest in Scotland, England's King George II, for example, commissioned a young engineer, William Roy to measure every inch of the Scottish Highlands in 1746. The resulting advantage this brought encouraged the government to send out Board of Ordnance teams to precisely survey the topography between London and the southern coast, creating the first one-inch map of Kent in 1801, the most detailed and accurate map produced to that date.

No matter how basic, all maps require at least one minor leap of imagination; until the invention of bulky models or computers, they forced the reader to *visualize* a larger three-dimensional place reduced to a two-dimensional medium.

Cartographers also utilize *symbols* to represent geographic features. Physical distance is invariably shrunk, even on a 3-D map. But railways began to be built in the age of science and reason. New cartographic, engraving, and printing techniques were giving rise to never before seen detail and beauty.

As torrents of track tore through the countryside, post–Industrial Revolution cities were oozing out over fields faster than cartographers could get their theodolites erected—far too quickly for meticulous surveyors to keep up. So the very first railroads were almost always added to the existing plates of topographic maps.

The Liverpool & Manchester Railway, for example, was crudely etched onto an earlier "Map of the County Palatine of Lancaster," published in 1830 (1). In the same era North America's first railways were opening, including, in 1833, the South Carolina Canal & Railroad Company, whose 136-mile line from Charleston to Hamburg was for a short time the longest track on the planet (2). In 1837 the New York and Harlem Railroad ran from what were then rural fields down onto lower Manhattan's streets, as a map drawn ten years later shows (3).

A railroad building frenzy engulfed much of the modern world in the middle of the nineteenth century. Between 1830 and 1850 over 6,000 miles of new railways were surveyed, mapped, approved, and built in Britain alone, and by 1850 America had mapped and built 9,000 miles, mainly in New England and the mid-Atlantic states.

The world's first real transcontinental railway linked the Pacific with the rest of the United States. Built by two different companies, it started from different places, meeting in the middle, at the infamous planting of the Golden Spike on Promontory Summit, Utah, in 1869 (4). Without accurate surveying and mapping the tracks would never have met up.

As the complexity of rail networks increased, so the methods each operator used to inform passengers of their superiority over a competitor also evolved. Maps and posters were the key to competing with the other operators and gaining new traffic. They became an art form in their own right wherever several operators covered a similar route.

In the 1800s, printing was achieved by carving away wood, leaving a raised surface to ink on (letterpress), or by marking metal plates that were soaked in acid to create grooves that held the ink (engraving). Fine detail, deep shading, and color were time-consuming processes, so early maps were often fairly crude affairs. New additions were usually overprinted onto existing maps or were etched onto old plates.

While handy for perhaps the first line or two, the technique quickly showed signs of overcrowding in the mesh of lines crossing North America and Europe. In the dense urban centers of Chicago, Boston, New York, Berlin, Paris, Vienna, London, Manchester, and West Yorkshire, mapmakers were severely challenged, but the advent of the Metropolitan Railway, with its even closer stations, would stretch legibility on traditional map designs just too far.

The problem was to present with clarity the multiplicity of densely concentrated routes and crossovers in the central area of a city, where stations are close together, while not wasting space in the suburbs where the stations were more spread out. The days of maps remaining true to geographic scale were numbered.

In addition, in the case of subways it was soon recognized that you do not necessarily need to know exactly which streets you are traveling under when you are many feet below them, just which station follows which.

A portentous 1874 map of London's Metropolitan Railway (5) removed almost all surface topography, including the street pattern. Similarly, by the late 1900s, Berlin's complex system of urban railways was depicted without any surface features (6). This style

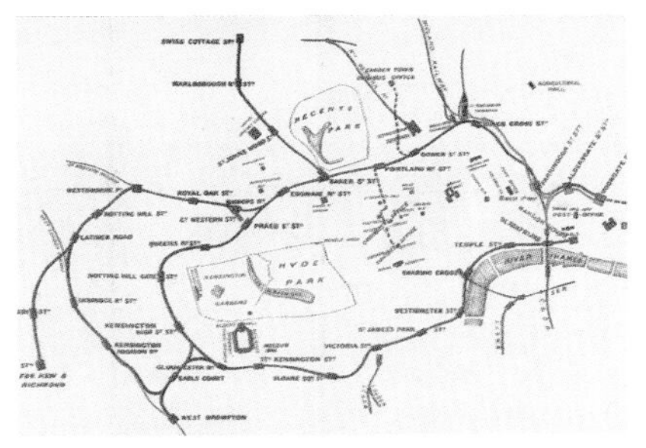

5. 1874: Most of London's geography erased.

was similar to that used by British Railways Clearing House in 1895 and John Airey some years earlier.

Remaining to scale and staying legible would have meant maps becoming far too unwieldy to carry around, so the next big jump was to distort the true nature of the distance between stations. This change was crucial for maps of the entire United States or lines like the Metropolitan with a close proximity of stations in central London but up to seven miles apart in the country.

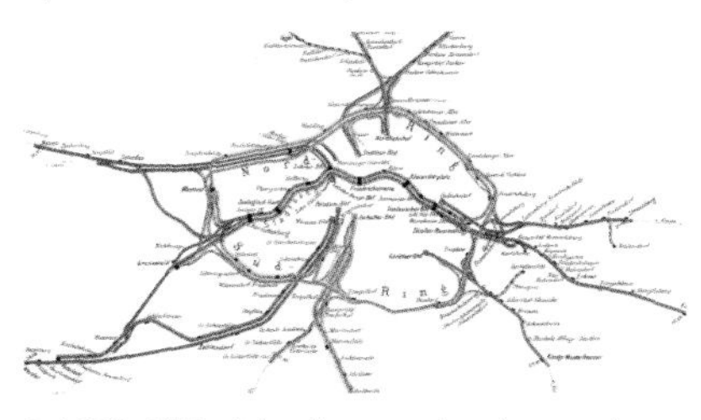

6. 1896: All Berlin's railways and no topography.

Signal box and railway gradient diagrams had already been standard in the industry for some years, but around 1904 the Central London Railway (now London's Central Line) produced a simplified diagram of its new service (7). Art itself became

7. 1904: CLR maps had a slight diagrammatic feel.

increasingly abstract during the early twentieth century, and this was reflected in commercial art (graphic design). The idea that one thing could stand for another was a great boost for mapmakers. Cubism and the work of artists like Mondrian and Kandinsky would have influenced cartographers to push the boundaries. Andrew Dow's book (see bibliography) has unearthed evidence that route diagrams within trains were taking abstraction to the extreme—possibly from as early as 1904, when the L&Y electrified its Southport line (8), and on a 1908 Bakerloo Line poster (see p. 20).

8. 1904: The Southport Line in-car diagram.

North American city grids enforce a more even distribution of stations (see p. 16), but the massive urban rail system of the Los Angeles Pacific RR was so dense (9), one route was stylized into the shape of a balloon (10). Boston's 1926 map (see p. 48) removed all streets and topographical features, distorted the true length of lines, and introduced another crucial feature seen previously only on the example from London in 1917 (11);

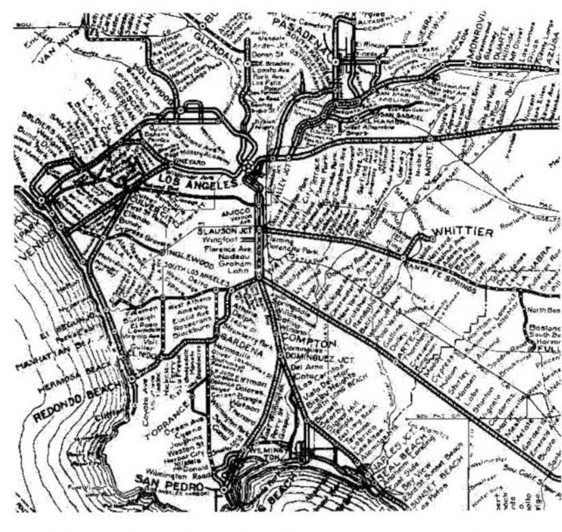

9. 1920s: Los Angeles Pacific streetcar network.

both smoothed out the bends and kinks of the tracks into neat straight or simple curved lines. This kind of simplification was just what legibility had been waiting for. An interesting variant, using only curves, appeared on a 1925 London in-car map (12).

10. 1926: A stylized Los Angeles Pacific route.

In 1929 an insightful London draughtsman, George Dow, began producing his first schematics for LNER suburban railways. They were arguably the first full system diagrams seen anywhere (13).

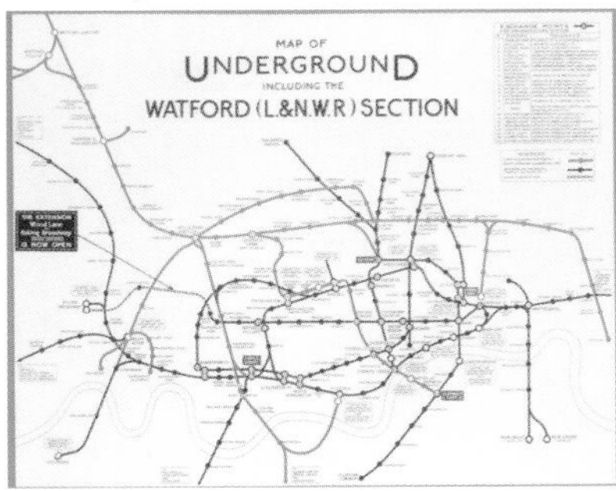

11. 1917: A much simplified diagram of London.

A 1931 diagram of Berlin's S-Bahn puts all diagonals at 45 degrees (see p. 12). Up until 1932 F. H. Stingemore's small, folded-card pocket maps for London Underground (see p. 21) lacked topographical features, so aiding clarity, but his lines appeared like a bag of interlaced colored wool.

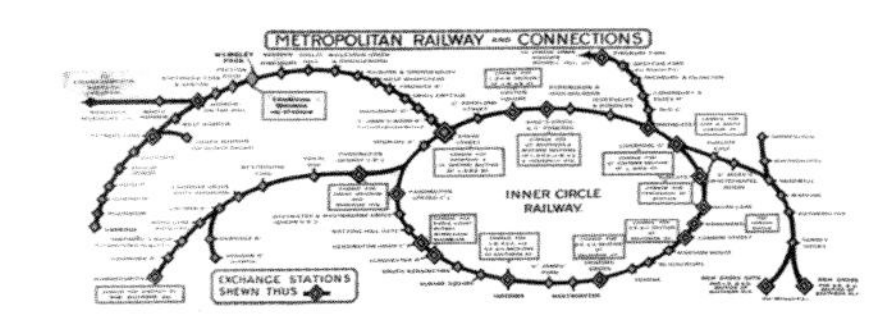

12. 1925: London Metropolitan in-car diagram.

Effectively combining all the in-car route maps in 1931, Harry Beck sketched a full system diagram making the lines conform to a set of rules utilizing horizontals, verticals, and sparing use of the diagonal at just one angle of 45 degrees (14). London Transport published Beck's version in 1933 (see pp. 21, 22). The way we looked at transit maps changed forever.

13. 1929: George Dow's LNER route diagram.

Most maps in this book are what we would now call schematics or diagrams because, essentially, in many cities the highly detailed topographic map has evolved into a simple route diagram as a more effective way to envision and thus navigate the labyrinth of urban rail tunnels right beneath our feet.

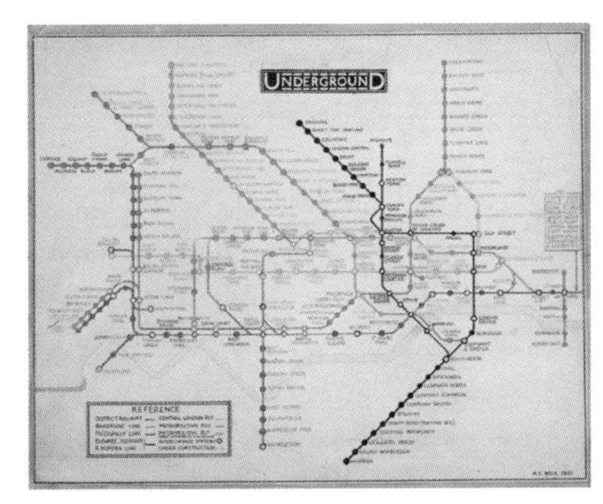

14. 1931: Harry Beck's presentation visual.

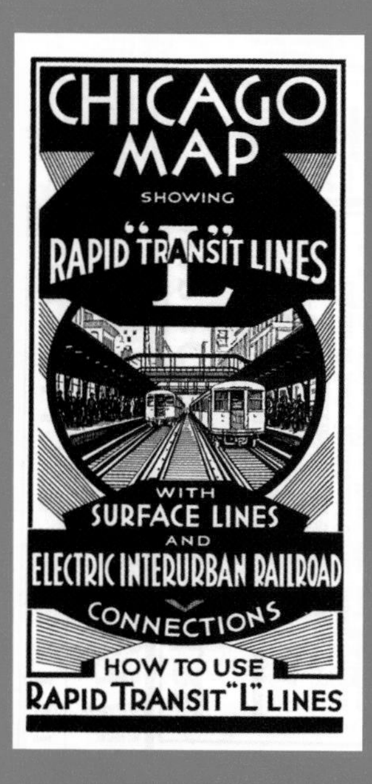
CHICAGO MAP
SHOWING
RAPID "TRANSIT LINES
"L"
WITH
SURFACE LINES
AND
ELECTRIC INTERURBAN RAILROAD
CONNECTIONS
HOW TO USE
RAPID TRANSIT "L" LINES

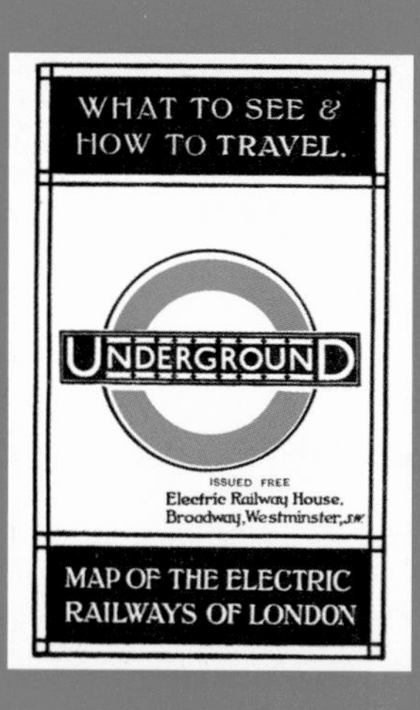
WHAT TO SEE &
HOW TO TRAVEL.
UNDERGROUND
ISSUED FREE
Electric Railway House,
Broadway, Westminster, SW
MAP OF THE ELECTRIC
RAILWAYS OF LONDON

Schnellbahnnetz S U
Stand 1.11.1990

Map Showing
CHICAGO'S
NEW SUBWAY
and
ELEVATED
RAILROAD
CONNECTIONS
RAPID TRANSIT LINES
SUBWAY ELEVATED

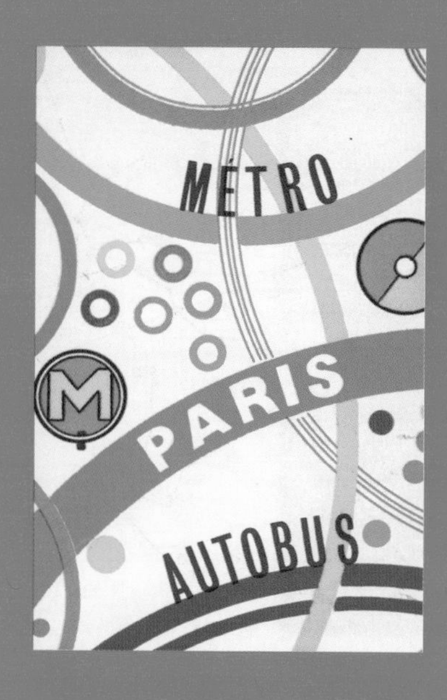
MÉTRO
M
PARIS
AUTOBUS

NEW YORK
SUBWAYS
UNION DIME SAVINGS BANK
6TH AVE., at 40TH ST., N.Y.C.

Paris

МЕТРОПОЛИТЕН

TOKYO
SUBWAY MAP
東洋唯一の地下鐵道
上野淺草間開通
TEITO RAPID TRANSIT AUTHORITY

Plano
Esquemático
de la Red
navegamadrid
www.metromadrid.es
Metro
CONSEJERÍA DE OBRAS PÚBLICAS,
URBANISMO Y TRANSPORTES
Comunidad de Madrid

SUBWAYS IN TŌKYŌ
TEITO RAPID TRANSIT AUTHORITY

Cities in Zone 1 have produced the greatest range of historical material. On this spread are examples from some of the beautiful and graphically diverse front covers that have been used for pocket maps of the urban-transit systems covered in this section. Many of these display what Frank Pick, creator of London Underground's influential graphic design, called "fitness for purpose," while others are purely decorative for art's own sake.

Zone 1

Berlin	012
Chicago	016
London	020
Madrid	024
Moscow	028
New York	032
Paris	036
Tokyo	040

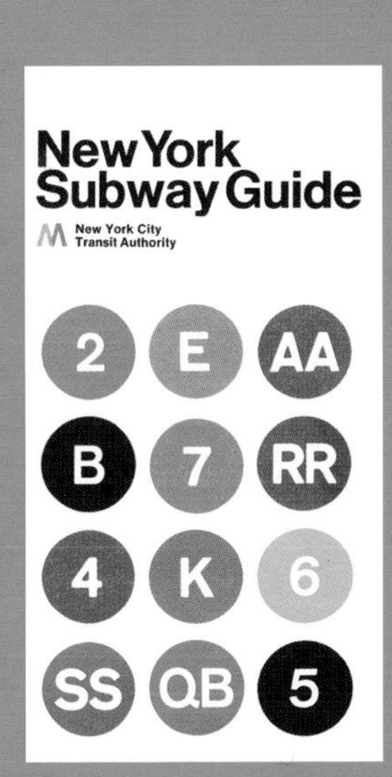

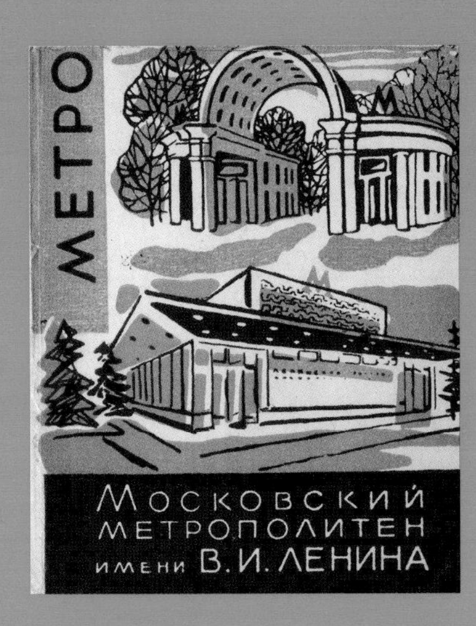

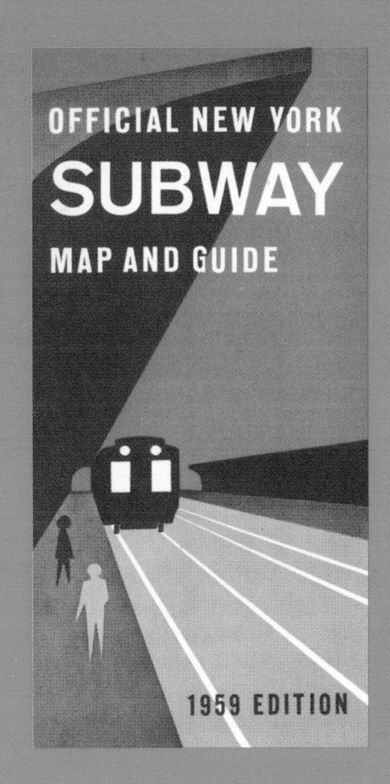

Berlin

Urban population: 3.5 million. Route length: 91.2 miles. Stations: 170. First section open: 1902. Underground: 90 percent.

Urban-transit maps echo the prevailing social and political trends of the societies they emanate from, but Berlin's U-Bahn and S-Bahn maps perhaps more than others give us a flavor of that city's unique life. The maps here illustrate the evolution of a network dating back over a century to when the great German capital dominated Europe, through the ravages of two world wars, and, during an inimitable division, now dismantled.

The roots of Berlin's mass transit are cited as making it one of the earliest urban-rail networks in Europe. A heavy-rail ring line encircled the city in the 1870s and by 1882 a central spine, the Stadtbahn (city line), had opened, giving it a lead in railway provision over other continental European capitals. A pioneering 1896 map, although not publicly available, was probably the earliest attempt to show the many *Vorortbahnen* (heavy-rail suburban lines) together by removing all topography (see p. 8).

The first line of the Untergrundbahn, from Knie (now known as Ernst-Reuter-Platz) to Warschauer Brücke (Warschauer Straße), with a branch to Potsdamer Platz (now U1 & U2), which opened in 1902, was mostly elevated above the streets. It became the backbone of the system we see today, though not all the lines envisaged were built, as can be seen from this 1910 plan of all urban rail, where up to nine were proposed (1). An innovative, public, color pocket map of 1914 (2) shows all urban-rail lines in existence, plus parks and major roads and an early use of the black circle with white-eye center device.

By 1922 new lines were opening rapidly and the maps gave emphasis to the growing U-Bahn (3). By 1926

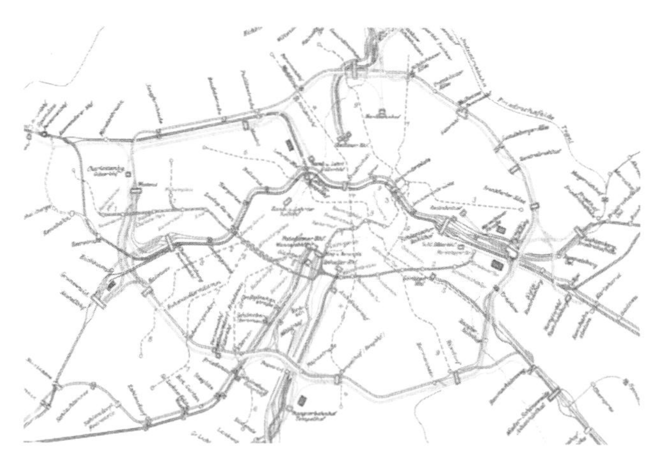

1. 1910: Detail from large poster, Eisenbahnnetz von Gross-Berlin.

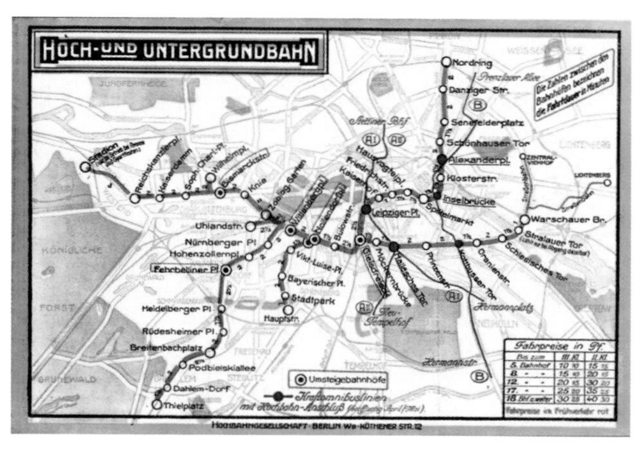

2. 1914: Color pocket map on card.

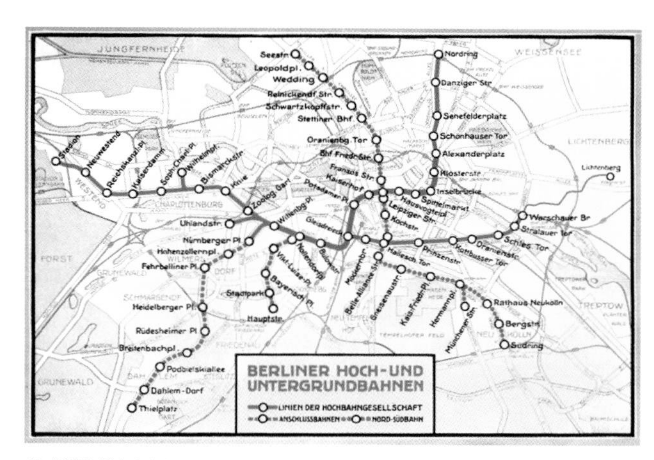

3. 1922: This joint map gave prominence to the Untergrundbahn.

4. 1926: This pocket map was one of the first to show just the U-Bahn.

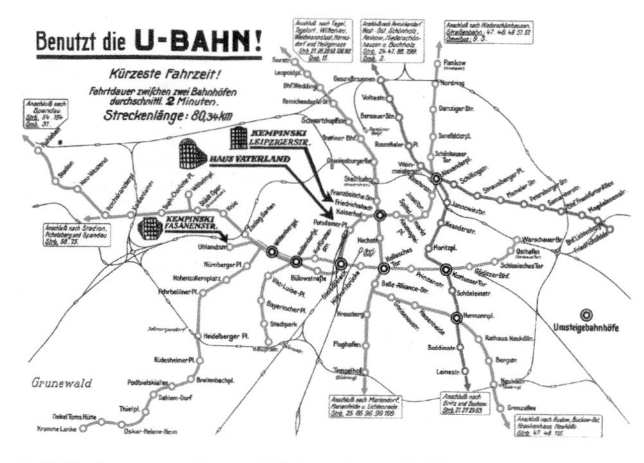

5. 1931: Pocket map showing S-Bahn in thinner black lines.

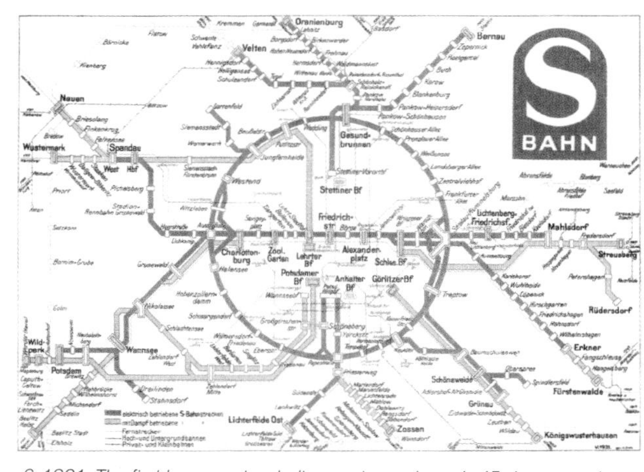

6. 1931: The first known network diagram to employ only 45-degree angles.

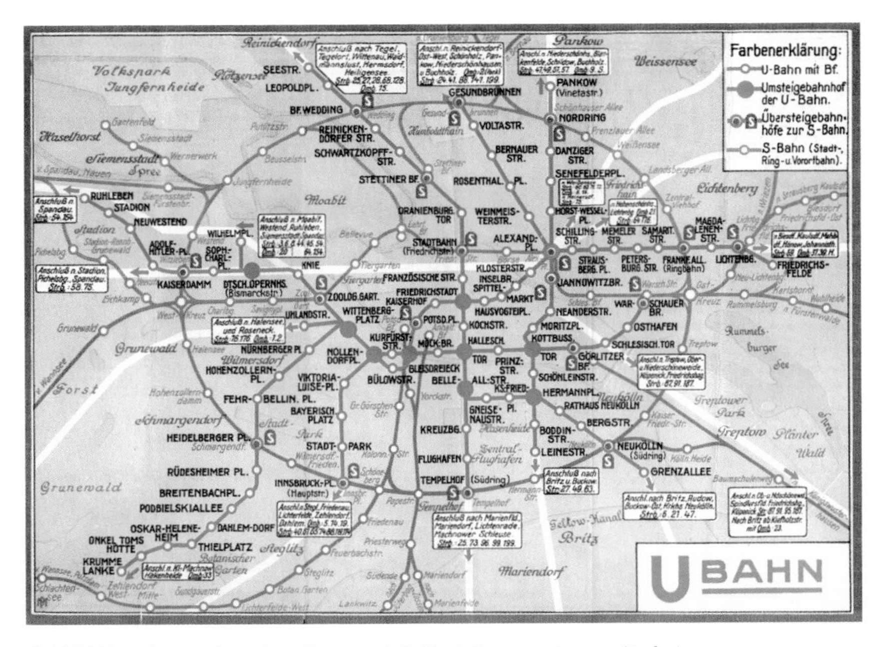

7. 1934: Pocket map using schematic concepts but including some topographic features.

it was appropriate to show the U-Bahn on its own, so this beautiful if poorly preserved example from that year knocks out all topography (4).

It also uses a little artistic license in straightening out the true geographic trajectory of some of the lines, making a clearer diagram. The 1931 version reverted to showing the other railroads but was among the first to use the shortened brand name U-Bahn (5), and interchanges also gained the double black circle symbol.

Due to their frequent running, segregation from other systems, and urban nature, the Stadtbahn, Ringbahn, and Vorortbahn were progressively electrified and collectively rebranded in 1930 as the S-Bahn, standing for Stadt (city) and Schnell (fast), which could now be considered a complete urban-rail network in its own right.

A huge leap in Berlin map design came in 1931 with the first S-Bahn diagram (6), which just predates Harry Beck's iconic London Underground diagram (see pp. 9, 21). It is a brave leap into pure geometry; the ring clearly stylized and enlarged in the center is the focus of the network, beating a similar ring design concept on Moscow diagrams by almost five decades. It is understood to be the very first urban-rail diagram to show all diagonals

8. 1940: Paper map due to wartime shortages.

at just the one simple angle of 45 degrees. Also, here stations are reduced to regimental, even spacing and the whole diagram employs fairly extreme distortion from geographic accuracy. The concept was extrapolated on the beautiful 1934 U-Bahn map (7), which elongates the S-Bahn ring into an elegant oval, straightens out lines, evens station spacing, and keeps station names from crossing over lines.

The large *U* seen outside stations was shown for the first time on maps from 1937, but neat design suffered

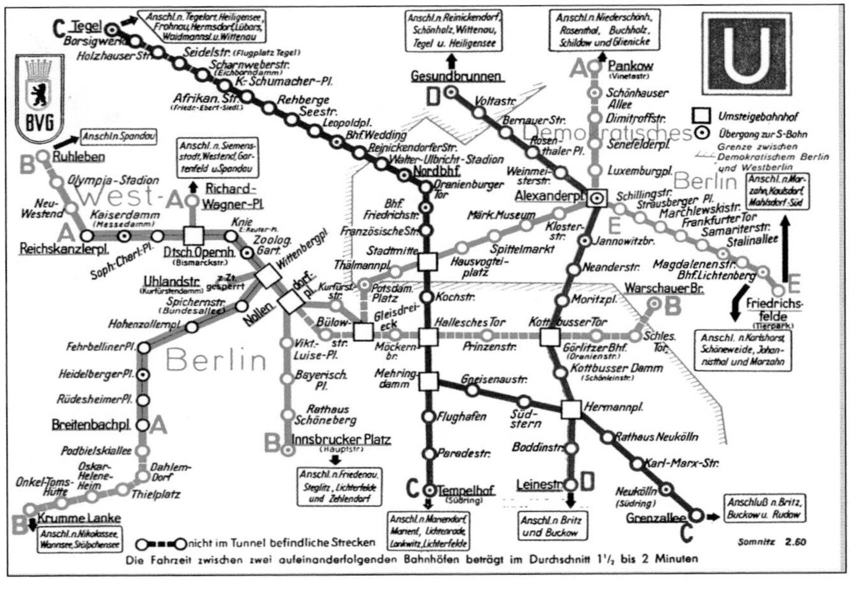

9. 1960: West Berlin pocket map—the last before the wall went up.

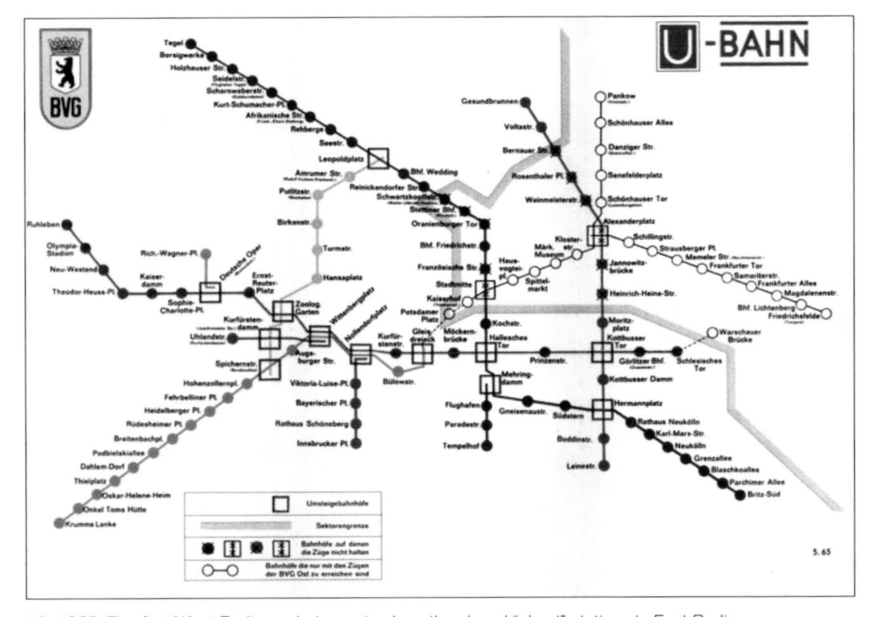

10. 1965: The first West Berlin pocket map to show the closed "ghost" stations in East Berlin.

Narrow-profile train at Schlesisches Tor.

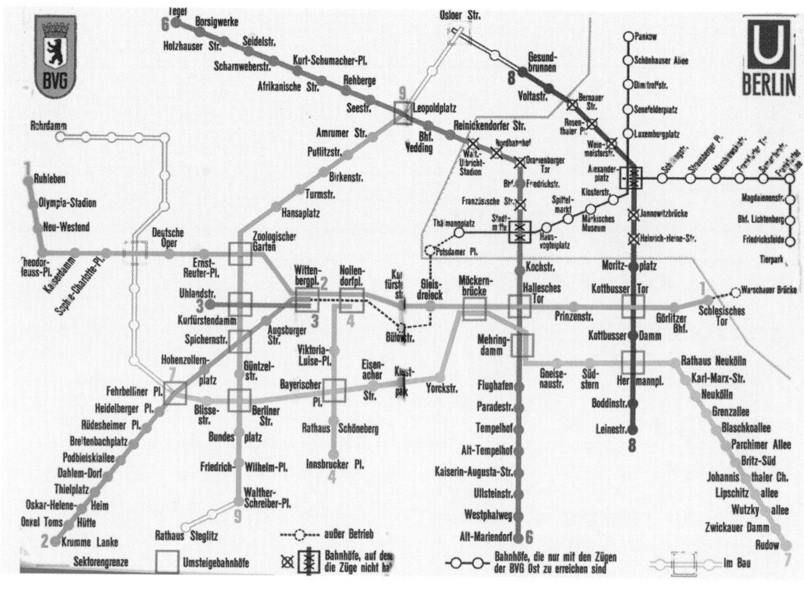

11. 1972: The large West Berlin in-car and station wall map with "ghost" stations x-ed out.

One-off: a gift to BVG from London Transport.

somewhat during the war. Nonetheless, the 1940 pocket diagram (8), while printed with just one color and on low-quality paper, was the first to show line names (A to E) and to attempt to distinguish between interchanges just within the U-Bahn system and those to other networks.

What happened to the U-Bahn after the Second World War, however, was unique in urban-rail operations. With the separation of the city aboveground came a subterranean division too. For the first few years the U-Bahn ran as normal, albeit while undergoing repairs from bomb damage, but as the Cold War deepened and the Berlin wall went up, urban-rail operations were heavily affected for the worse.

Running beneath East Berlin from the Western sector, lines D (now U8) and C (now U6) were shown on the 1960 diagram (9), but by the time the wall was erected in 1961, formerly joined-up lines (like that which has now become U2) were abruptly terminated on the Eastern side. Some stations, like Schlesisches Tor to Warschauer Br, were cut off completely from the network for thirty-odd years. Elsewhere, services ran from the West straight through the sealed-off, ghostly stations beneath the East. This is seen in the x-ed out stations on the small 1965 West Berlin pocket diagram (10), which uses yet more geographic distortion to create a pleasing, cleaner balance and some white space. It was the first to show the ghost stations with a large black X through them. Another feature here is the increasing diagrammatic direction, though some neatness is lost on the 1972 in-car poster (11). By about 1978, all diagonals reverted back to 45 degrees to incorporate some new extensions on the Western side.

A glance at two of the last preunification maps from East Berlin shows how both sides kept a unified approach to system logos, but observe how clever graphic design was used to marginalize (12) and all but obliterate (13) West Berlin, a trick used on many East Berlin maps to varying degrees of success. In the latter diagram, though, the stylized S-Bahn ring makes a welcome return.

It was the long-boarded-up station of Jannowitzbrücke that was the first to reopen, just two days after the

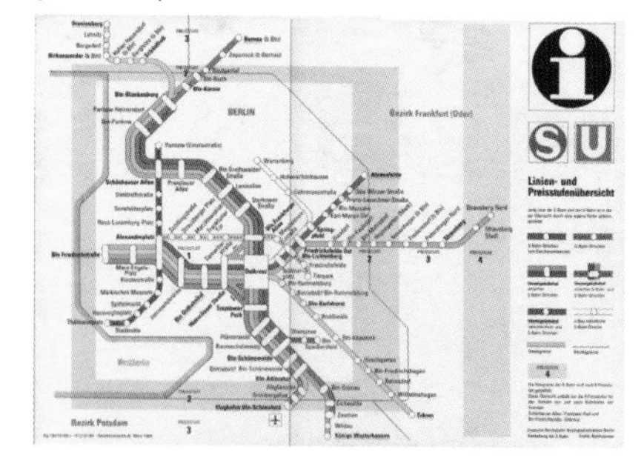

12. 1984: Diagram of East Berlin–only services.

13. 1988: Last East Berlin pocket map graphically shrinks the Western sector.

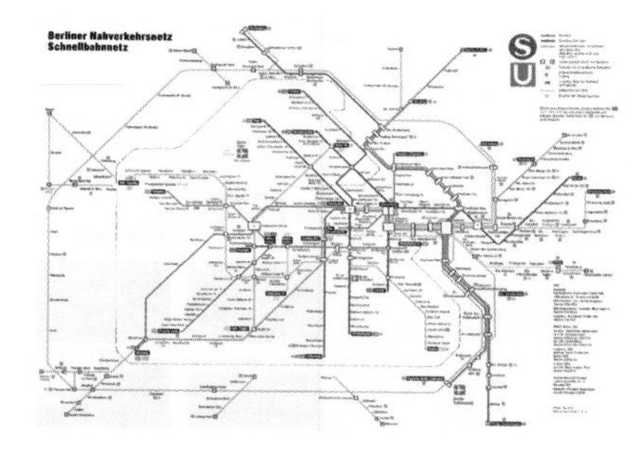
14. 1990: First to show a reunified network.

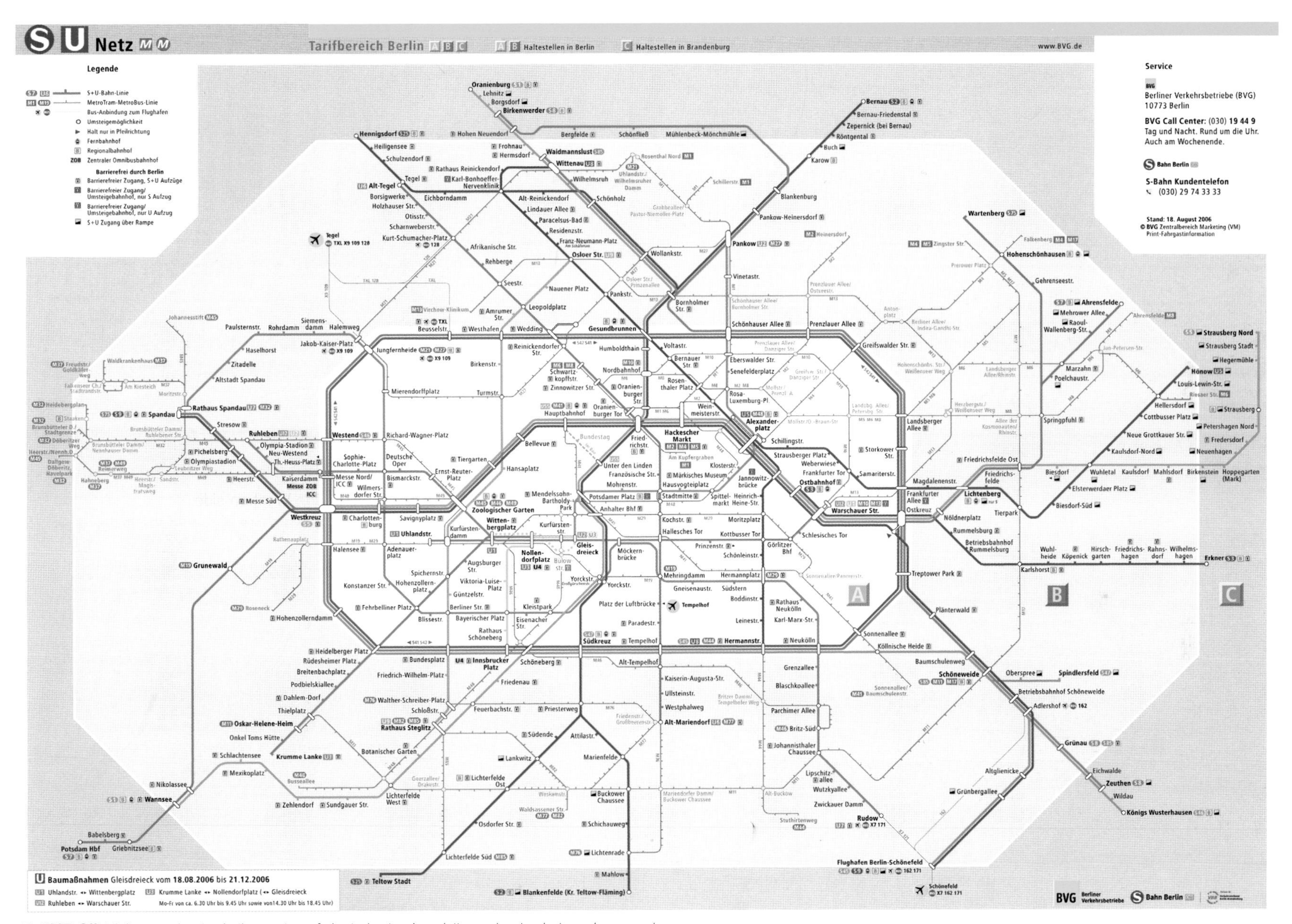

15. 2007: Official diagram showing both a century of physical network evolution and major design enhancements.

wall came down in November 1989. From that date on design concepts were assimilated from both the former East and West for the healing of the fragmented system, which benefited greatly from the involvement of German typeface guru Eric Spiekermann, who was also responsible for kick-starting a much-needed overhaul of the fragmented network signage.

The earliest version of a newly unified network map, though hurriedly assembled, was proudly published in April 1990 (14). This has evolved into the outstanding diagram we see today (15), whose graceful balance and impeccable neatness developed in tandem with what is now one of the world's most effective integrated urban mass-rapid-transit systems.

Chicago

Urban population: 7.5 million. Route length: 107.5 miles. Stations: 151. First section open: 1892. Underground: 95 percent.

Chicago is one of America's greatest cities, a claim that goes back to the middle of the 1800s, so naturally it was among the earliest U.S. urban-rail pioneers and became the world's largest rail hub. It broke new ground in mass transit and consequently the city's maps have often been smart and innovative. It had one of the most extensive streetcar systems and boasted some early elevated railways (Chicago South Side opened in 1892). These structures on stilts tower over the city streets, keeping travelers moving above the crowded downtown thoroughfares. However, unlike New York, which tore down most of its Els (albeit mostly to replace them with subways), Chicago kept many of its rails in the sky, bequeathing one of the city's most enduring landmarks, the now infamous downtown "L" Loop.

Opened in 1897, this was one of the features that helped Chicago stand out in the field of transportation excellence.

A large 1898 foldout map (1) of the Metropolitan West Side El includes this beautifully drawn detail of the Loop (2).

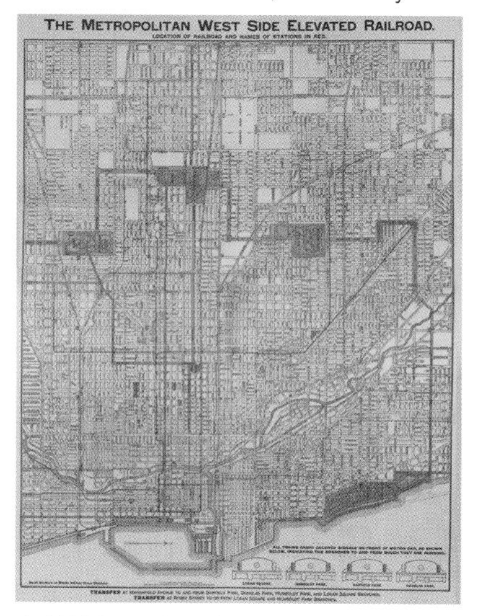

1. 1898: Large foldout map.

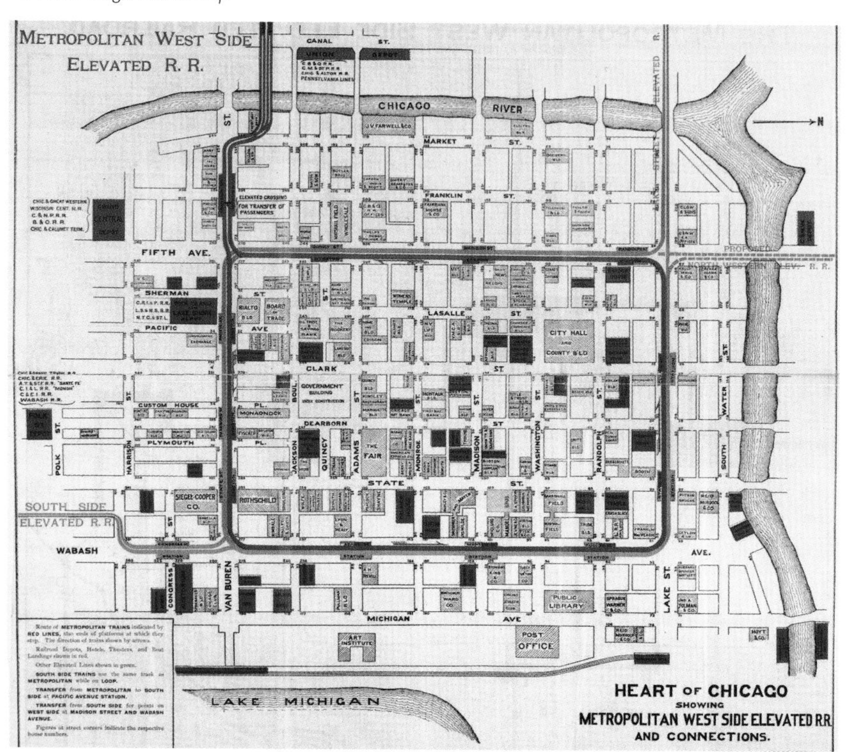

2. 1898: Detail of downtown Loop from the back of the large Metropolitan West Side Elevated foldout map.

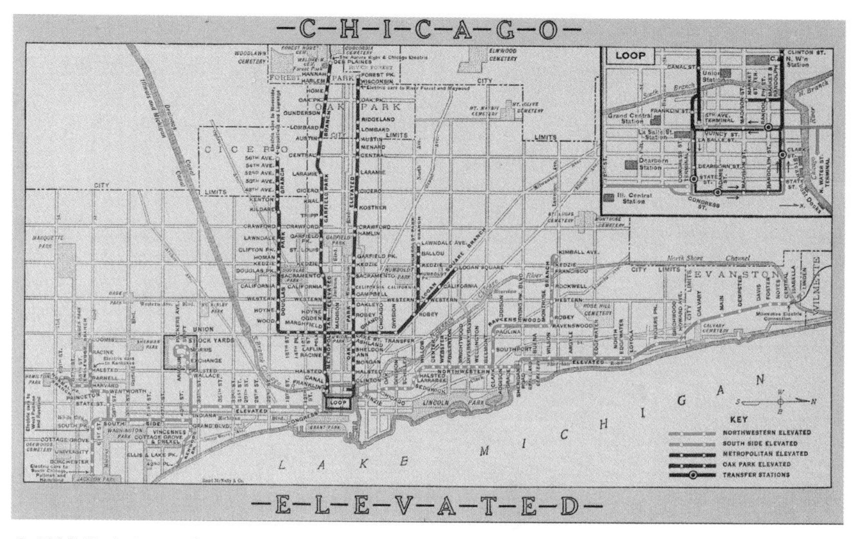

3. 1913: Pocket map often reproduced as a postcard. Much of the current system is already in place.

With the city's population topping a million, the late 1800s saw Chicago at the forefront of innovations in electrification, signaling, and car design. For this Chicago owes much to the financier Charles Tyson Yerkes, who by the late 1890s had skipped town following a few financial irregularities! He landed in Britain, where he became father to London's deep tube network. To this day American influences are all over it, one of the most permanent being the habit of calling "carriages" cars.

For many years Chicago's railroads were printed over street maps and depicted Lake Michigan in the lower half, inaccurately portraying the shoreline as running east–west. Although the maps remained strictly topographical, by showing less detail the beautiful 1913 pocket map (3) is somehow clearer. Note also early use of the open circle with bull's-eye center on the Loop detail of interchange stations.

As if predicting the austere times to come, the 1926 map dropped back to monochrome, but by 1934, despite the Great Depression, heavy stylization work appeared on the central Loop area (4).

Chicago joined the subway league with the opening in 1943 of a five-mile line below State Street and another parallel north–south line known as the Dearborn Street Subway, which finally opened in 1951.

Around the time of the first subway, most maps were skewed to show the lake on the right, thus allowing the coast to run north–south like it really does, as in the large 1946 foldout map (5). This publication was the first to show the subway and the last to be issued by Chicago Rapid Transit before the formation in 1947 of the Chicago Transit Authority (CTA).

Each subway has the unique feature

Contemporary Roosevelt station on the Red Line.

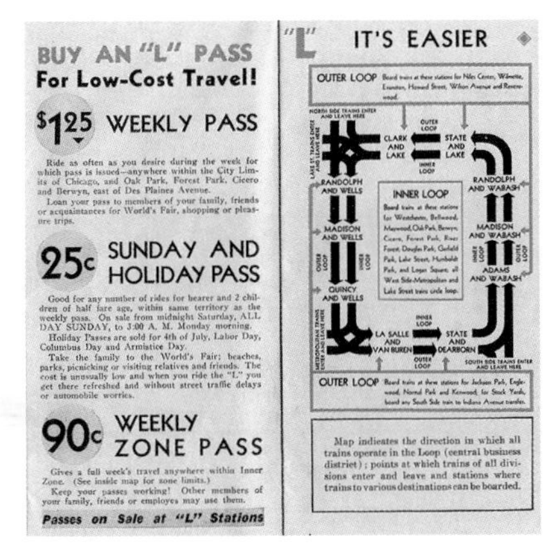

4. 1934: Simplification of the Loop detail.

of a single platform that runs the entire length of the tunnel, from Lake to Jackson on the Red Line and Washington to Jackson on the Blue Line. Passengers can enter at stairwells that don't have stations and can walk the full length, but trains stop only at named stops.

While the automobile reigned supreme across America, even Chicago abandoned some Ls in the 1950s and '60s. However, the city continued to invest in public rail transit; in the wake of major freeway construction the median strip was occasionally used to relocate and extend existing rail lines. This led to the happy state that provided direct mass-transit access to both of the city's main airports. The CTA also rescued a stub of what was once the most famous of interurban rail lines, the North Shore, marketed now as Skokie Swift.

Due to the authority's meticulous attention to detail, and the importance attached to the unified system, all publicity material was in the form of large foldout maps showing the entire metropolitan area. It wasn't until 1965 that the first stand-alone schematic materialized, showing just the urban-rail lines (the Ls, the Loop, and the

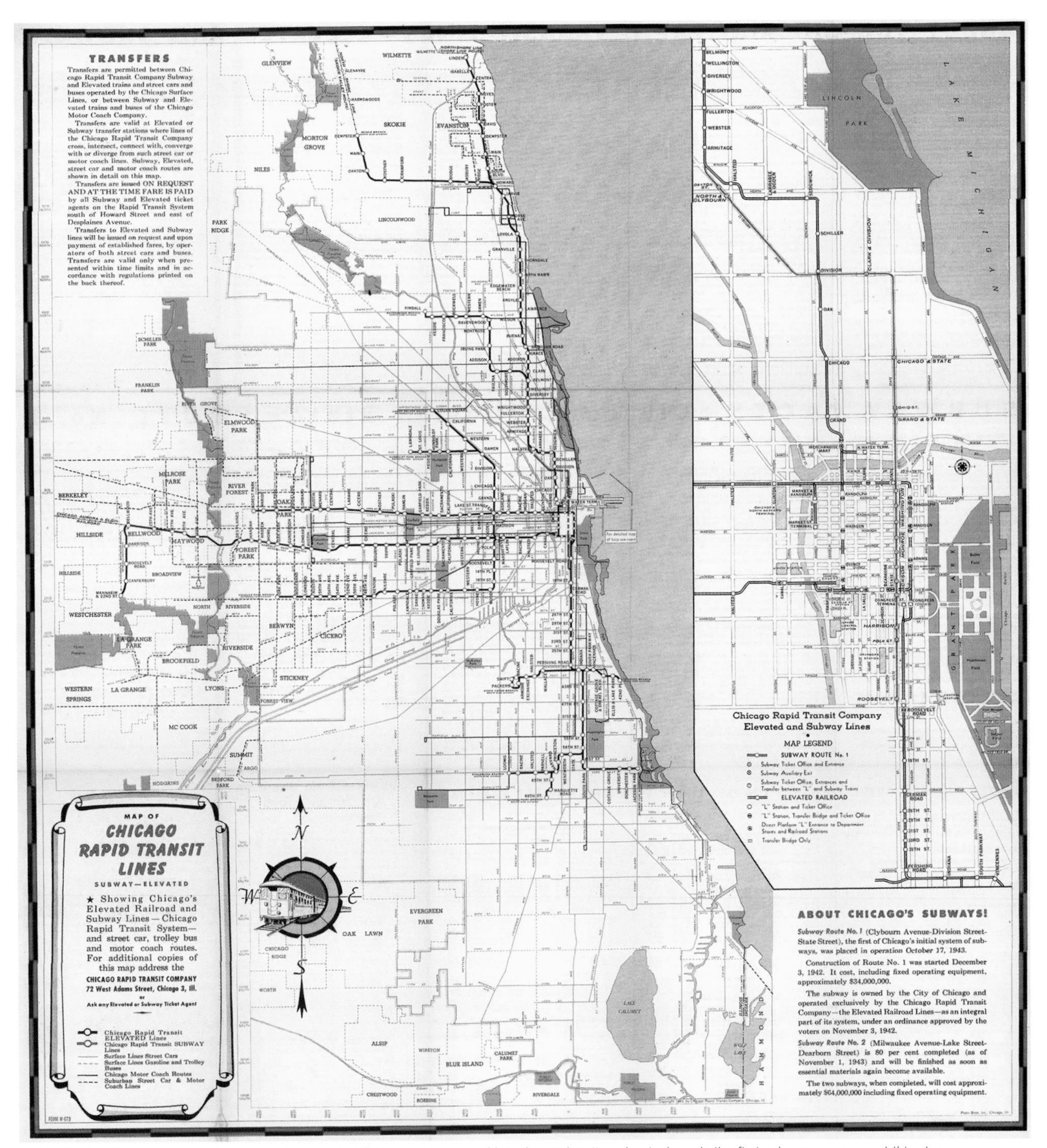

5. 1946: The CRT's last effort, the beautifully executed foldout system map. Note the explanation about why only the first subway was open at this stage.

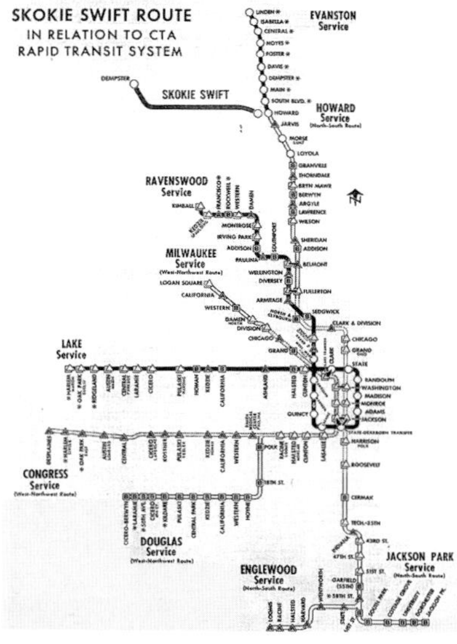

6. 1965: First schematic for the rail network.

subways) with no streets or bus routes (6), appearing all blue by 1970 and colored by 1991 (7).

For most of their busy working lives the lines had names like Congress (now the Blue Line), Dan Ryan (now the Red Line) and Ravenswood (now the Brown Line), coined mostly after their original termini. In 1993, following a major rebranding and station re-signing by the CTA, which coincided with a swap of some Loop and subway services, all became known by their designated colors. The simplified line-color naming scheme was carried through to some station signage and to the diagrams, which began to truly distort the geography to fit in all the routes, using standardized curves.

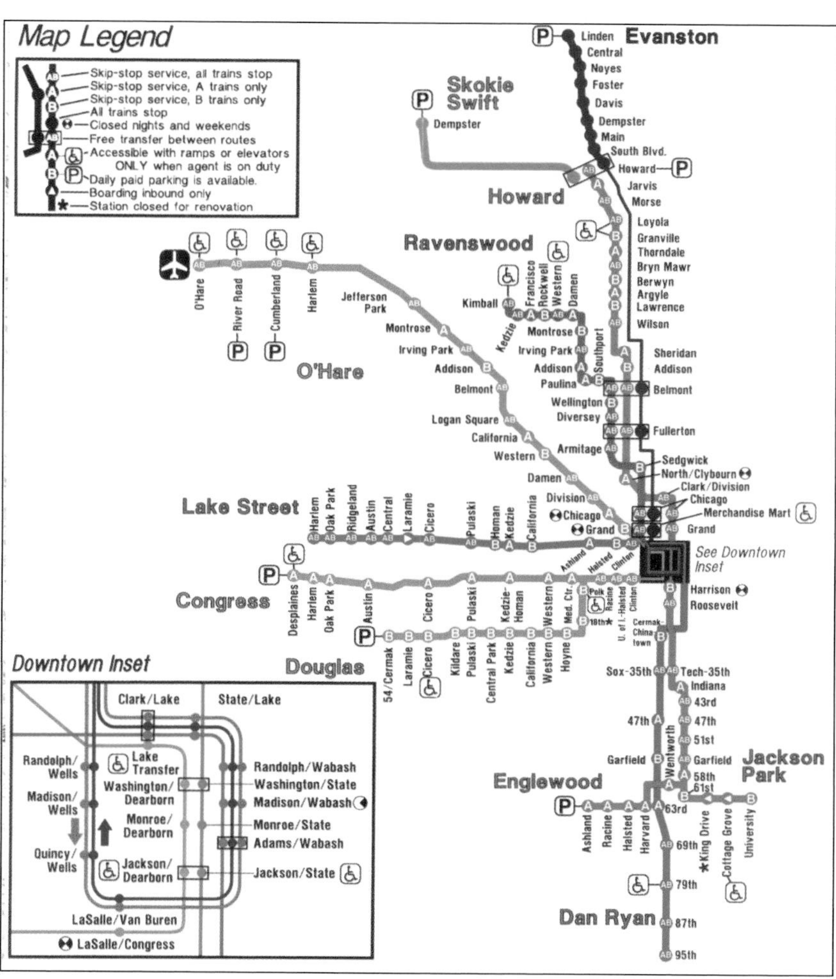

7. 1991: The introduction of color-aided legibility.

Train on the downtown Loop.

The in-car diagram retains the early-twentieth-century geographic twist to fit more easily in the space available inside the trains (8). The version currently in use (9), with its open circles and clean lines, stands on the shoulders of its predecessors and epitomizes the clarity and stylishness of graphic design proudly used across the entire CTA rail system.

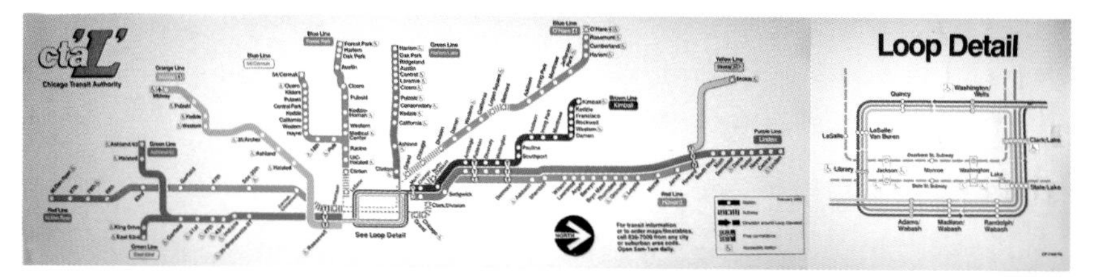

8. 2005: In-car route diagrams are backlit and show north to the right, echoing earlier maps.

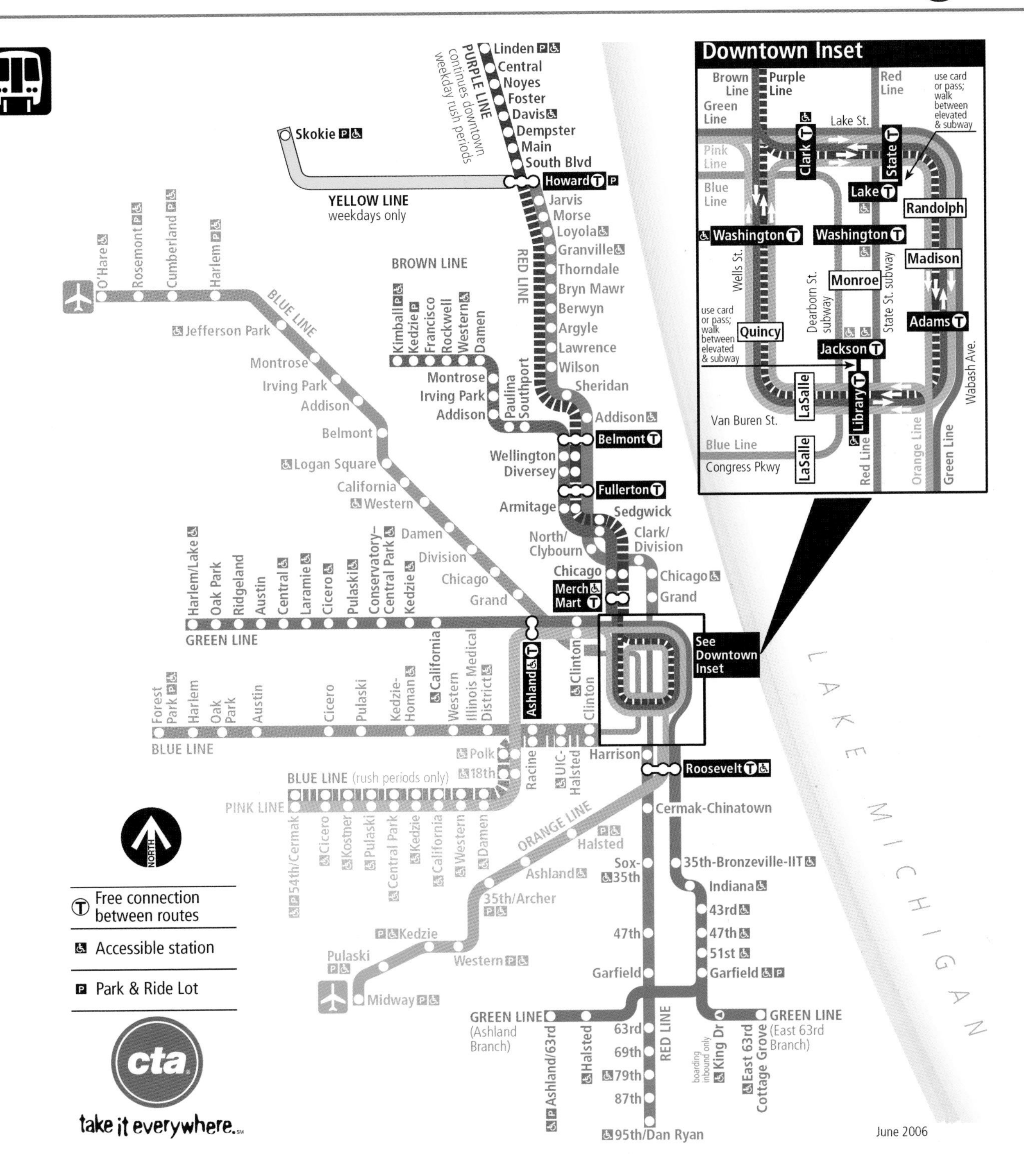
Downtown Inset
Brown Line
Purple Line
Green Line
Red Line
use card or pass; walk between elevated & subway
Lake St.
Clark
State
Pink Line
Blue Line
Lake
Randolph
Washington
Washington
Wells St.
Madison
Dearborn St. subway
Monroe
State St. subway
use card or pass; walk between elevated & subway
Quincy
Adams
Jackson
Wabash Ave.
LaSalle
Library
Van Buren St.
Blue Line
LaSalle
Congress Pkwy
Red Line
Orange Line
Green Line
Linden
Central
Noyes
Foster
Davis
Dempster
Main
South Blvd
PURPLE LINE
continues downtown weekday rush periods
Skokie
YELLOW LINE
weekdays only
Howard
Jarvis
Morse
Loyola
Granville
Thorndale
Bryn Mawr
Berwyn
Argyle
Lawrence
Wilson
Sheridan
RED LINE
BROWN LINE
Kimball
Kedzie
Francisco
Rockwell
Western
Damen
Montrose
Irving Park
Addison
Paulina
Southport
Addison
Belmont
Wellington
Diversey
Fullerton
Armitage
Sedgwick
North/ Clybourn
Clark/ Division
Chicago
Merch Mart
Chicago
Grand
O'Hare
Rosemont
Cumberland
Harlem
BLUE LINE
Jefferson Park
Montrose
Irving Park
Addison
Belmont
Logan Square
California
Western
Damen
Division
Chicago
Grand
Harlem/Lake
Oak Park
Ridgeland
Austin
Central
Laramie
Cicero
Pulaski
Conservatory-Central Park
Kedzie
GREEN LINE
Ashland
Clinton
Clinton
California
Forest Park
Harlem
Oak Park
Austin
Cicero
Pulaski
Kedzie-Homan
Western
Illinois Medical District
BLUE LINE
See Downtown Inset
LAKE MICHIGAN
Polk
18th
Racine
UIC-Halsted
Harrison
Roosevelt
BLUE LINE (rush periods only)
PINK LINE
54th/Cermak
Cicero
Kostner
Pulaski
Central Park
Kedzie
California
Western
Damen
ORANGE LINE
Cermak-Chinatown
Halsted
Ashland
Sox-35th
35th-Bronzeville-IIT
Indiana
35th/Archer
43rd
Kedzie
47th
47th
51st
Pulaski
Western
Garfield
Garfield
Midway
GREEN LINE (Ashland Branch)
Ashland/63rd
Halsted
63rd
69th
79th
87th
95th/Dan Ryan
RED LINE
boarding inbound only
King Dr
East 63rd
Cottage Grove
GREEN LINE (East 63rd Branch)
NORTH
Free connection between routes
Accessible station
Park & Ride Lot
cta
take it everywhere.
June 2006

9.

Maps reproduced by kind permission of and copyright CTA 2007. Antique maps: Graham Garfield, Mike Ashworth Collection, author. Photo: Brian Patton.

London

Urban population: 8 million. Route length: 272.5 miles. Stations: 306. First section open: 1863. Underground: 40 percent.

More than any other transit map, the London Underground diagram forms a mental map of the city for both residents and visitors. There are few other cartographic images more ingrained into the very psyche of a population. Despite being just colored lines on paper, this diagram has become a cultural icon; it *is* London.

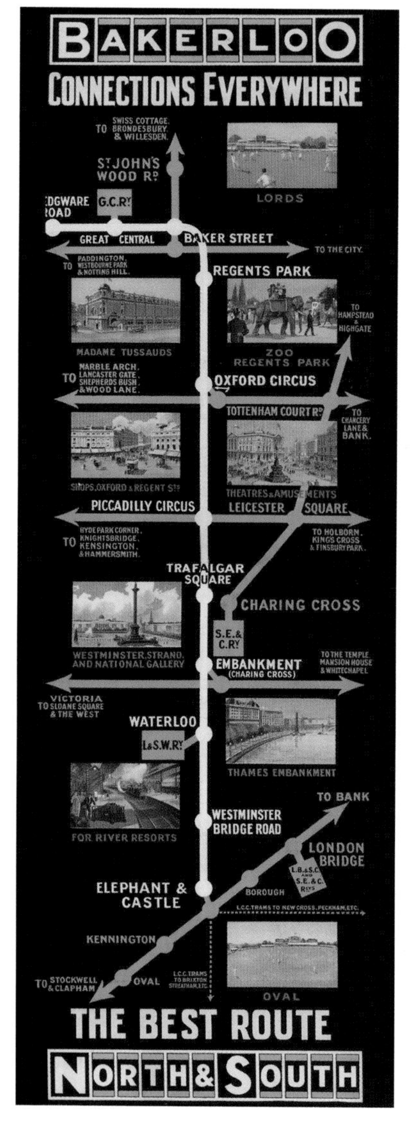

1. 1908: Very stylized feel for central London.

For the first few decades following the 1863 opening of the Metropolitan Railway, tracks were printed over existing street maps, but as the system grew in complexity, many maps became increasingly illegible. Cartographers were forced to experiment, resulting in significant improvements (see pp. 8–9), but the arrival of tube lines directly under London's hectic streets called for a radical rethink.

By the early 1900s, a fragmented system was unified as Underground Electric Railways of London, using the shorthand "UNDERGROUND" as a trademark from 1908. A portentous poster of that year (1), extolling the virtues of the newly opened Bakerloo Tube, dispensed with true geography in favor of simplifying the routes into straight lines intersecting each other at right angles. Also in 1908, an elongated route map inside District Line trains (2) started a worldwide trend of in-car diagrams, seen to this day. From the same year, the Map of the Tubes of London, issued by a department store, allows the rail routes to stand out by removing the streets (3).

A series of small pocket maps like the 1911 version (4) cemented the use of different colors for each line and was freely available from ticket offices. It also utilized a feature that would become one of the archetypal elements of the classic urban-rail diagram: the open circle with white bull's-eye center symbol at interchange stations, and the white-line connectors—devices now found the world over.

In 1916 the Johnston typeface was used for the first time on an in-car route diagram for the Piccadilly Line (5), but it was the popularity of portable pocket maps that encouraged the Underground group to start using commercial artists (today's

2. 1908: The earliest known use of a diagram inside a mass-transit train, from London's District Line.

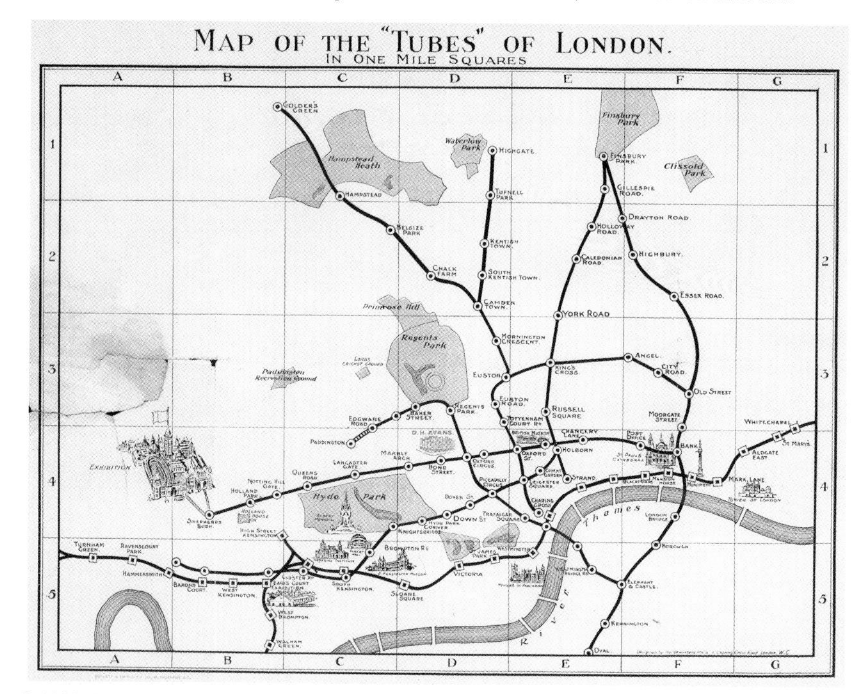

3. 1908: Though not an official UERL publication, this map has some diagrammatic elements.

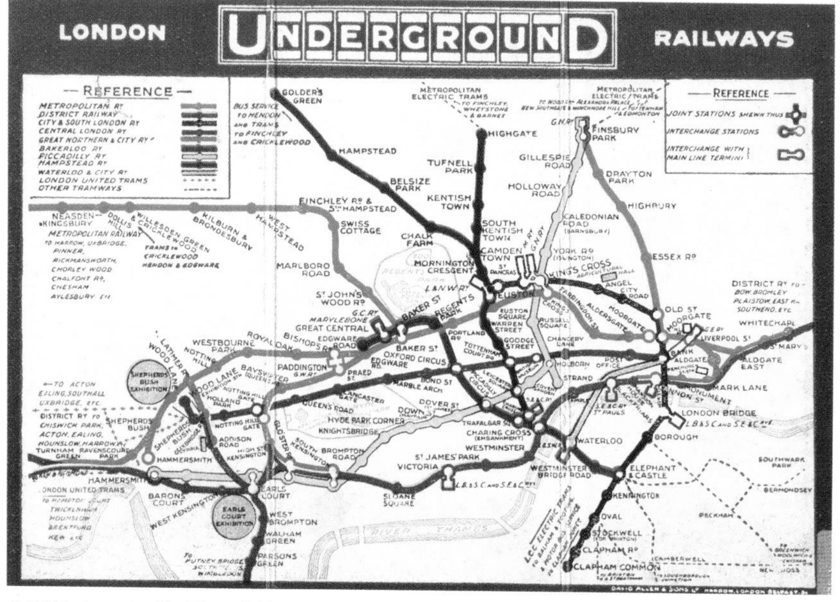

4. 1911: An early offical-issue folding pocket map on card, with colors designating different lines.

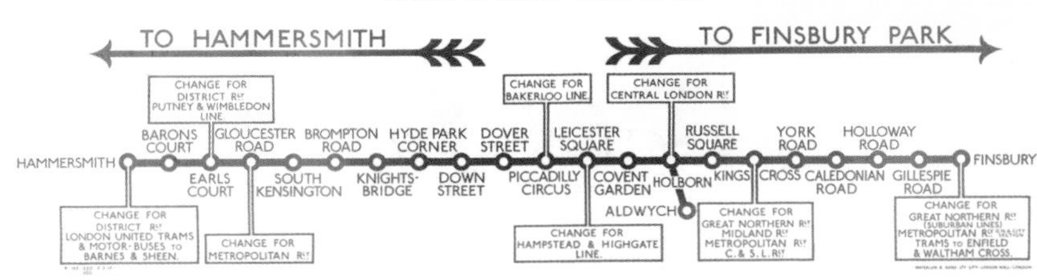

5. 1916: A Piccadilly Line in-car diagram was the first use of Edward Johnston's sans-serif face on a map.

graphic designers). One of the first was by calligrapher MacDonald Gill, whose cursive station names and elaborate borders of 1920 (6) held

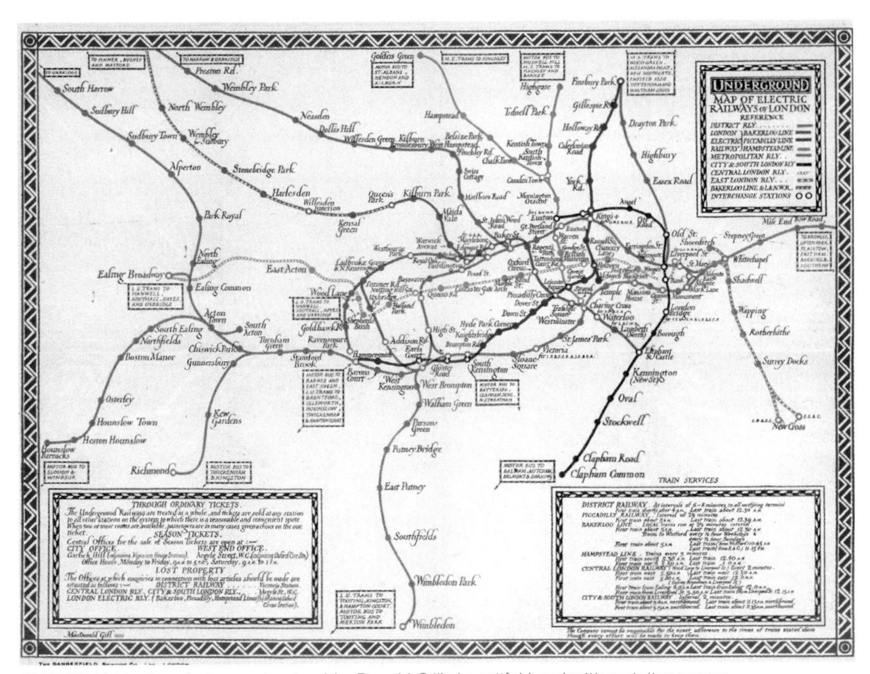

6. 1920: Foldout pocket map showing MacDonald Gill's beautiful handwritten station names.

sway until artist Fred Stingemore was commissioned to improve legibility.

Aping Gill, he left out all topography and used artistic license to squeeze in some station names, and even bent some of the lines (Bakerloo and Hampstead particularly), but the density of stops in the central area and his retention of topographic reference to tunnels meandering beneath the surface hampered the map's effectiveness (7). Compare these to the sea change wrought by the work of Harry Beck, a twenty-nine-year-old draftsman whose first sketch in 1931 (8) evolved into the masterpiece that was diffidently issued by skeptical managers in 1933 (9).

Recent research (notably by Andrew Dow and Max Roberts, see bibliography) and many illustrations shown here (and on pp. 8, 9, 12, 32 and 36) demonstrate diagrammatic elements evolving not just in London but also Berlin, Paris, and the United

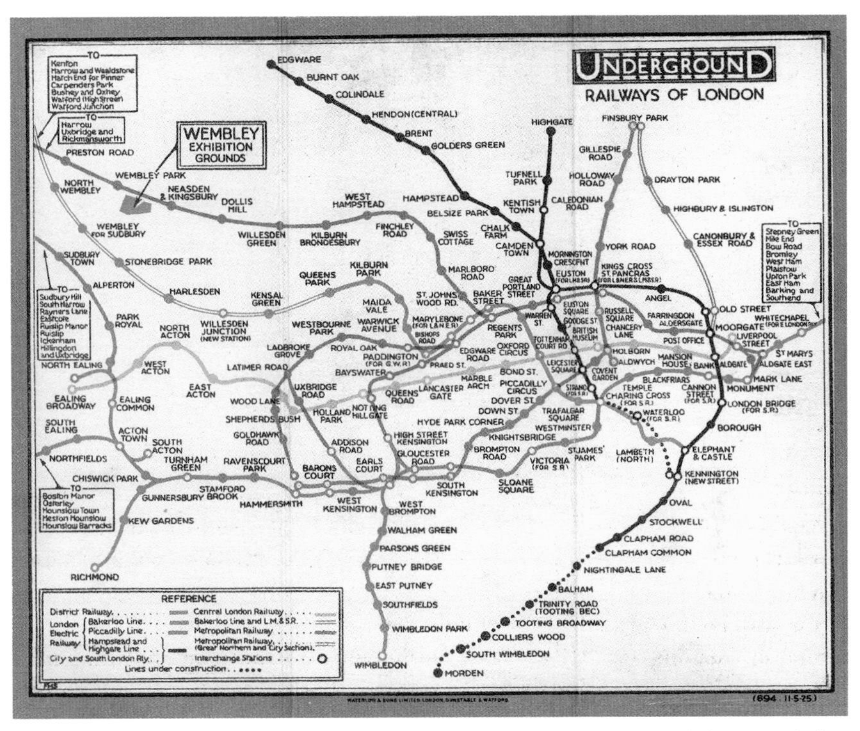

7. 1925: Stingemore's pocket map. Even without topography some central area names feel too squashed up.

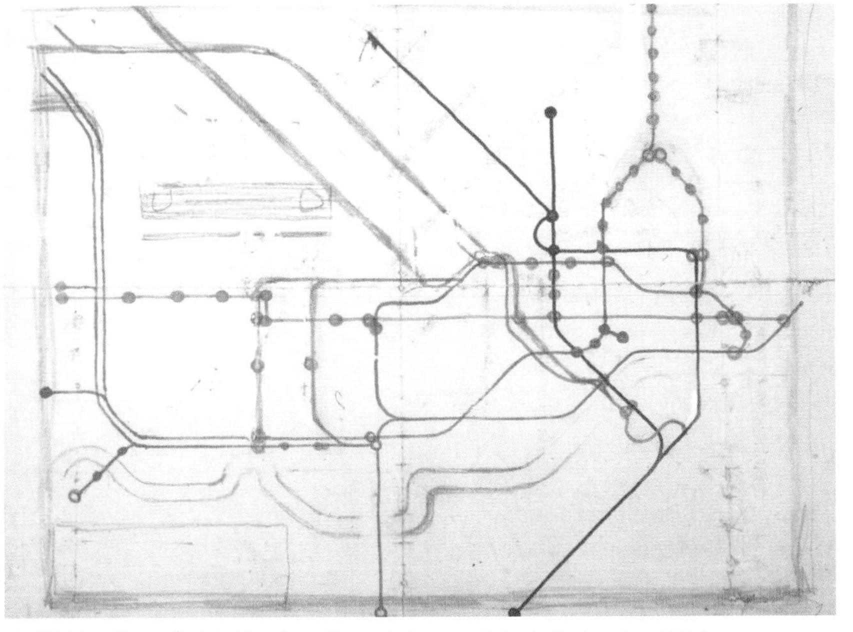

8. 1931: Beck's very first sketch enlarged the central area and standardized angles at 45 degrees.

London

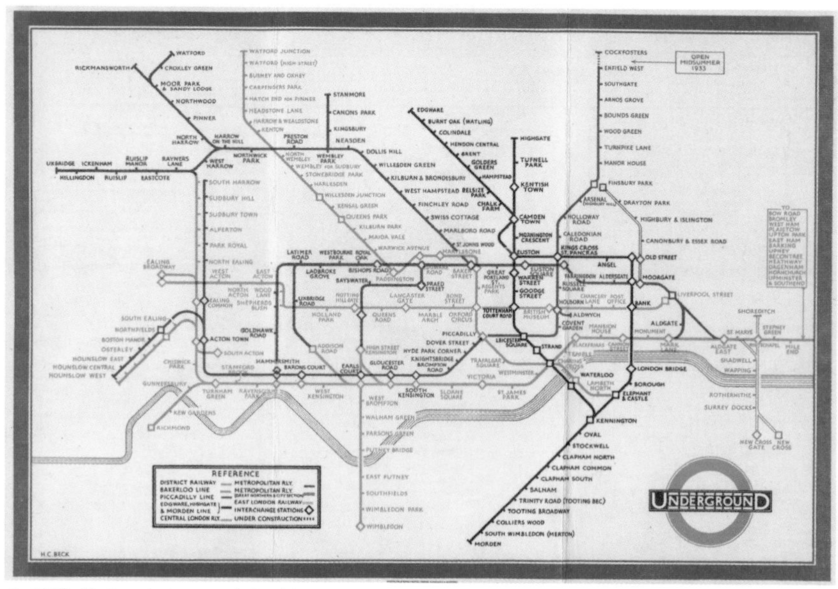

9. 1933: First pocket map with Beck's design, a concept now synonymous with London itself.

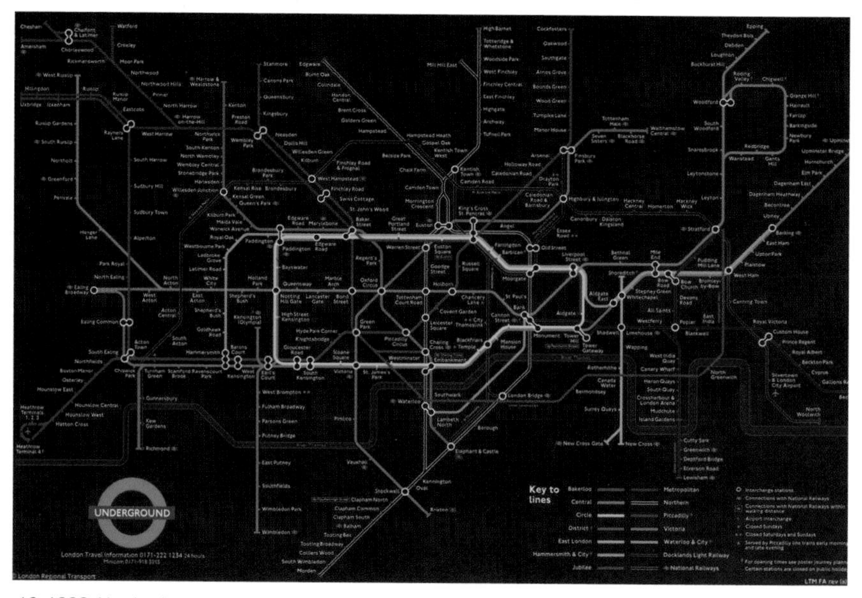

10. 1998: Used only on souvenirs, this black-background map was never intended for general use.

States. However, Beck's multicolor, single-angle, full system diagram, with an artificially enlarged central area and based on station relationships rather than geographic scale, gained instant public approval.

In describing the affiliation between lines, interchanges, and stations, it gives all the information one needs when traveling underground—which station follows which and where lines meet. By using an octagonal grid, it allows lines to intersect only at right angles.

Beck explained that the "physical distance from one station to the next is immaterial" and thus made space for the typesetting of the station names by compacting the suburbs and expanding the central area. All topographical detail is suppressed, but London's sense as a great city on either side of its major waterway is preserved by the inclusion of the River Thames.

The use of color for line identity and the incorporation of the exquisite Edward Johnston sans-serif font, first seen in the system in 1916, made this arguably the finest urban-rail diagram in the world.

It has been copied, parodied (even by Beck himself), and turned into a work of art (Simon Patterson's *The Great Bear*, 1992, now in the Tate Gallery). It has inspired cartographers and academics around the globe; often cited as the epitome of information graphics, it remains an undisputed design classic.

What Beck did was, with hindsight, surprisingly straightforward. Like a seamstress tightening a slack thread, he pulled his predecessor's wobbly Central Line into a simple horizontal. He smoothed the wiggly District branch to Wimbledon into a single vertical column and neatened the meandering Bakerloo, Metropolitan, and Hampstead railways into fixed 45-degree diagonals. Now a taut and perfectly woven geometric spider's web, the complete diagram resembled precision wiring rather than a plate of spaghetti!

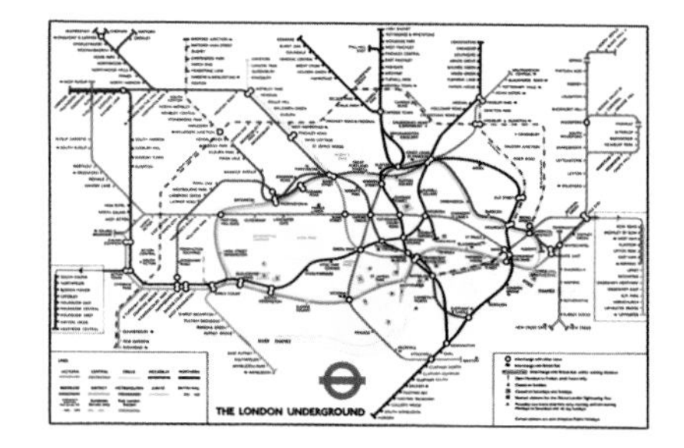

11. 1983: © Paul Mijksenaar. A clever hybrid.

Though cognizant of electrical diagrams, Beck later claimed he was more inspired by those of the sewage system. He sculpted each necessary curve into identical urban trajectories and used smart rudimentary ticks for stations rather than the more clumsy blobs on Stingemore's version (or his own presentation visual, p. 9).

He introduced gracefulness and modernity where before there had been unsatisfactory and cramped irregularity. This formerly intertwined rabbit warren became a logical system. Its inspirational stroke of genius was at once austere in an almost mechanical manner, yet balanced, clean, neat, and unambiguous.

The diagram has been refined both by its inventor and subsequent designers, including Paul E. Garbutt and Alan Foale of LS London, who both contributed to this book.

To illustrate its inherent flexibility, look from Beck's 1933 diagram (9) to the 1998 black version (10) and then at the 2007 London system map (13). The modern designs accommodate over 110 more stations and six more lines. Even the playful frontispiece of this book (see p. 1) is an adaptation of Beck's concept.

There have been attempts to move on from the 1933 vision, such as Paul Mijksenaar's valiant efforts using a more geographically accurate central

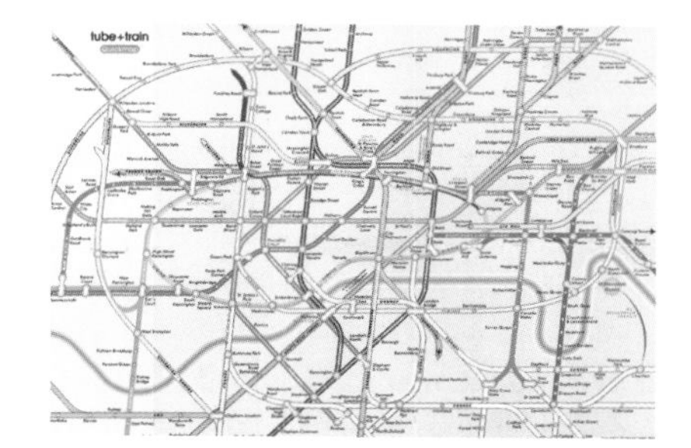

12. 2000: © QuickMap. A leaning concept.

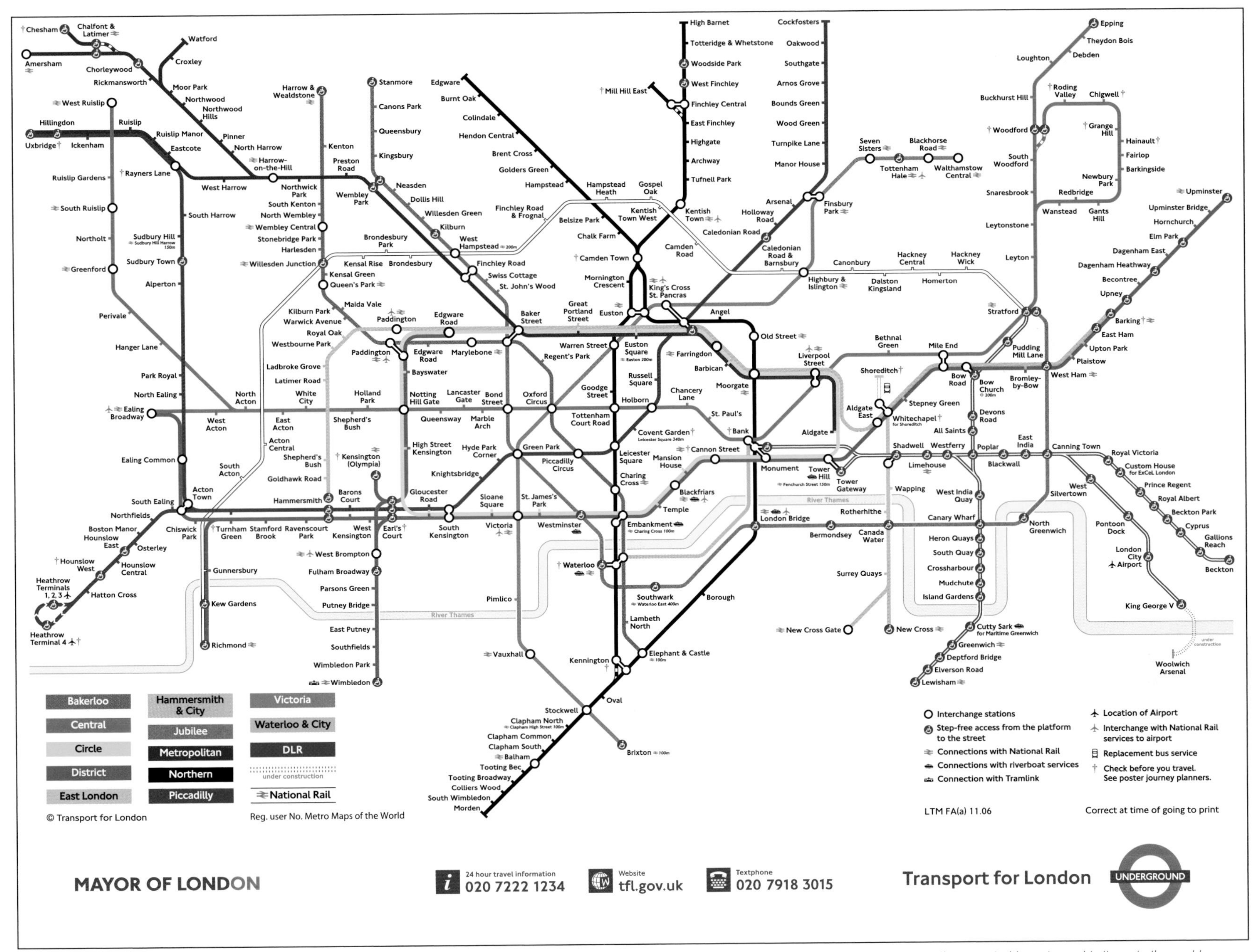

13. 2007: London's current official diagram. Wall posters of it refer back to the key role played by Harry Beck in developing what has become one of the most instantly recognizable cartographic items in the world.

area and diagrammatic rules in the outskirts (11) and the leaning diagram of all London rail networks produced by QuickMap (12)—but nothing has yet surpassed the original.

The history of London Underground is more than adequately documented elsewhere (see bibliography), but beyond pioneering cartographic, engineering, and technological solutions to mass rapid transit, London can also cite an astonishing architectural and artistic heritage. Advertising, publicity, signage, and structures are bound by the use of the roundel logo and the Johnston typeface. This powerful partnership was arguably the world's first widespread corporate identity. The person who commissioned these elements, former publicity manager Frank Pick, deserves as much credit as Beck because the logo and the typeface, like the map, have themselves become shorthand for London.

Madrid

Urban population: 6 million. Route length: 175.8 miles. Stations: 242. First section open: 1919. Underground: 92 percent.

The Spanish capital was first promised an underground railway as far back as 1892, but it was not until 1916 that a concession was approved to start work on Line 1 of the Madrid Metro to run between Sol and Cuatro Caminos. Although little printed material survives from the opening in 1919 by Compañía Metropolitano Alfonso XIII, the system, city, and regional governments now produce a plethora of excellent publicity, including many high-quality maps and diagrams.

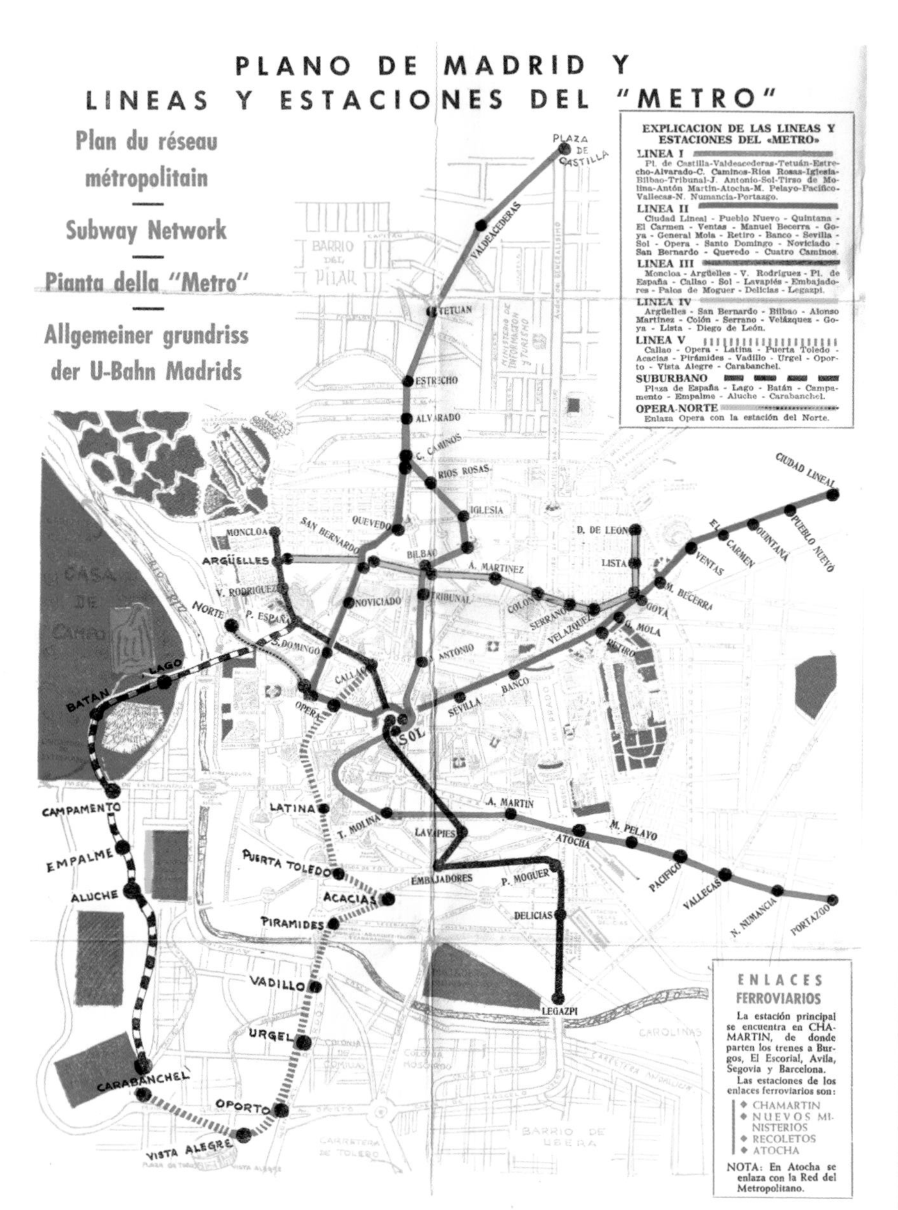

1. 1952: Despite a geographic feel, this early multilingual tourist guide contains much distortion.

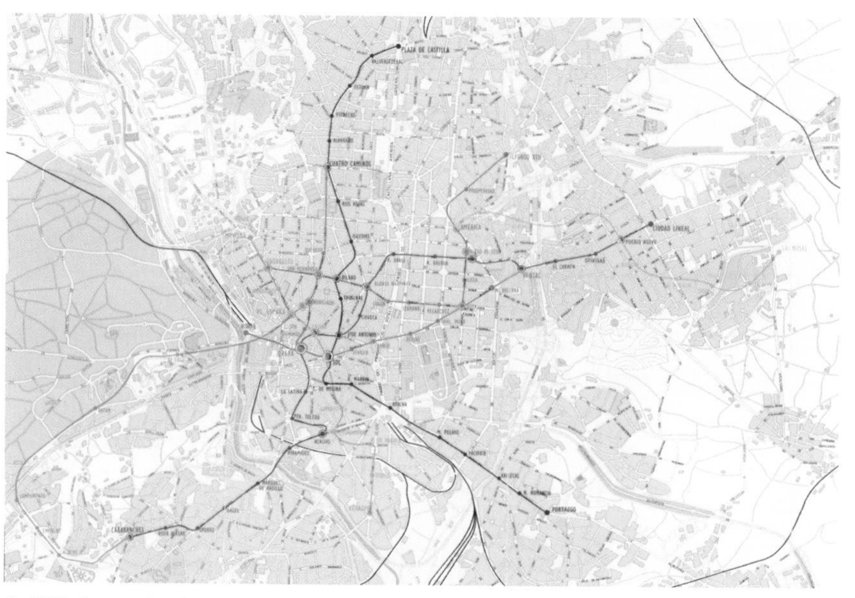

2. 1976: Geographic plan with split-color interchange symbols.

3. 1981: The first classic diagram was released as a pocket map on a folded card.

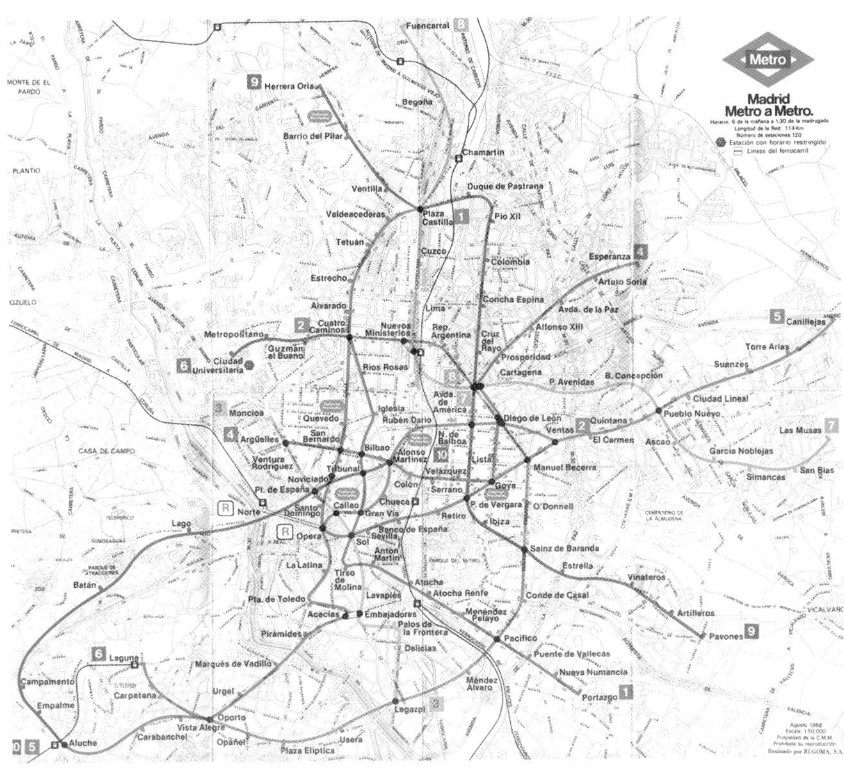

4. 1988: Detailed topographic Metro de Madrid foldout pocket map.

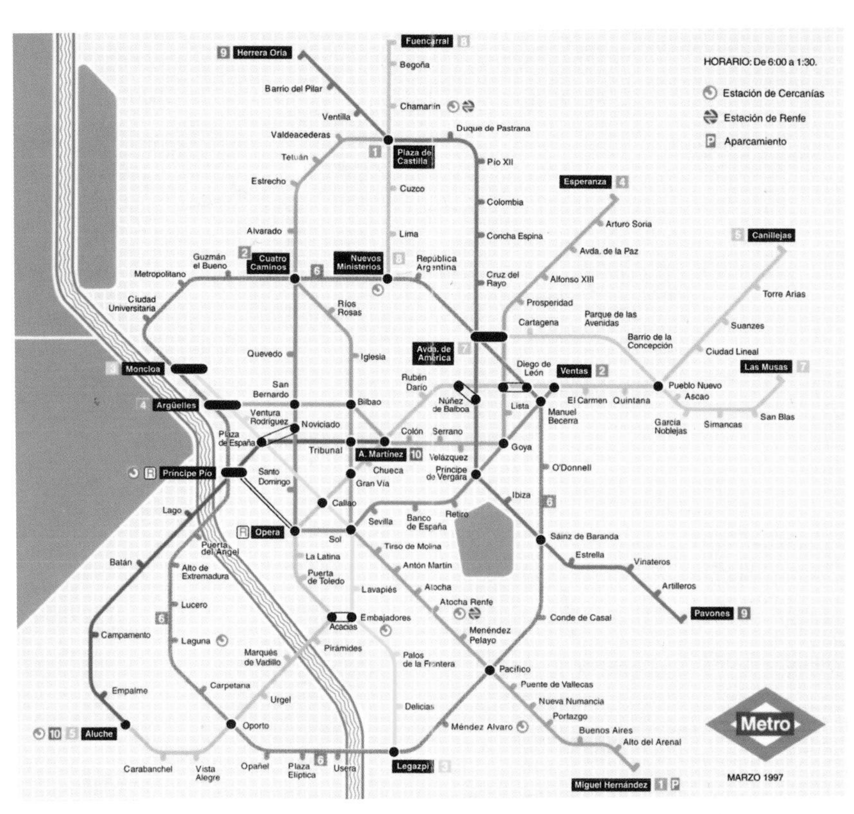

5. 1997: This pocket map on a folded card demonstrates how quickly the system was growing.

Since early on, Metro lines were shown in relation to the streets above—for example, this colorful snapshot of the Metro system in 1952 (1) taken from the Collado city tourist guide. A rare 1976 version (2) appears to be printed over with what looks like virtually every street in the city. This practice was the norm for many years, and Barcelona Metro maps showing selected streets are still produced today. This version from 1988 (4) is typical of that geographic heritage.

A proper diagram, complete with 45-degree angles and minus all topography, appeared in 1981 (3) after some far-reaching extensions. Some surface features like the major parks and river Manzanares were reintroduced in stylized form by the 1997 pocket map (5) but did not last long. There are a couple of distinctly Gallic influences visible here—for example, terminal stations highlighted in white on black.

Since the establishment of the Consorcio Regional de Transportes de Madrid (CTM) in 1993, the Metro has expanded massively in several phases. The 1995–99 extensions (6) were well publicized, and diagrammatic emphasis was given to the completion of the full-circle line, here seen as a central focus on the diagram in a style similar to the Moscow maps. However, with the opening in 2003 of the next massive project, the whole diagram had to change proportions to portrait format to accomodate the new, self-contained circle line 12, called MetroSur (7). This monumental construction effort was a $1.6-billion, 25.5-mile, fully underground loop, whose twenty-eight stations link a belt of five new towns to the south of Madrid, and which has interchanges to both the Metro and mainline suburban lines, the Cercanías.

The CTM's latest schematic (8) retains these proportions while accomodating yet more extensions—notably to Line 10 in the north, 9b in the east, and three light Metro lines.

The current system diagram (9), much like Barcelona's, retains the feel of a classic urban-rail diagram, despite the angles not being at 45 degrees but at a more gentle 35 degrees instead. It uses strong, bright colors for the lines and, one angle for diagonals, expands the center, equally spaces the stations, and features the archetypal black circle with white-eye center symbol at interchange stations, plus white-line connectors. Tints on the current diagram are used for the stations on the southeastern Line 9 extension, which are situated outside the main fare scheme and given a boxlike

6. 1999: The four-year expansion plan complete with emphasis on the completed circular Line 6.

structure—shades of the old Paris Métro's treatment of the Sceaux line or its outer RER services.

Madrid numbers, rather than names, its lines. It also utilizes what has become Spanish shorthand for a Metro system, the red diamond with a blue bar, which can be traced back as far as the early 1930s. Such a powerful icon, which some might argue owes more than a little to the London Metropolitan Railway's 1920s diamond signage, can also be seen elsewhere in Spain, such as on the Barcelona Metro and its suburban Ferrocarrils de la Generalitat system, although Bilbao, Valencia, and

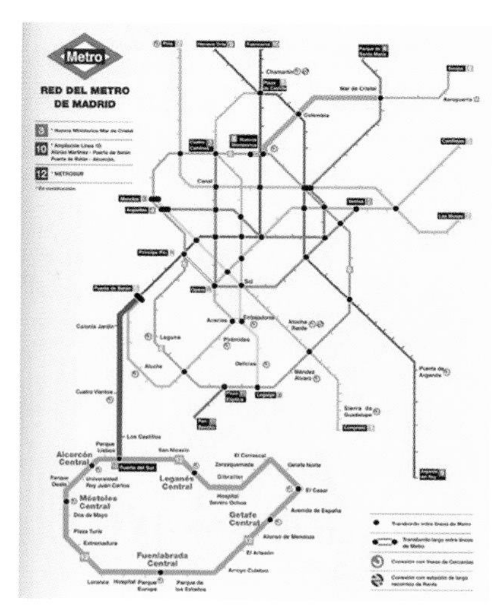

7. 2000: MetroSur would force a new shape.

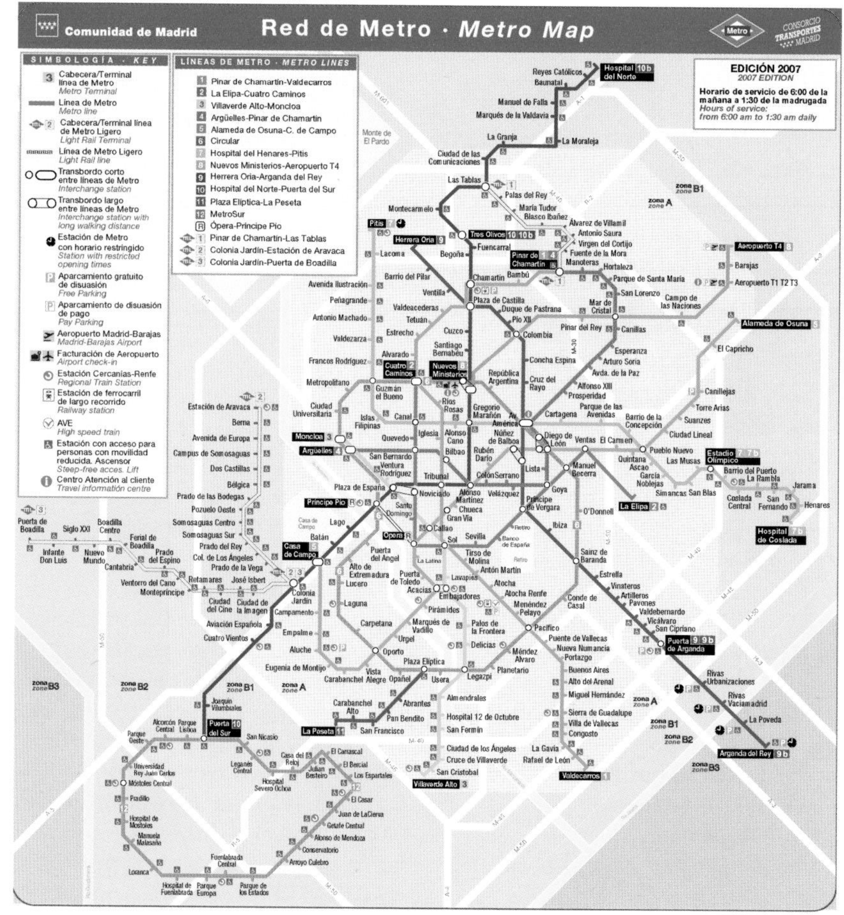

8. 2007: The CTM system map sticks with 45-degree angles but includes some major highways.

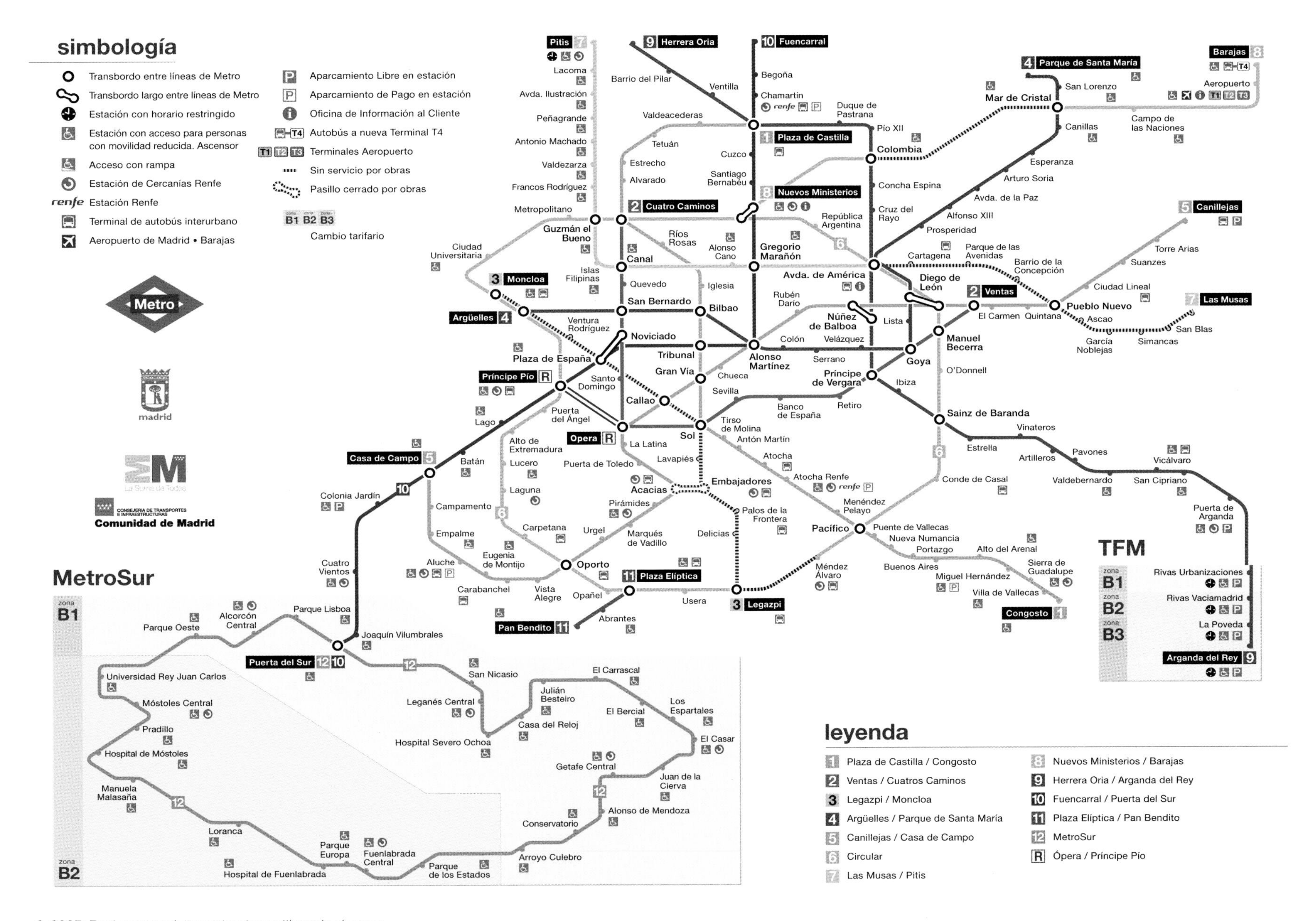

9. 2007: Easily accomodating extensions with angle changes.

(forthcoming) Sevilla have opted for more chic and contemporary logos.

The Spanish eye for design, in both architecture and corporate identity, has become more evident as the Madrid Metro has seen astonishing growth, much of it in the last decade and with more to come. Work is already well under way on MetroNorte (10b), making it the fastest-growing system in Europe, and Madrid now has Europe's second-longest route mile-length.

The quality of stations, trains, and graphic design is exemplary—this is what urban mass rapid transit should be about in the twenty-first century.

Moscow

Urban population: 10.5 million. Route length: 173.2 miles. Stations: 172. First section open: 1935. Underground: 96 percent.

The Moscow Metro has some of the most beautiful and ornate structures ever created belowground. It is by far the world's busiest Metro system, carrying over 7 million people a day through its cavernous cathedral-like central stations of the 1930s to its modern and more utilitarian suburban extensions of later decades. Apart from the stations, and more relevant to this book, Moscow also has the most diverse collection of map designs of any system. Barely the only device that has remained constant in the frequent alterations of graphic direction has been the simple *M* of the Metro logo, seen on everything from tickets to station entrances.

Many of the cartographic ideas utilized have displayed quite bold and innovative concepts—although on closer inspection some of these may well have been influenced by earlier designs from further afield. In spite of the ostentatious decor lavished on many stations, numerous examples of Moscow's braver cartographic efforts were hampered by years of restrictions on paper, printing, and use of color brought about by rationing and war. The first examples were mostly monochrome, and the earliest known map showed the new Metro with major streets in 1935 (1), although even here we see use of the black circle with white-eye center symbol.

An in-car diagram shows travel time between stations (2). The first stripping out of all topography is seen in a rare and flimsy American sketch (3) believed to stem from a tourist publication of 1935 celebrating the system's opening. The second line opened in 1938. A highly topographic map from this year has a somewhat 3-D feel, thanks to the drawings of each station entrance (4). A monochrome diagram of 1938 (5), from a newspaper, was the first to show part of the proposed route of the new circle line.

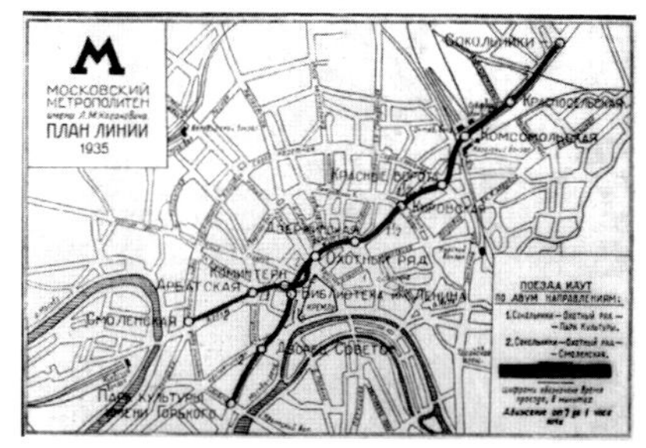

1. 1935: The backbone of the system, Line 1.

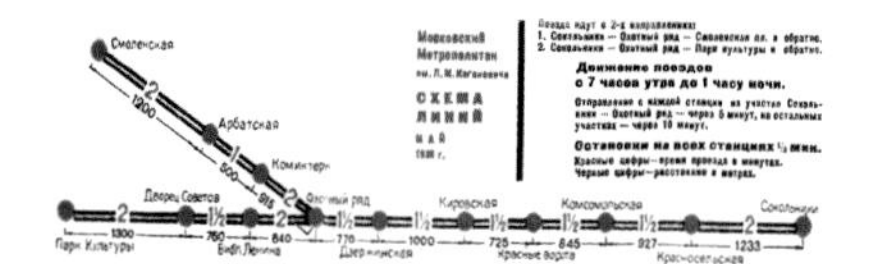

2. 1935: One of the first in-car diagrams.

СХЕМА ЛИНИЙ МОСКОВСКОГО МЕТРОПОЛИТЕНА ИМЕНИ Л. М. КАГАНОВИЧА 1-й и 2-й ОЧЕРЕДЕЙ

4. 1938: Sketches of each elaborate station building appear alongside the route.

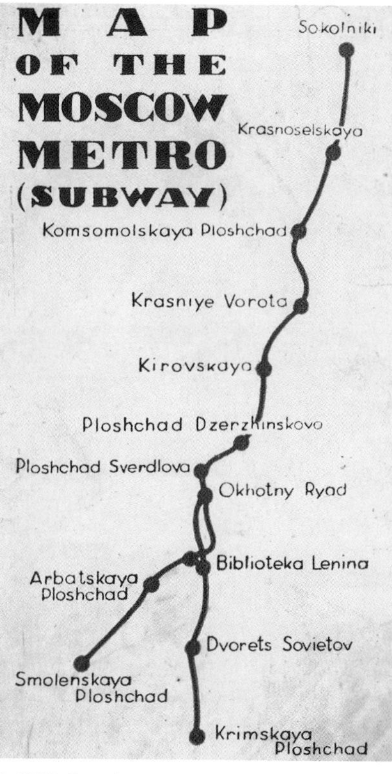

3. 1935: One of earliest known English maps .

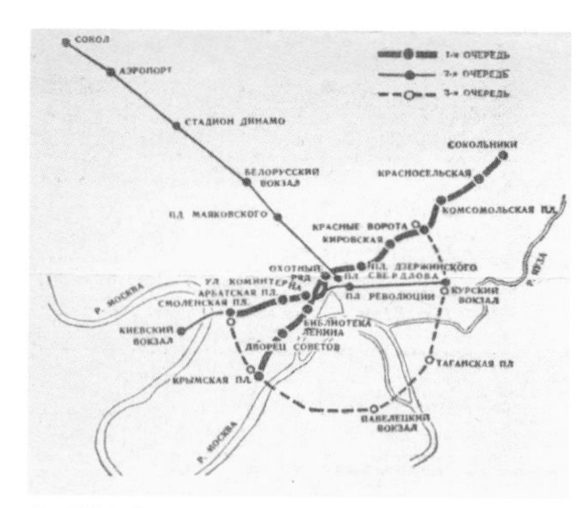

5. 1938: Showing the first half of the circle line.

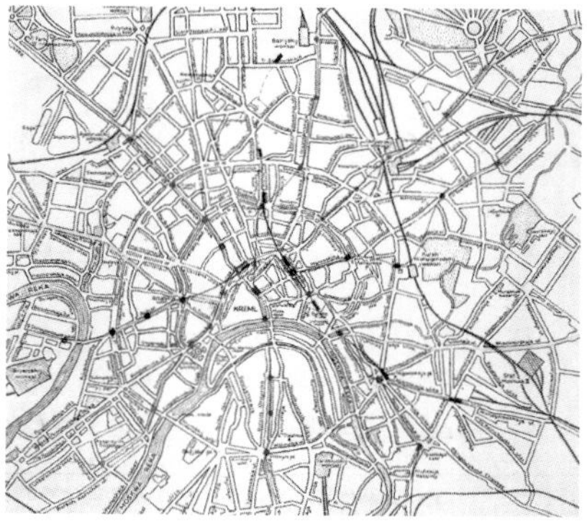

6. 1939: Stingemore's version of the entire network.

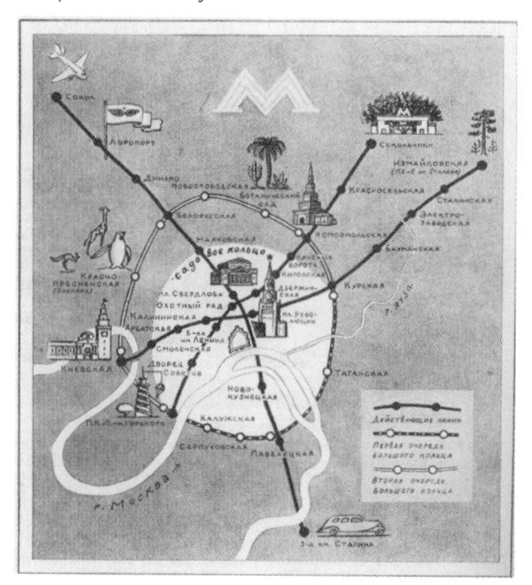

7. 1947: Palms, kangaroos, penguins and cacti!

8. 1947: Clearly aimed at day-trippers!

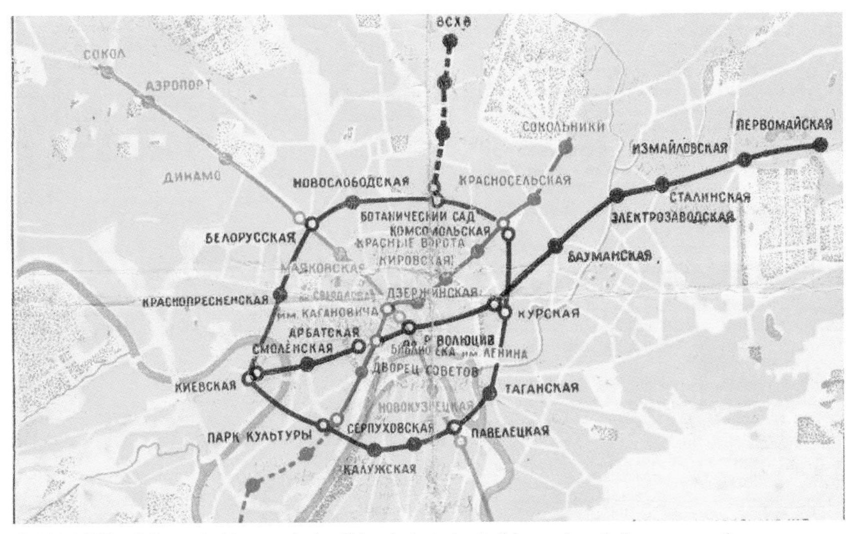

9. Mid-1950s: Color-coded lines and simplifying trajectories in this small pocket map on card.

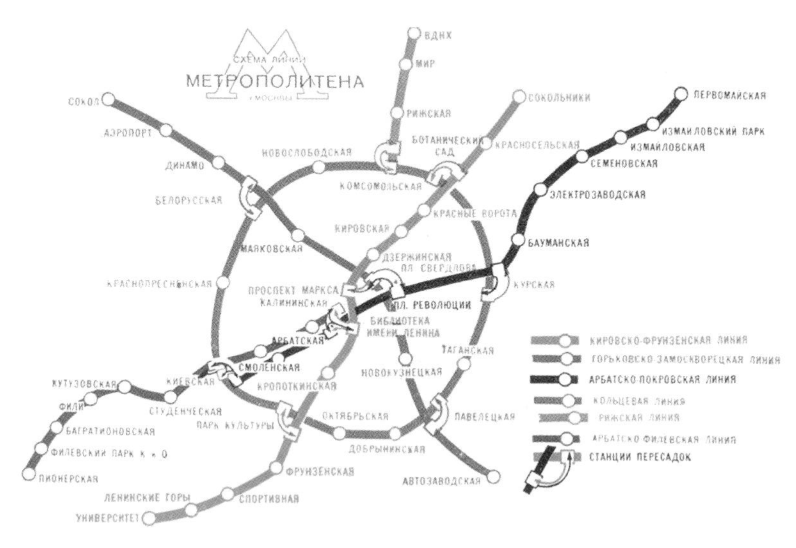

11. 1962: One of many different methods used to depict interchanges—here with curved arrows.

10. 1958: The buildings depicted are not monuments or landmarks but the grand station entrances.

The construction of the system, built partly by "volunteers" under Communist conditions and scarcity of materials, was helped by the involvement of London Underground, which drew up plans for stations and rolling stock.

This led to grand plans of an initial network being drawn by F. H. Stingemore, one of the pre-1933 designers of the London diagram (6).

By 1947 construction on the first section of the Garden Line belt around the boulevards was under way and shown in full on a 1947 monochrome map complete with exquisite sketches of places to visit (7). A color poster from the same year introduces yet more abstraction—few

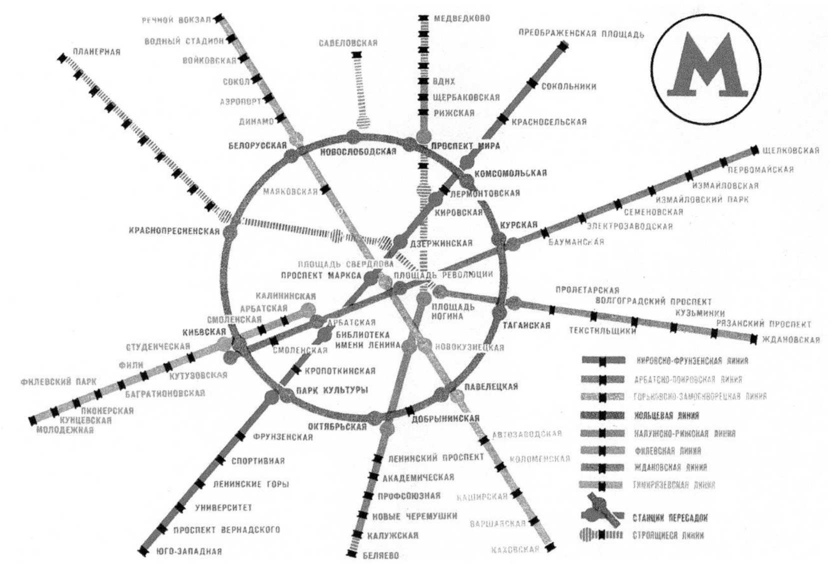

12. 1970: This sudden leap into extreme geometry was Moscow's first true diagram.

Moscow

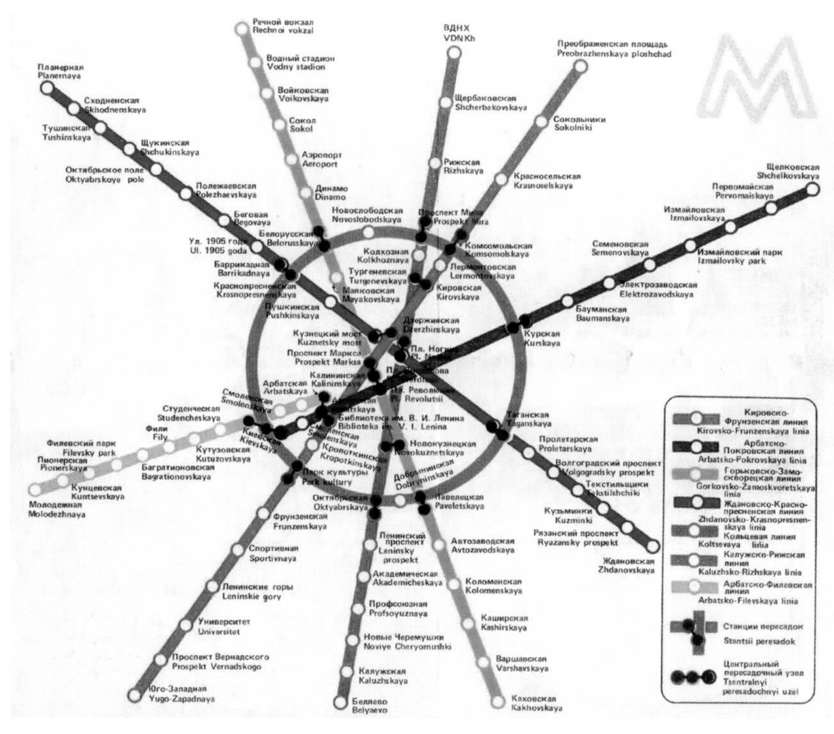

13. 1977: Like knitting needles spearing a ball of wool, this could be the most abstract diagram ever!

stations are named (8). Both have some distortion to fit all the lines in, as well as evenly spaced stations in the center, and one has the return of the black circle with white bull's-eye symbol. The mid-1950s version (9) was the first full-color pocket map.

Justifiably, Moscow is the only system to proudly display its elegant station entrances on maps, and an example from 1958 exemplifies the trait (10).

A typical Moscow interchange has several names, but during the early 1960s a new icon for them emerged (11): an open-ended cube with one arm stretching into its neighbor and terminating in a minute arrow. Although innovative, this proved impractical and was later dropped. Major changes that appeared on the 1970 diagram (12) were to set the tone to the present day. First, a geometric circle focused the eye perfectly; second, all the lines were completely straightened out. (Lines 1–4 had no bends at all!) This was the biggest improvement the Moscow map ever underwent, and most diagrams since then have been based on its simplicity.

A 1977 version is probably one of the most extreme departures into pure abstraction of any urban-rail map ever produced (13). It consists of only straight lines and a circle. That's it. No bends, no topography, *no use,* some might even say!

In 1979 the designers had another minirevolution. Two new features were introduced: the "oval-ization" of the circle and the innovative "beading" of outer stations, in which the line itself is removed but the station locations are retained as dots (14). The 1983 pocket map (15) turned the oval on its side and also segmented the interchange circles (an idea first used in 1980).

It was not until 1988 that the standard 45-degree angle was tried (16), a feature that has remained till the present day. This central area plan from 1991 even goes as far as showing corridors connecting interchanges (17)!

With the introduction in 1992 of black-rimmed open circles with white-line connectors came a look toward a more "westernized" feel for the Moscow Metro (18). The year 1998 saw another idea: a beautiful 3-D tubular design (19) on which the beading reappeared and continued until 2003.

The current diagram (20), drawn by Artemy Lebdev's studio, represents the culmination of seventy years of innovation in map design on one of the world's most impressive and effective mass transit systems.

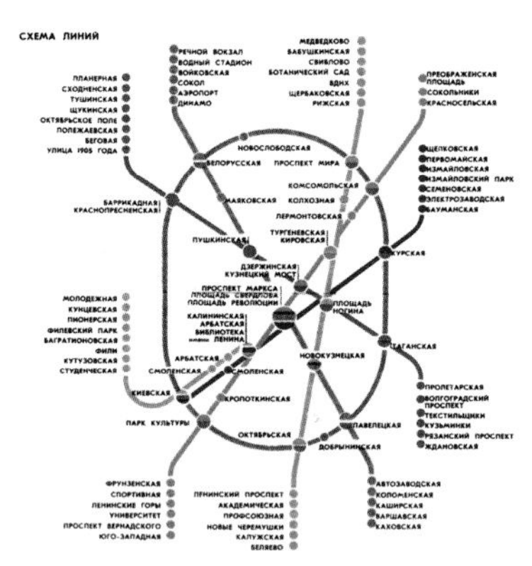

14. 1979: First "beading" of outer stations.

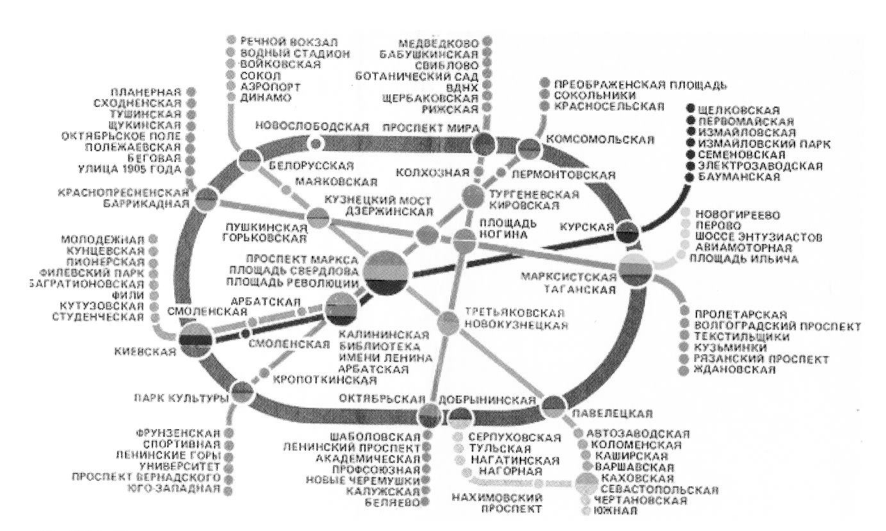

15. 1983: The segmented interchange spheres and landscape format did not last long.

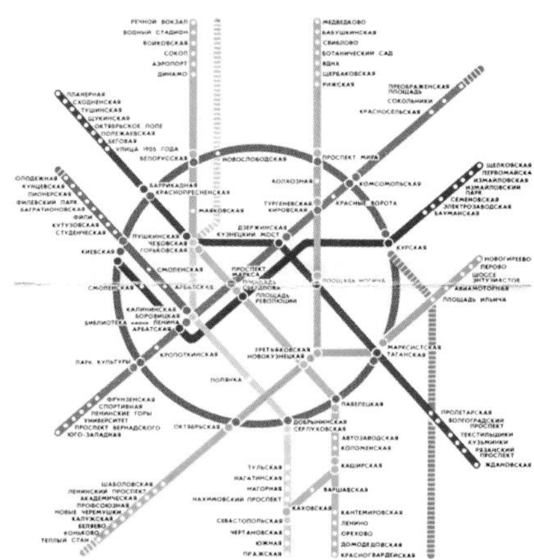

16. 1988: The arrival of 45-degree angles.

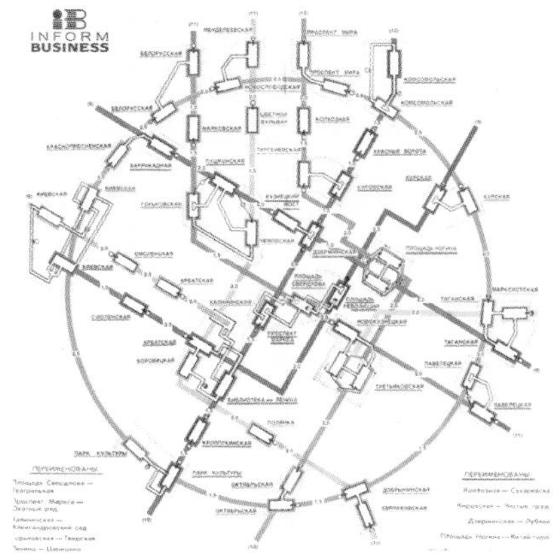

17. 1991: Coal mine, bunker, or mass-transit map?

18. 1992: A more "Western" feel.

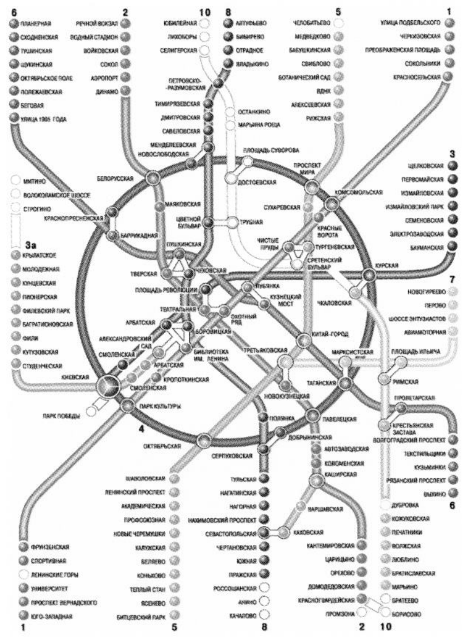

19. 1998: Beading returns with 3-D tubular lines.

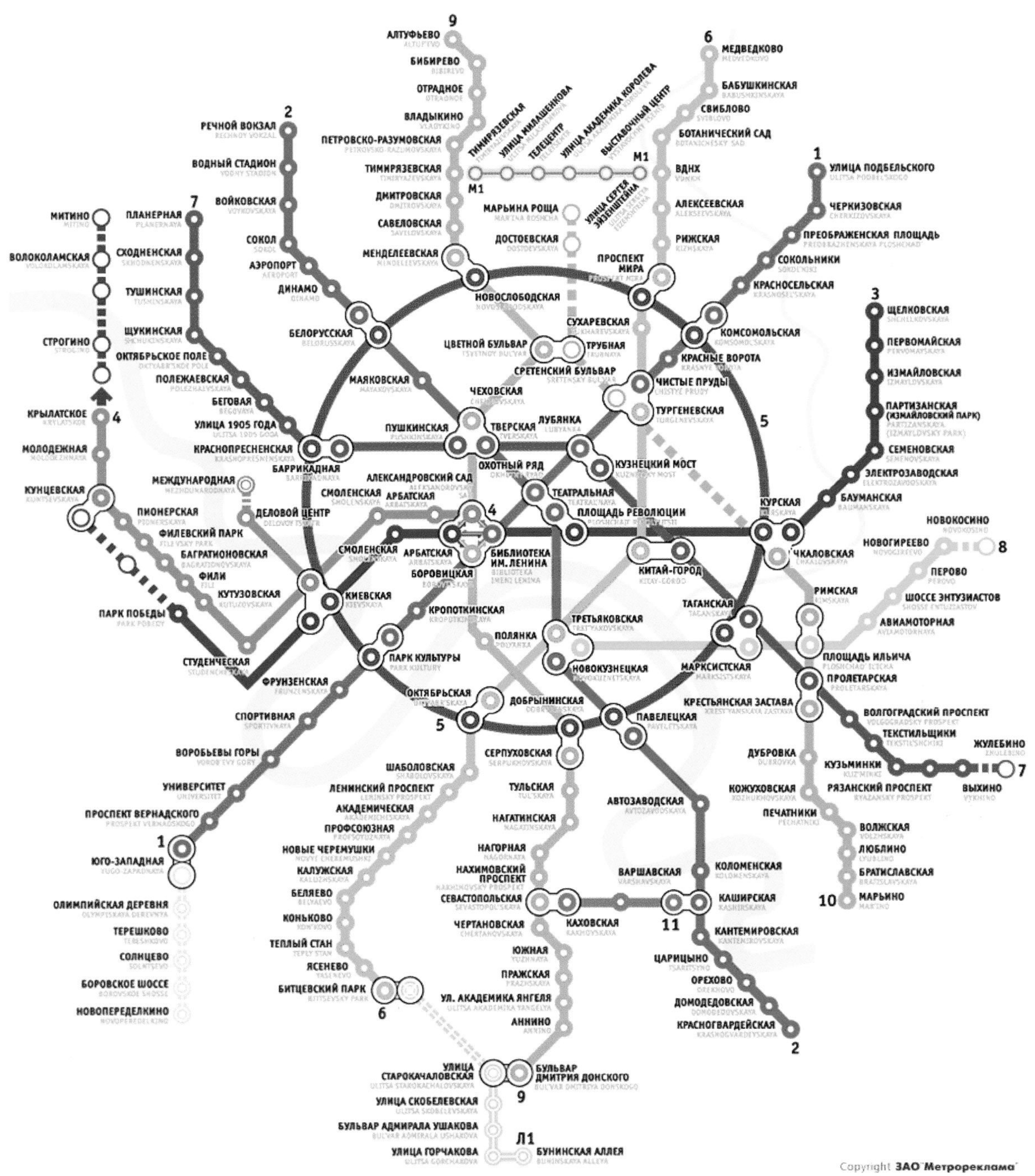

20. 2007: Striking contemporary official diagram courtesy of Artemy Lebedev (www.metro.ru).

New York

Urban population: 14.6 million. Route length: 228.6 miles. Stations: 468. First section open: 1904 (1868). Underground: 60 percent.

The world's largest subway has produced a rich and diverse range of map styles. This is due to the challenge presented by its sheer scale and complexity and its heritage of three separate operators. Even since unification in 1940, there have been several experiments with signage and various attempts at an ideal solution for the map, some better than others.

By 1898, when the City of Greater New York was created, many elevated railroads (aka Els) already existed and several tunneled rail attempts had been made. Though the subway was built with public money, its original operators were privately owned; the Interborough Rapid Transit Company, the IRT, opened the first nine miles in 1904. A beautiful map from the following year (1) prophetically claims it to be the "Greatest Underground Railway in the World" and shows the extent of the existing Els. Brooklyn Rapid Transit Company (BRT) ran the Els on the Long Island side, where it opened its first subway section in 1908 and then crossed into Manhattan in 1911 (2). The city awarded BRT contracts to run some new subways and it in turn created the New York Municipal Railway, which operated the Broadway Line (3) among others. Following bankruptcy in 1924, it became Brooklyn Manhattan Transit, the BMT, and produced a very clear early diagram (5).

The IRT meanwhile produced a neat schematic in 1924 (4). Following an outcry over the large profits being made, a publicly owned and run system was inaugrated. Its first line, the Eighth Avenue opened in 1932. By 1938 the Independent City Owned Rapid Transit Railway had a sizable network (6). The IND, as it became known, had proposed a second phase of construction but the arrival of the Great Depression stopped all that.

It was the Board of Transport (which operated the IND) that took over the IRT and BMT lines in 1940. Until then the only maps produced of all lines together were made for private companies, banks, and stores by designers like Andrew Hagstrom. His 1948 map was badged as being from the Board of Transport and gave equal weight to all lines (7).

After 1952 the rebranded Transit Authority made contact with graphic designer George Salomon, who wanted to tidy up the signage and attempt a unified corporate image. He introduced the Akzidenz-Grotesk font to the system on his impressive first diagram dated 1958, which included much geographic distortion and 45-degree diagonals (8). The schematic style, which continued for over twenty years, was revised in 1966 by Stanley

Goldstein, using concepts submitted by Raleigh D'Adamo in a 1964 contest to design a new map (9), and revised again in 1969 (10), but it was the 1972 diagram (11) by Massimo Vignelli that became a modern graphic design icon.

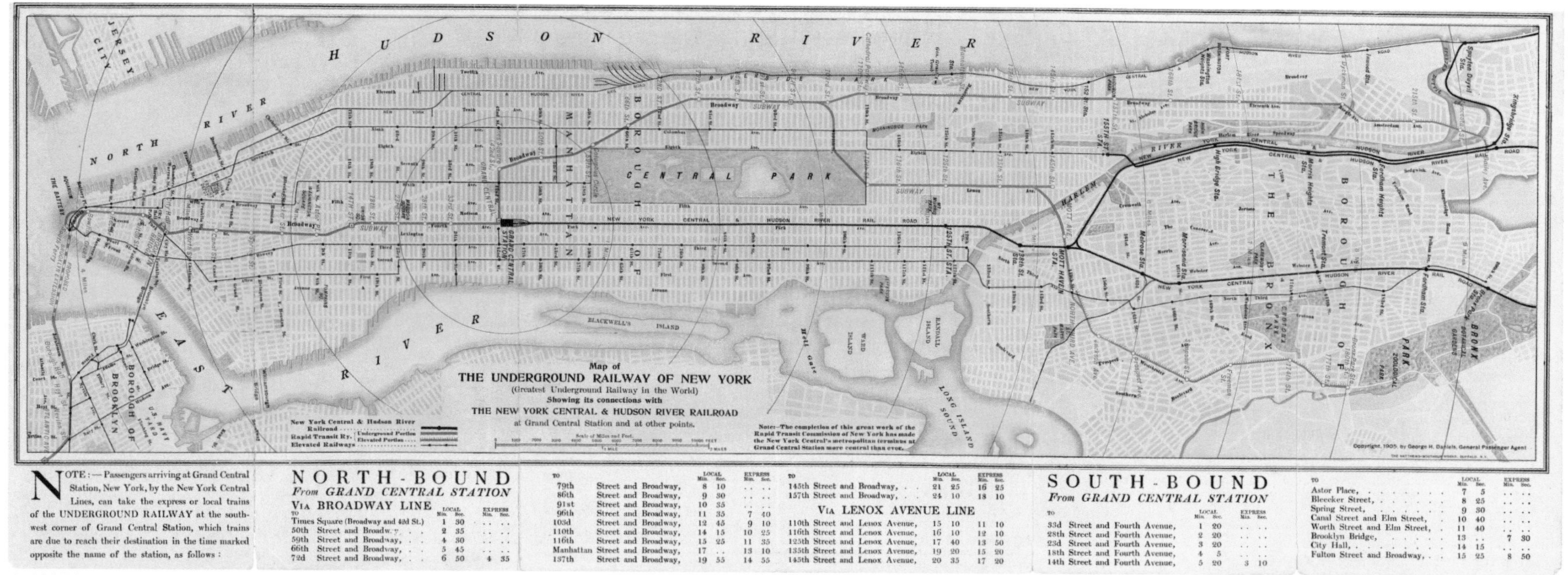

1. 1905: Unusually describing the IRT subway (in red) as an "Underground Railway," this highly detailed map was probably used inside cars. Manhattan stayed tilted at this incline on subway maps for twenty years.

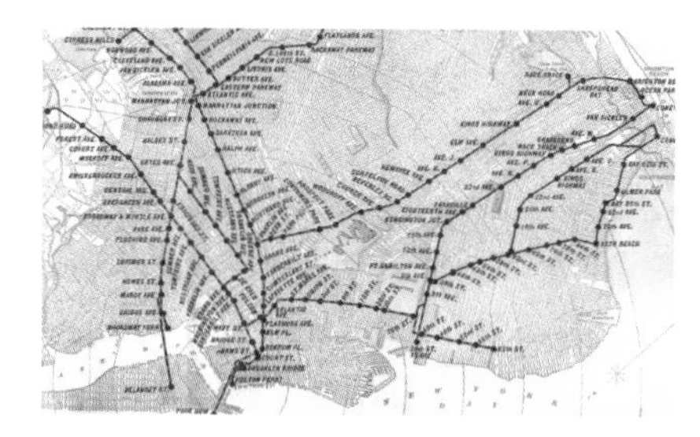

2. 1911: Detail from BRT map.

3. 1919: The MR: dispensing with street clutter.

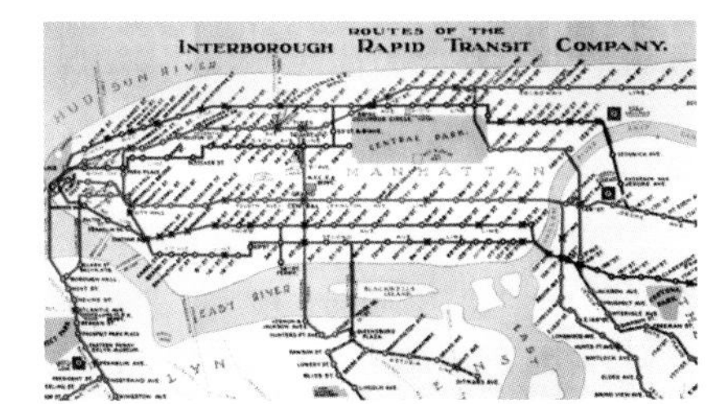

4. 1924: The IRT's subways and Els.

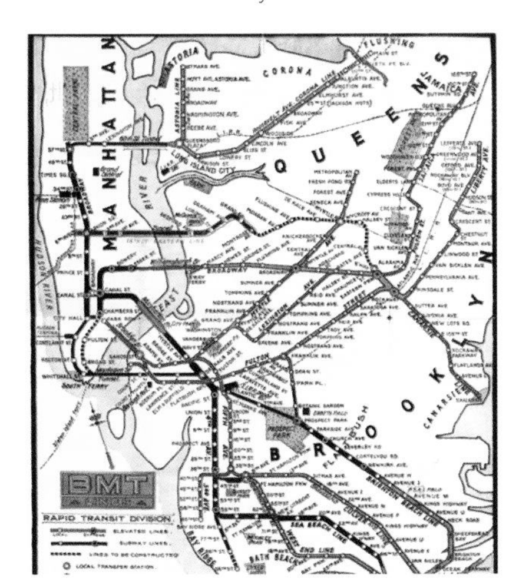

5. 1924: BMT twisted the axis the "right way" up.

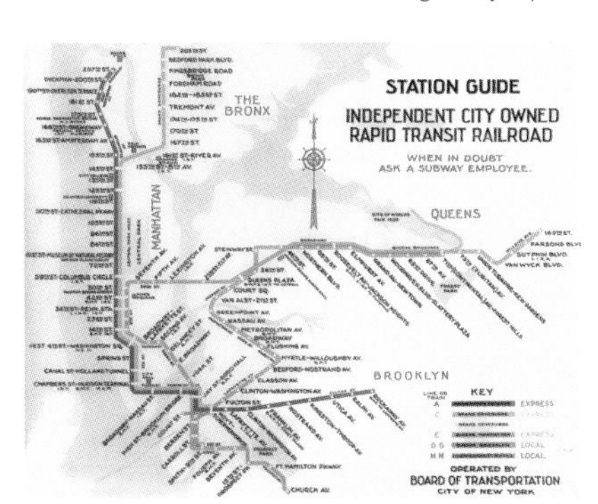

6. 1938: The Independent plan, clear and bright.

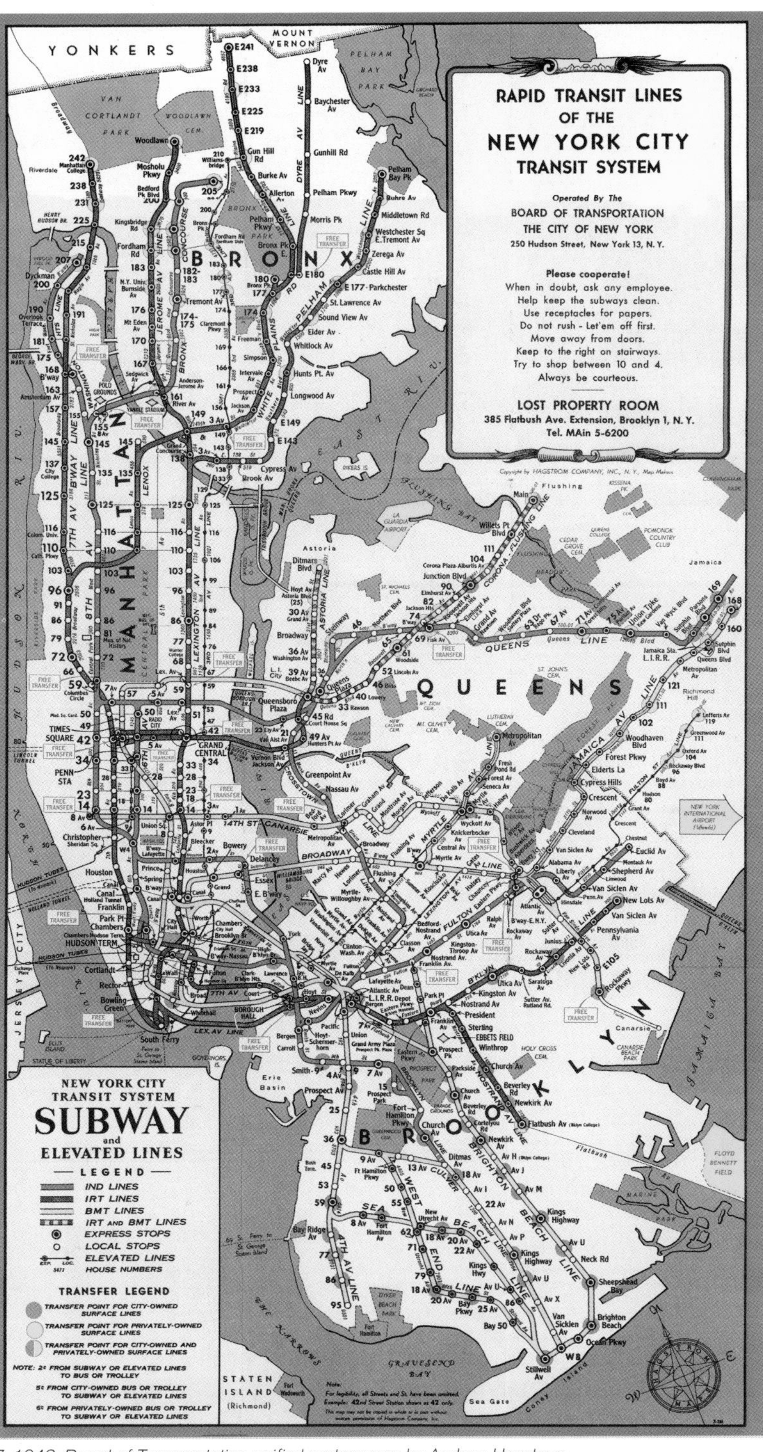

7. 1948: Board of Transportation unified system map by Andrew Hagstrom.

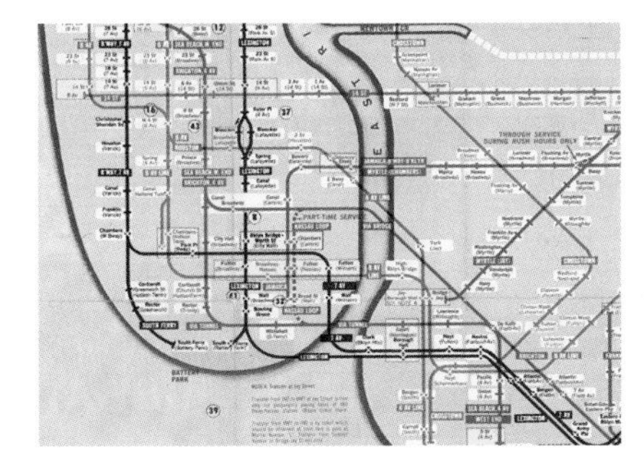

8. 1958: George Salomon's first schematic.

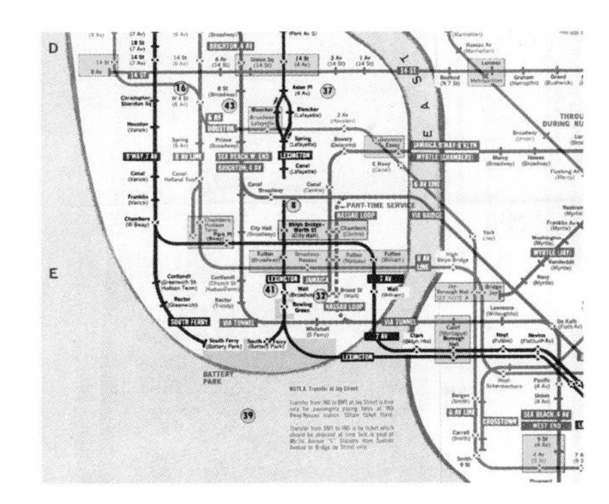

9. 1966: Goldstein/D'Adamo version.

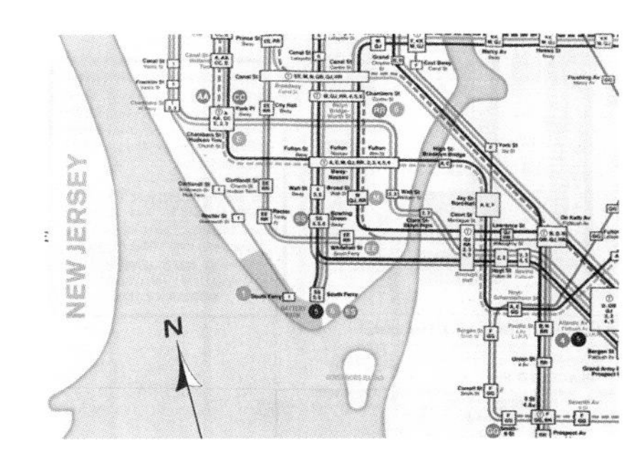

10. 1969: Revision showing more "services."

New York

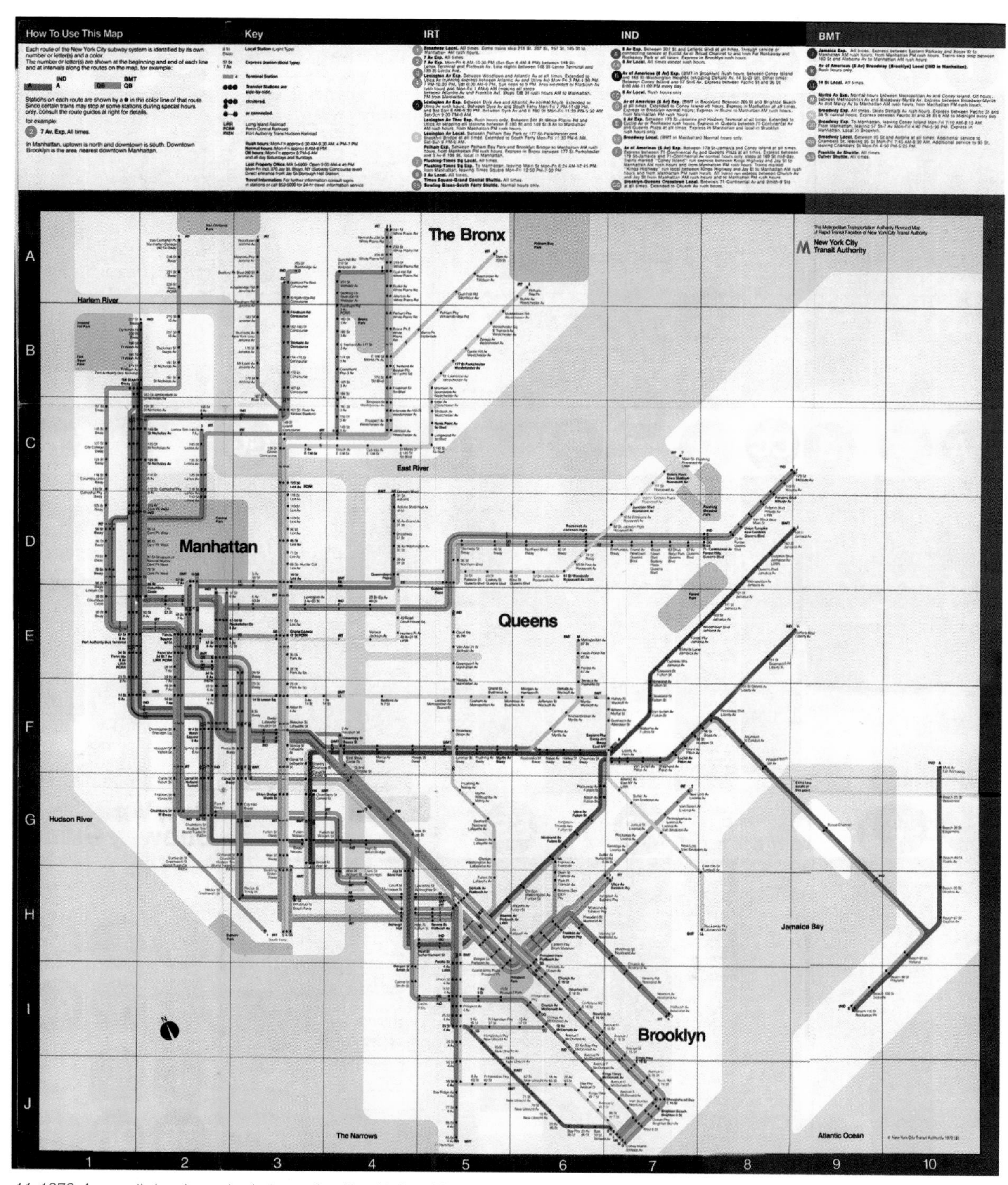

11. 1972: Apparently loved more by designers than New Yorkers, Massimo Vignelli's diagram undeniably evokes the spirit of the age.

Developing the themes of 45-degree diagonals, Vignelli's Unimark company used different colors for each service and endorsed blanket use of the Akzidenz font, including an entire graphics package for the signage. This was the closest New York got to the simplicity of other urban rail diagrams. (In homage to Salomon and Vignelli, all headlines in this book are set in the Akzidenz-Grotesk Medium font.)

Many New Yorkers complained that the attention to geometry made Vignelli's map confusing to use; for example, 50th Street–Broadway station appears west of 50th Street and 8th Avenue when in reality it's east. So in 1975 the MTA formed a committee, finally chaired by John Tauranac, to revise the map, and the new design was executed by Mike Hertz. The result, a major shift back to topography, was published in 1979 and has stabilized the design ever since.

However, the switch away from geometry made New York the only large system not to use a traditional diagram, which often surprises visitors unfolding their unwieldy, cluttered MTA maps. There are increasing calls for New York to reintroduce a smaller, simple diagram comparable to those of London, Moscow, Tokyo, or Paris.

New ideas for evolving a schematic version circulate on the Internet. Joseph Brennan from Columbia University is inspired by classic urban-transit diagram designs, (12) while Eddie Jabbours' "Kick map" cleverly manages to combine clear topography with schematic rules to show every service, while retaining key streets and even districts (13). It suggests that a hybrid based on diagrammatic concepts with some useful geography could work as well for New York as it does elsewhere.

As this debate rages, the current plan (14) remains broadly geographic.

12. 2002: Joseph Brennan's neat schematic suggestion.

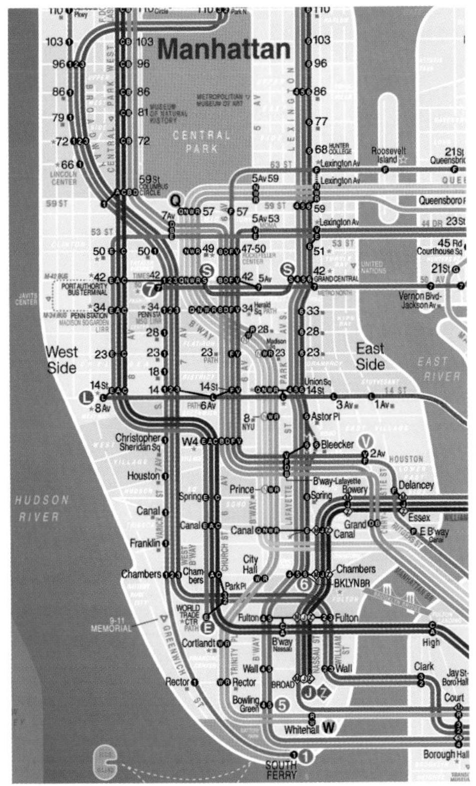

13. 2007: Kick Design Inc.'s clear hybrid concept.

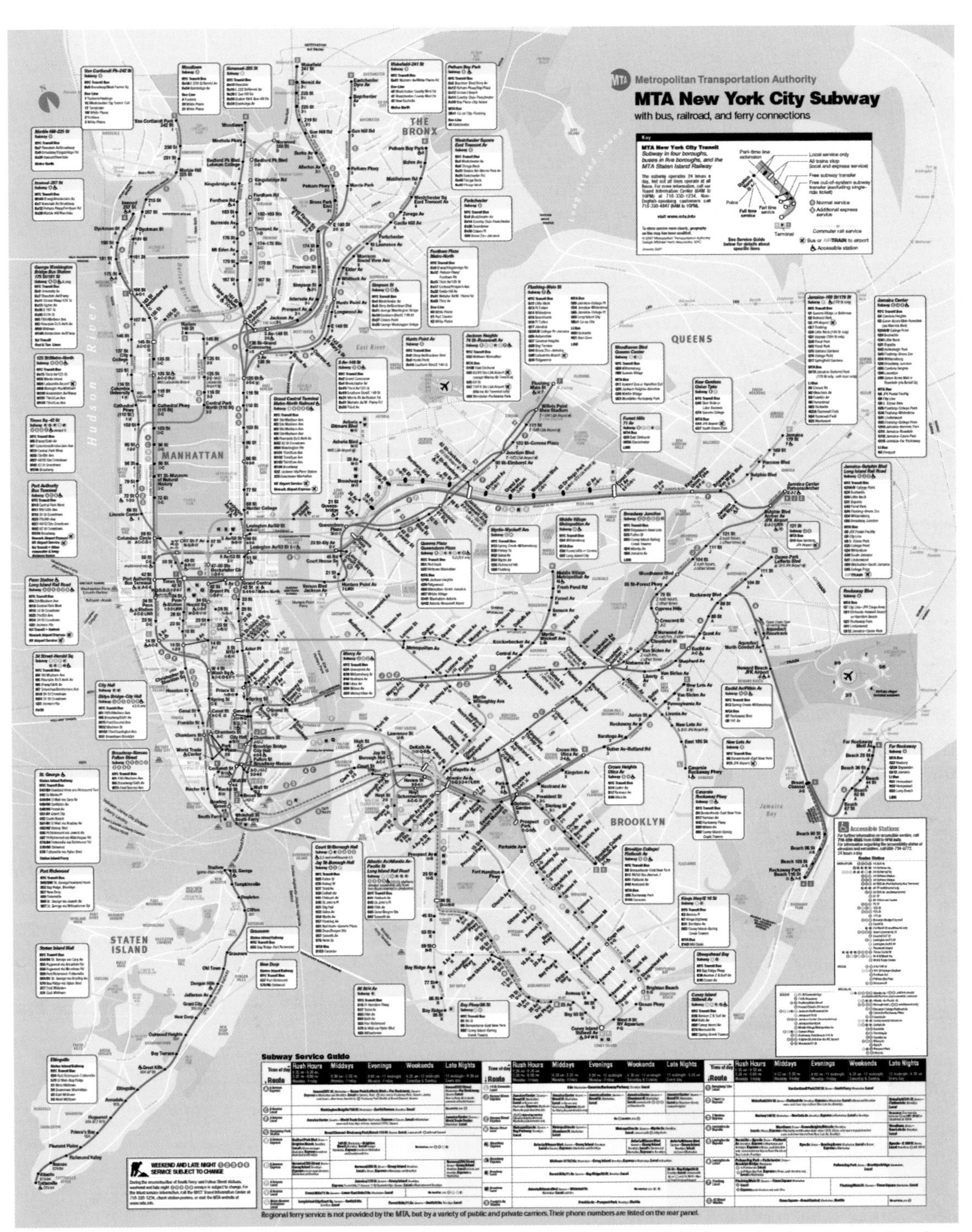

14. 2007. Contemporary wall poster and foldout map shows a huge amount of detail. However, there is a less cluttered version available online.

Maps reproduced by kind permission of and copyright MTA 2007. Antique maps: Mike Ashworth Collection, New York Transit Museum. Photos: Author.

Paris

Urban population: 11.1 million. Route length: 133.7 miles. Stations: 381. First section open: 1900. Underground: 95 percent.

Europe's tourist capital is blessed with one of the world's most effective urban-rail systems and has had a myriad of different Métro map styles since Line 1 opened in 1900. They now come in standard sizes, from the schematic Paris Poche (pocket map), minus all topography but the river Seine and the city boundary (6.5 x 6.5 inches, on paper, folded to smaller than credit card size), to large foldout guides (25 x 15 inches), which are highly detailed geographic maps. In stations there are larger wall posters, with yet more streets and details shown. The operator, RATP, licenses the official diagram to hotels, shops, etc. Unofficial maps are available in hotel lobbies, tourist guides, or street atlases.

Despite the architectural heritage of the Métro's art nouveau station entrances and much engineering innovation, official maps were not produced until 1922, so all early ones were made by private companies. This 1900 sketch was published in *Le Petit Journal* (2).

Much of the system's elegance is owed to Hector Guimard, the architect employed by the first system operator, the Compagnie du Chemin de Fer

1. 1900: Guimard entrance at Porte Dauphine.

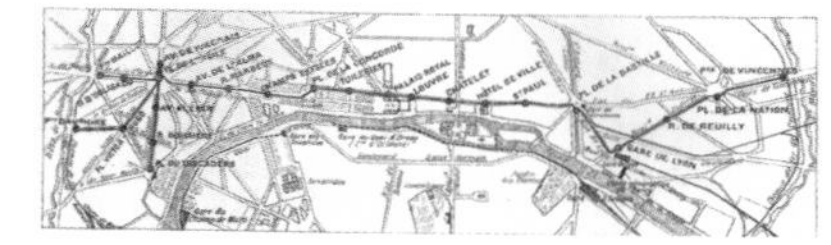

2. 1900: The first sections of line to open.

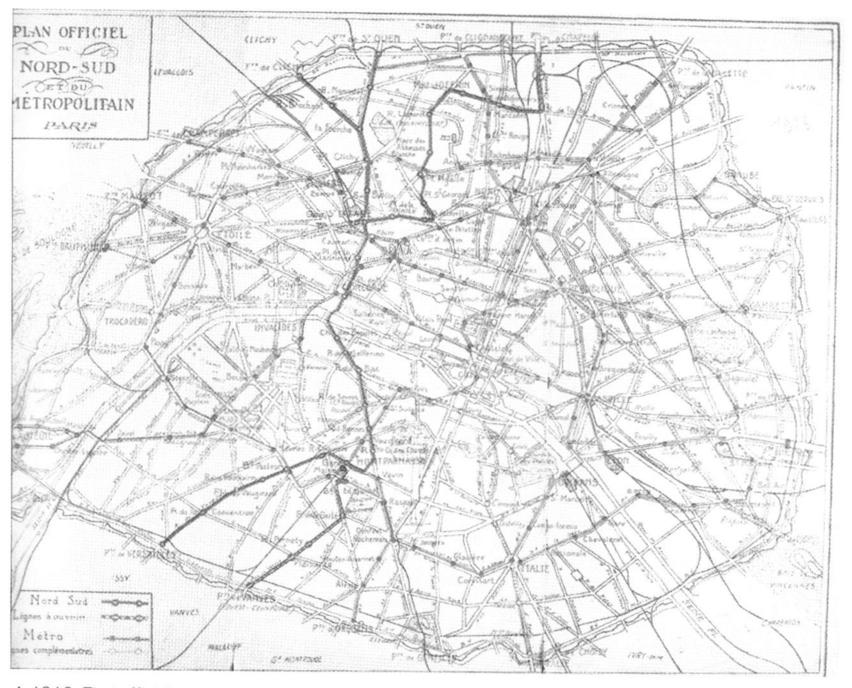

4. 1910: First official map produced by the Nord–Sud company showed all lines in operation.

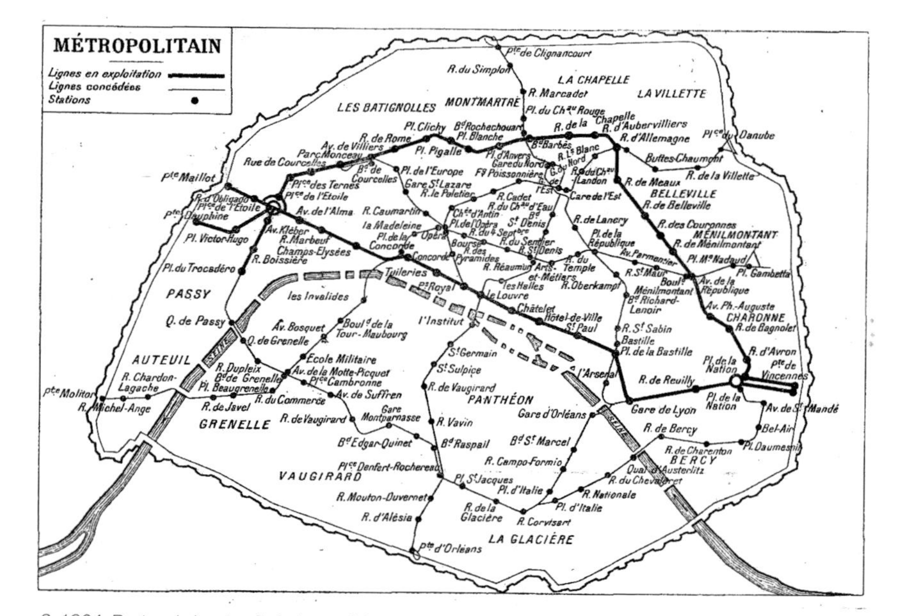

3. 1904: Postcard showing first phase of the proposed network.

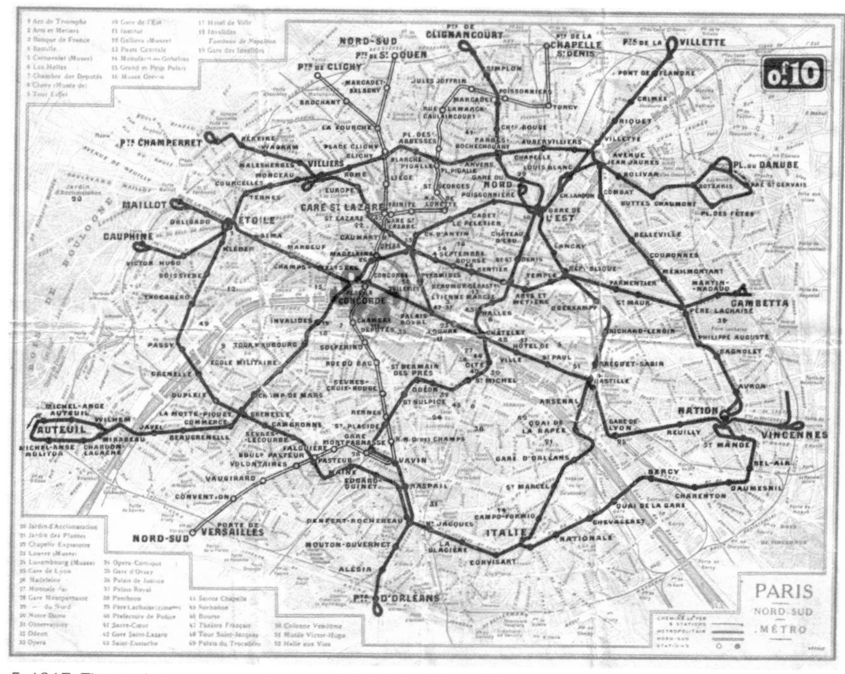

5. 1917: This pocket map on card was so detailed that it would have been hard to read in poor light.

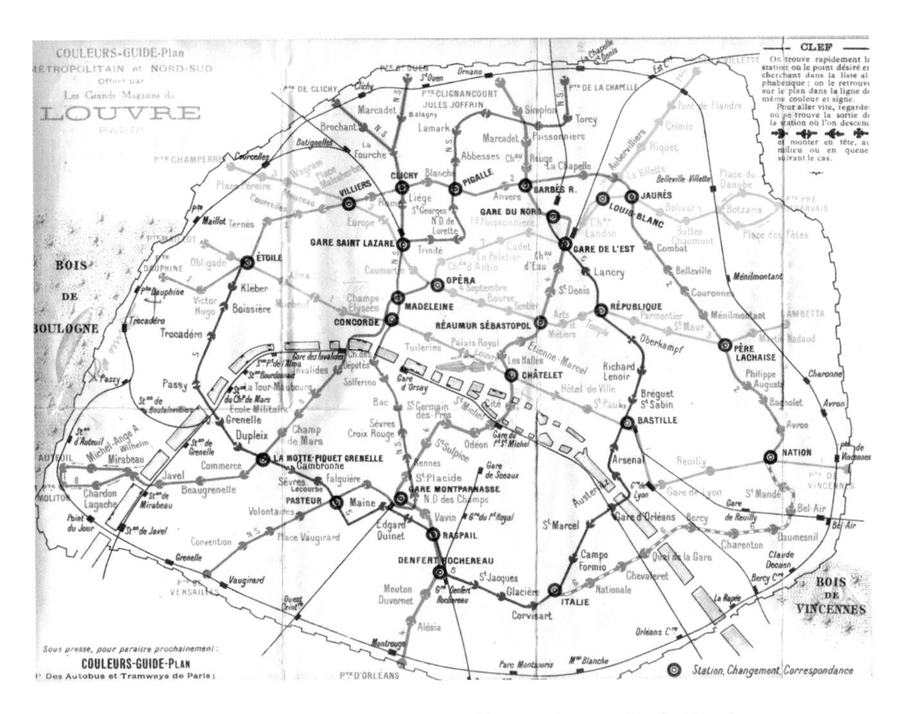

6. 1916: Couleurs Guide Plan, available from the Louvre Museum shop, was the first in color.

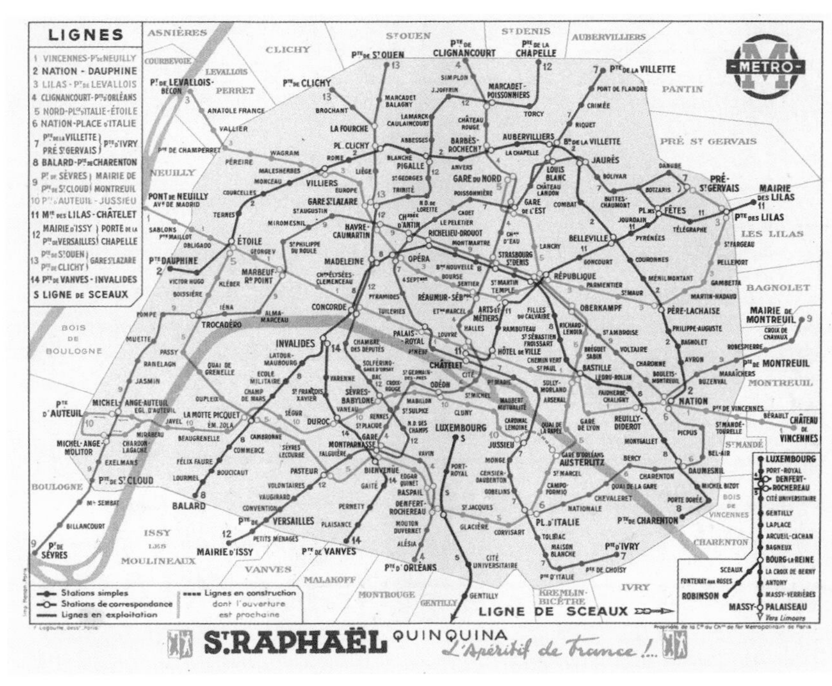

8. 1939: Early F. Lagoutte design for the CMP.

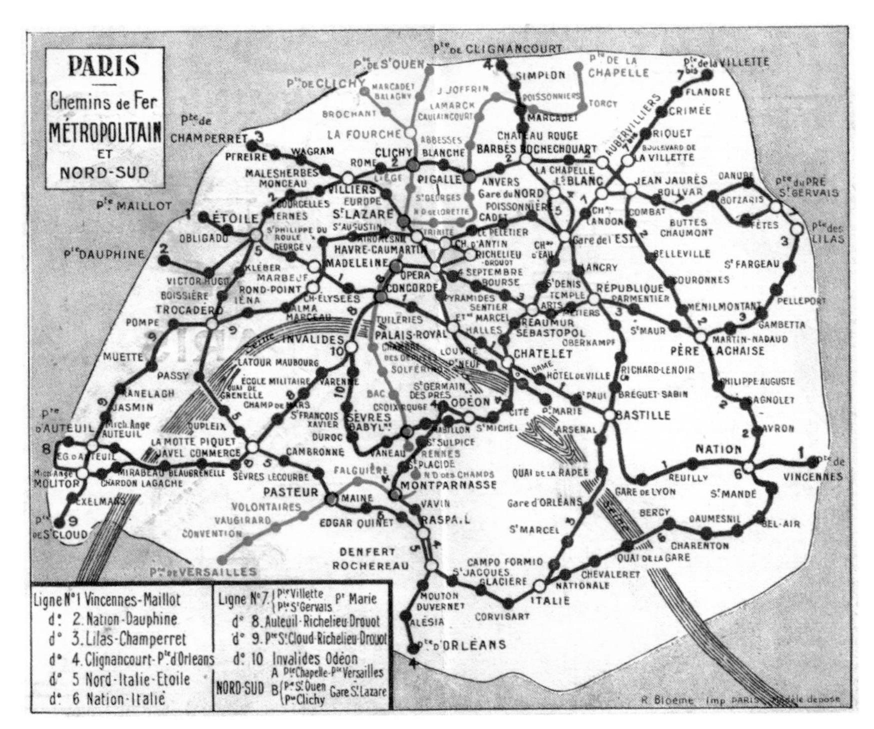

7. 1929: The first CMP pocket card folder removed all topography but the river Seine.

An in-car route diagram for Ligne 9 dating from the late 1940s.

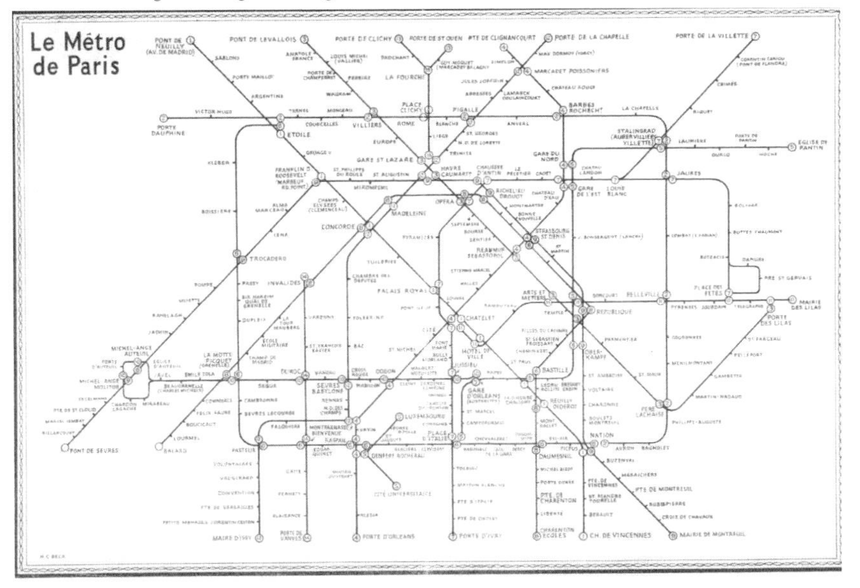

9. 1946: Harry Beck's ingenious attempt at diagrammatic simplification was rejected by the RATP.

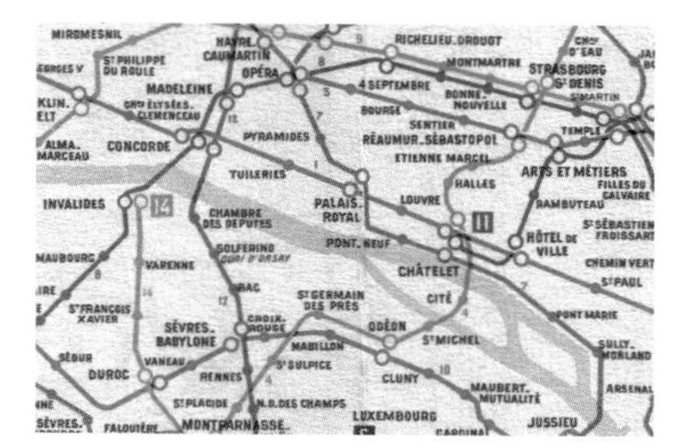

10. 1961: First Georges Redon version.

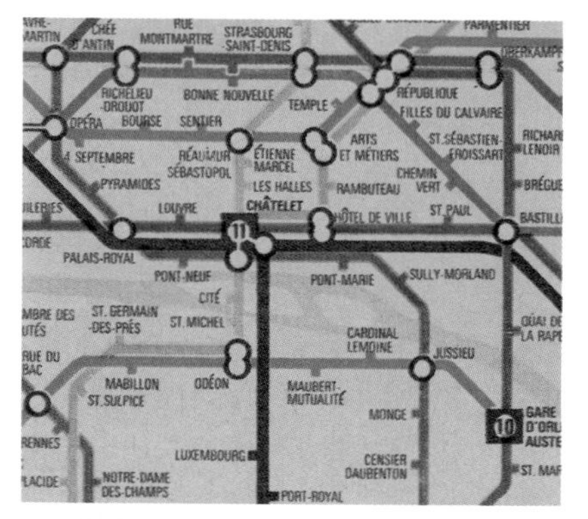

11. 1978: Ken Lewis in London style.

12. 1983: Constantin Spandonide's neat attempt.

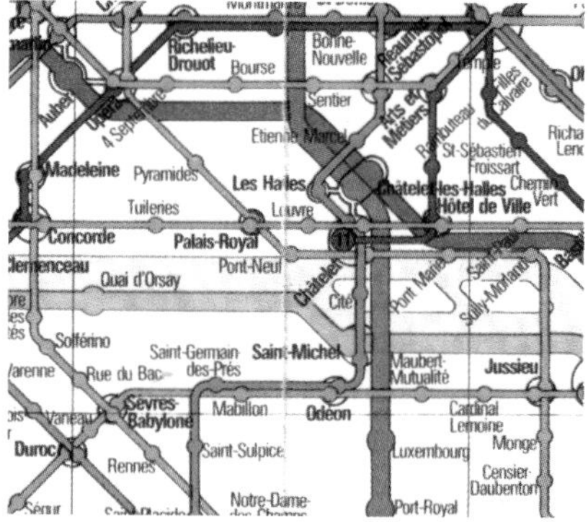

13. 1983: In-house RATP design by Rouxel.

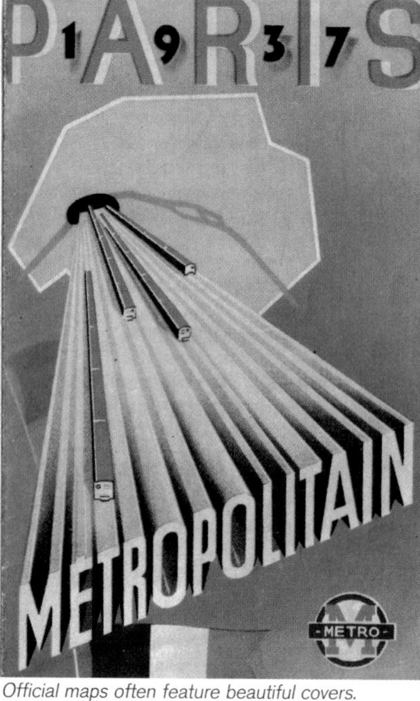

Official maps often feature beautiful covers.

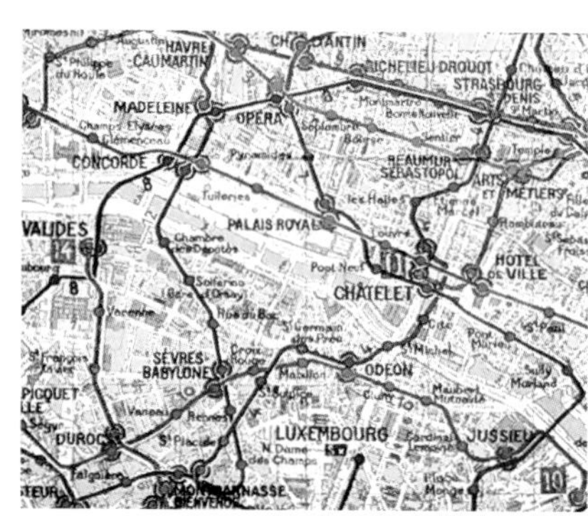

14. 1964: Station wall map.

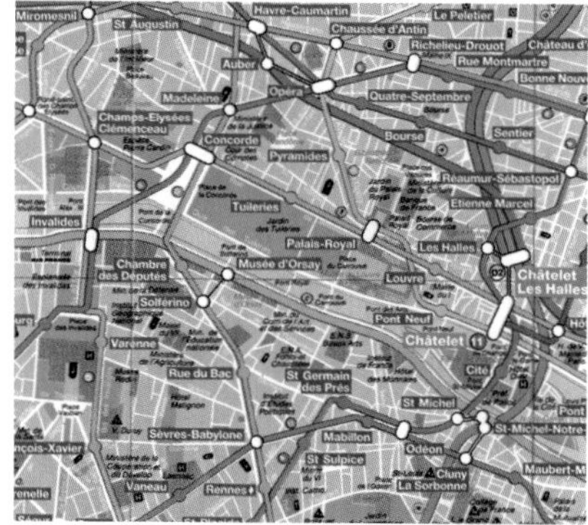

15. 1998: Station wall map, also in tourist guides.

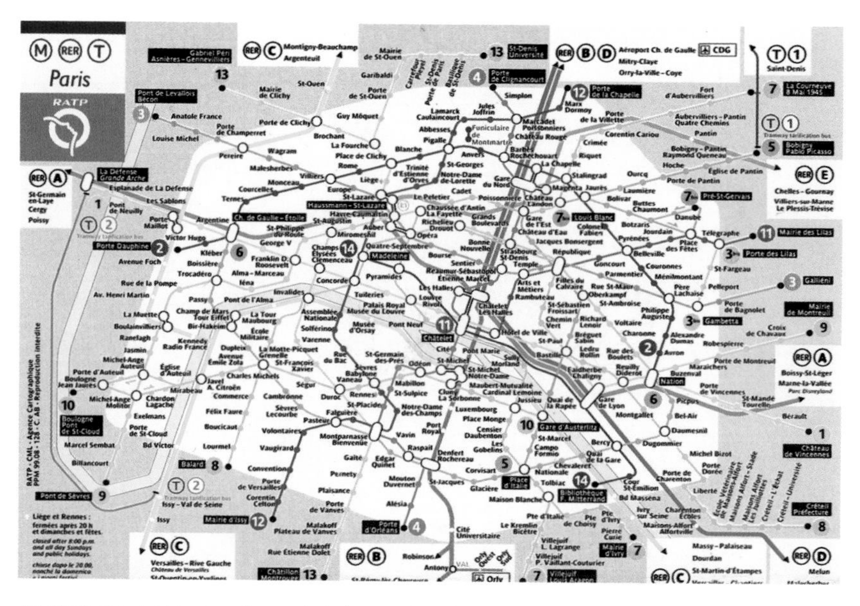

16. 1999: This right-leaning diagram lasted over a decade on all maps issued to the public.

Métropolitain de Paris, the CMP. Guimard came up with the "Pavilion" station entrances, now regarded as design classics. On these were the classic Métropolitain signs in an art nouveau style by font maker Georges Auriol, which along with the Eiffel Tower, are now logos for Paris itself.

A 1904 postcard (3) showed the first six lines, which were mostly completed in just ten years. A private company was permitted to construct a new north–south route that opened in 1910. Their first official map showed the CMP lines as well as those of the Nord-Sud (4) and squeezed the entire city onto one map, a trait still common on Métro platform wall maps today.

Around the First World War, a few neat pocket maps appeared (5). An early attempt to iron out some kinks and remove the street clutter above came on an unofficial 1916 Louvre bookshop map (6). Further improvement arrived with the 1929 CMP pocket card map (7). A Mr. F. Lagoutte took over design in 1937 (8) and added the contemporary Métro logo, which, with its bar/circle and red-blue mix, resembled London's roundel. His design set the tone for many years but fell short of standardizing angles.

For such a tightly knit web of lines, where stations are typically a third of a mile apart, a crystal clear diagram would have been helpful, so in 1946, after what is believed to be over a decade of work, Harry Beck, without so much as a commission in the air, had created a simplified Paris map (9). Although apparently presented to the RATP, they stuck with their more wavy version.

Artist Georges Redon took over in the early 1960s (10). Another British designer, Ken Lewis (11), attempted to add the 45-degree concept with great success but little interest from RATP. Greek-French architect Constantin Spandonide also had his 1983 schematic idea (12) rejected. However, in the same year RATP designer Patrice Rouxel attempted his first schematic, which did not last long.

Station wall maps (14, 15) have

continued to be fully geographic even to the present day and are very useful for orientation aboveground.

Throughout the 1980s and '90s, though, almost all official diagrams were gradually conforming to schematic standards while adhering to the slight rightward lean of the city's streets at surface level. Tilt the book down to the left so the north–south RER spine on the 1999 diagram (16) becomes vertical and, if you ignore the black type, the whole thing seems more legible. However, with a 70-degree rhomboid slant, horizontal station names can be tricky to read since most are forced to crash over lines.

With the production, in 2001, of its new official 45-degree-based Métro schematic, still in use today (17), RATP designers seem to have acknowledged that diagrams can be useful, as Beck, Lewis, Spandonide, and others have argued for over half a century.

The corporate identity of the Métro as a whole suffered from a variety of logos and typefaces over the years. A special 1973 variation by Adrian Frutiger of his capital letter–only font Univers, and more recently Neue Helvetica, were both superseded in 1996 by Parisine. Conceived by French type designer Jean-François Porchez, it can now be seen system-wide, giving a more welcoming feel to the "city of light."

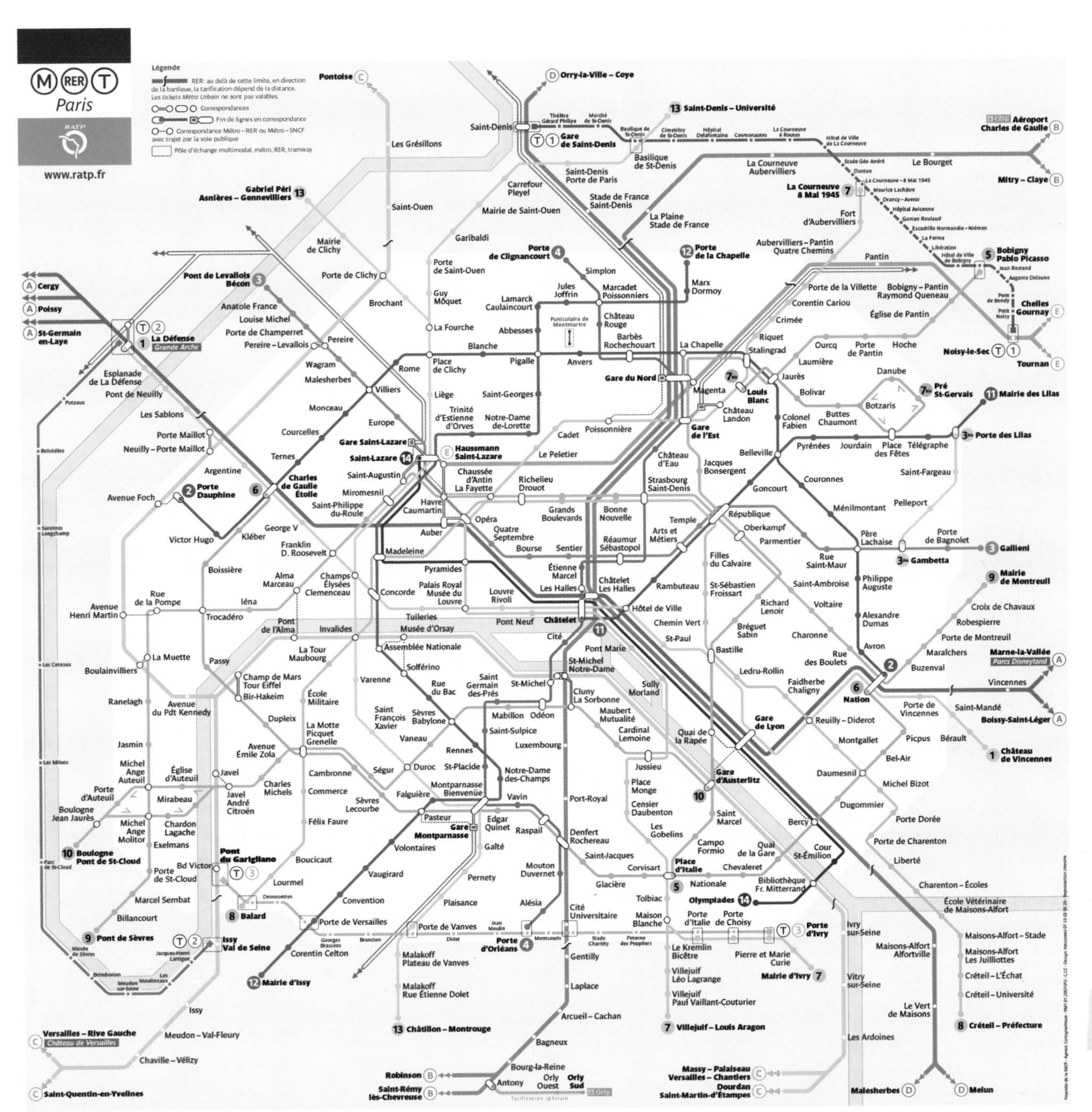

17. 2007: Produced for RATP by BDC Conseil.

Tokyo

Urban population: 8 million (35 million megalopolis). Route length: 181.6 miles. Stations: 202. First section open: 1927. Underground: 45 percent.

For the uninitiated Westerner, the dazzling Japanese capital can appear unfathomably complex. In truth, the urban-rail maps shown here, designed to make the system easier to use, might feel equally baffling! This isn't because of the language or because there are two subway operators; quite simply, there is a gargantuan urban-rail system in this, the world's biggest megalopolitan area. It's an astonishing urban expanse of, depending where the line is drawn, up to 5,200 square miles and home to anything from 8 million (the "twenty-three special wards") to 35 million.

Apart from the world's biggest network of heavy-rail commuter lines and the main JR (Japan Rail) network, the Tokyo Metropolitan Government Transportation Bureau, or TOEI, manages four Metro lines while Tokyo Metro Corporation (formerly Teito Rapid Transit Authority, the TRTA) runs eight lines known as the Eidan Subway. Established in 1941, the Eidan now operates what was Tokyo's first urban-rail line, which opened in 1927, although the original proposals for an underground here date back to 1912.

The size of the population explains the incredible passenger numbers carried and why there are numerous plans for extensions. The first diagram available here is the Greater Tokyo rail systems map of 1965, with its excellent circular focus of the busy JR Yamanote

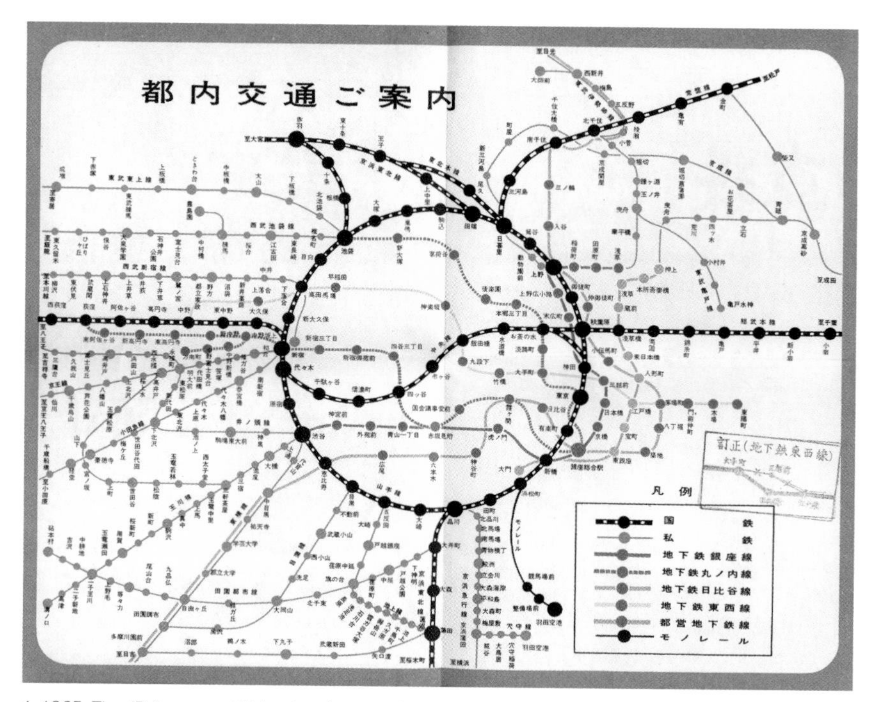

1. 1965: The JR line around Tokyo is schematically extrapolated into a perfect circle.

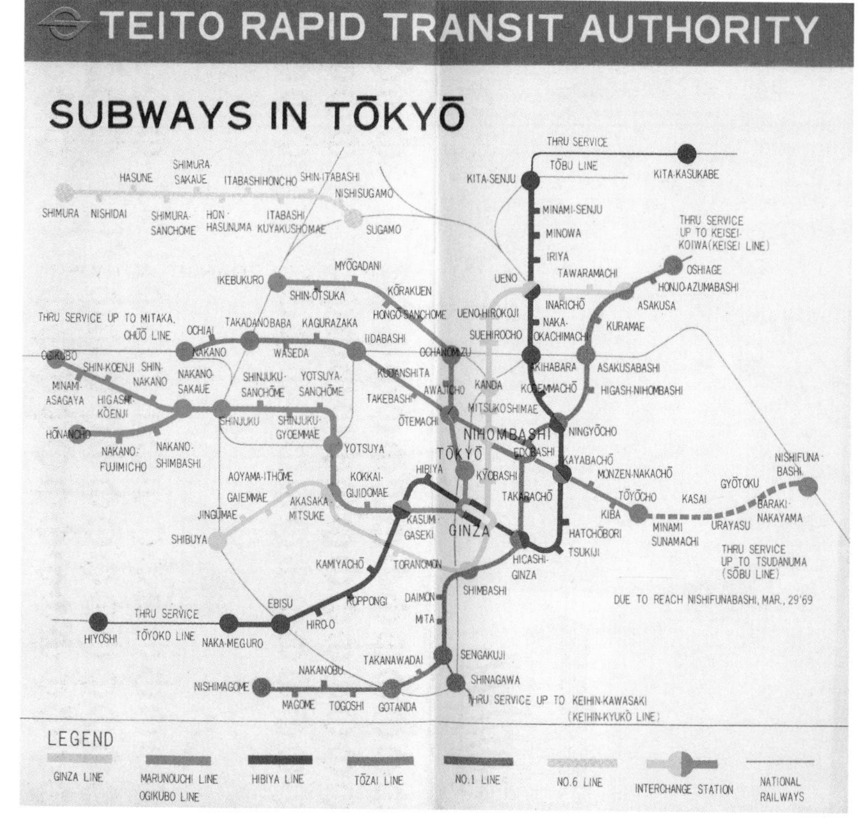

2. 1969: Early TRTA pocket card map for tourists—one of earliest to be printed in English.

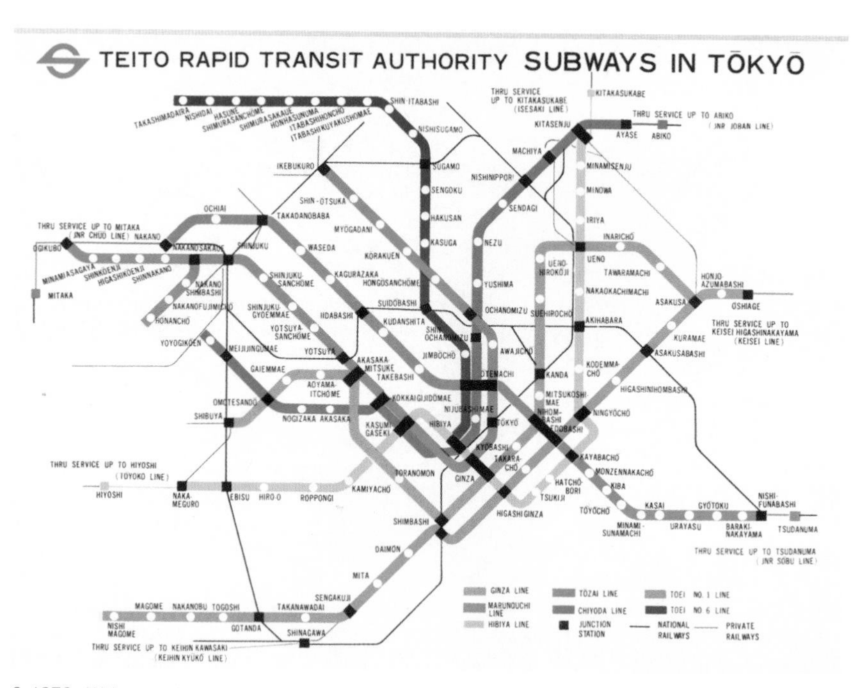

3. 1972: With several new lines open, a Western diagrammatic approach was taken for this pocket map.

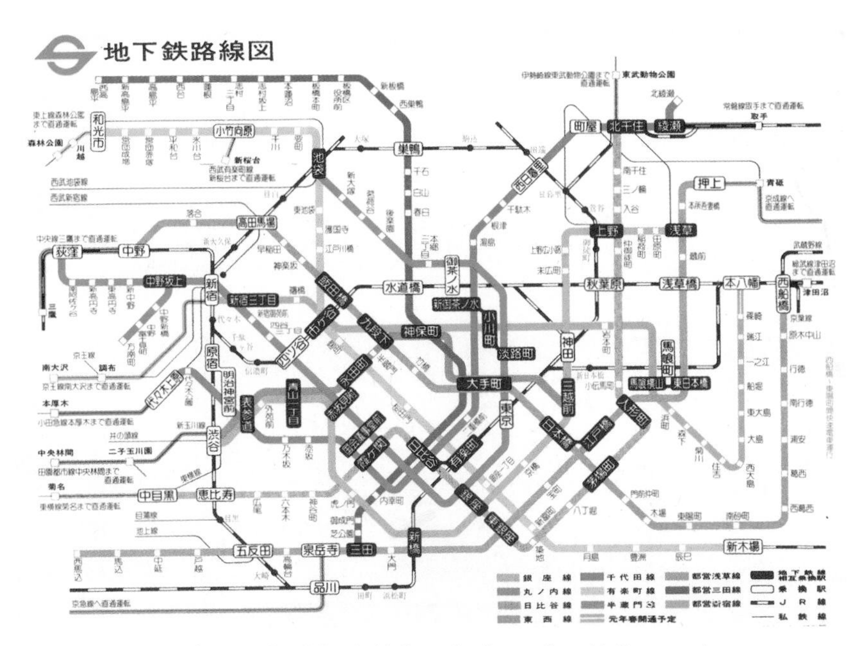

4. *1989: The practice of naming the station inside its marker has continued to the present.*

Line (1). The 1969 guide (2), while spacing the stations fairly equally and cutting out all topographic features, makes little attempt at standardizing angles or curves.

Tokyo had to adopt a diagrammatic form early on because there would simply be no other way to show such

5. *1997: Showing stairways at interchanges.*

interwoven lines covering an area so compact. The 1972 diagram (3) introduces the classic 45-degree-angle form with open circles for stations and is a good deal easier to read than some of its successors.

Much standardization on the 1989 diagram (4) helps to accommodate many new lines and introduces the practice of putting the station's name inside its marker—in this version reversed out of black on interchanges. In Japan, exotic treatment can be given to commonplace items; in this remarkable example from a restaurant placemat (5) the location of *stairs* inside the stations is attempted! No prizes for guessing this was not an official TRTA or TOEI publication!

A 1997 ticket (6) shows a layout change, with more horizontals to allow for a landscape view and to fit in new lines. The 45-degree diagonals and even the interchange stations using a variant of the "black circle with white center" symbol, shown here as black-outlined oblongs with white centers, are still much in evidence. A small handout diagram from 2000 (7), which also has the Osaka subway printed on the reverse, shows the whole system squashed using 50-degree angles. Japan's commitment to multilingualism, particularly with American English, has led to station signage and announcements in at least two languages.

The subway map itself has been reproduced officially and otherwise on many products and souvenirs. On pages 42 and 43 are the most prolific diagrams in recent circulation. There is one from each of the two main operators. The 2002 TRTA schematic (8) and the TOEI (10) both heavily distort geography. The TOEI diagram of

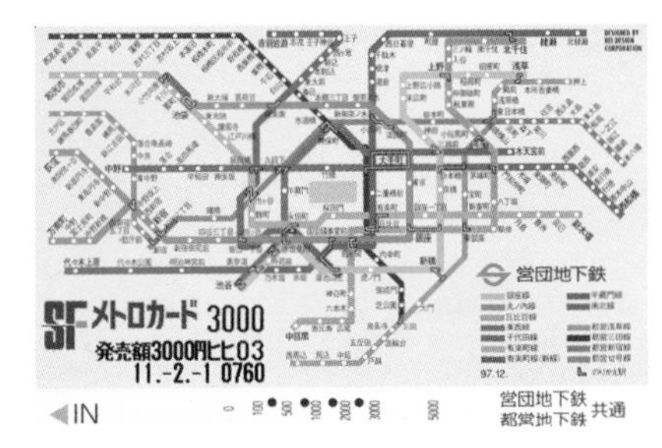

6. *1997: The entire system on the back of a ticket.*

only its own lines (9) feels almost like a beautiful Japanese kanji, or logogram, character.

TRTA was privatized and rebranded Tokyo Metro in 2004. Logos were changed, and this was followed by a new initiative to make the system easier for visitors to use. Every station was numbered (a common feature on East Asian metros) and the Japanese-named lines were also allocated a Latin-

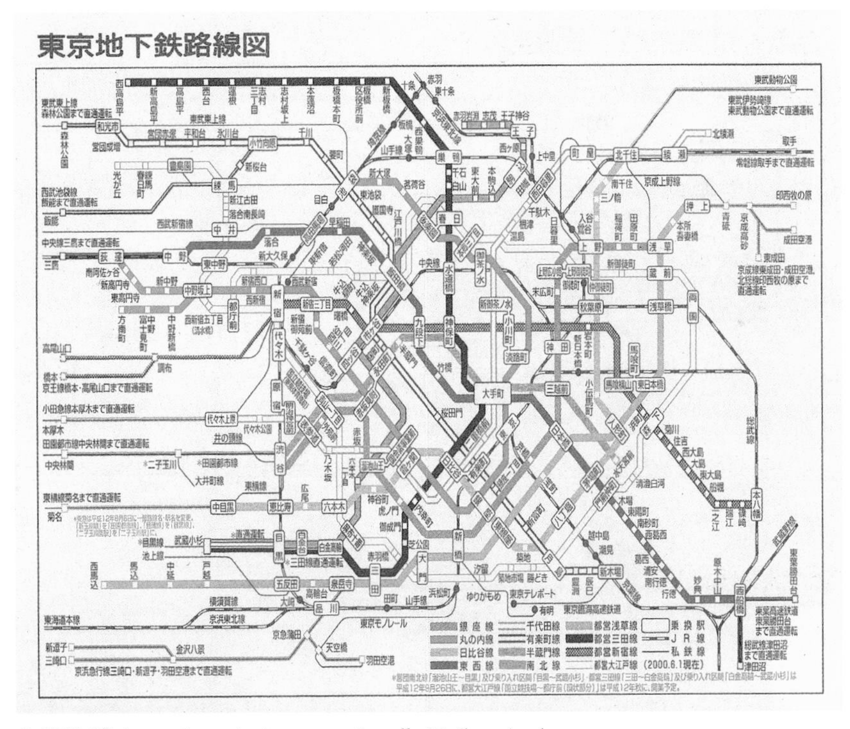

7. *2000: 50-degree diagonals give a concertina effect to the network.*

Tokyo

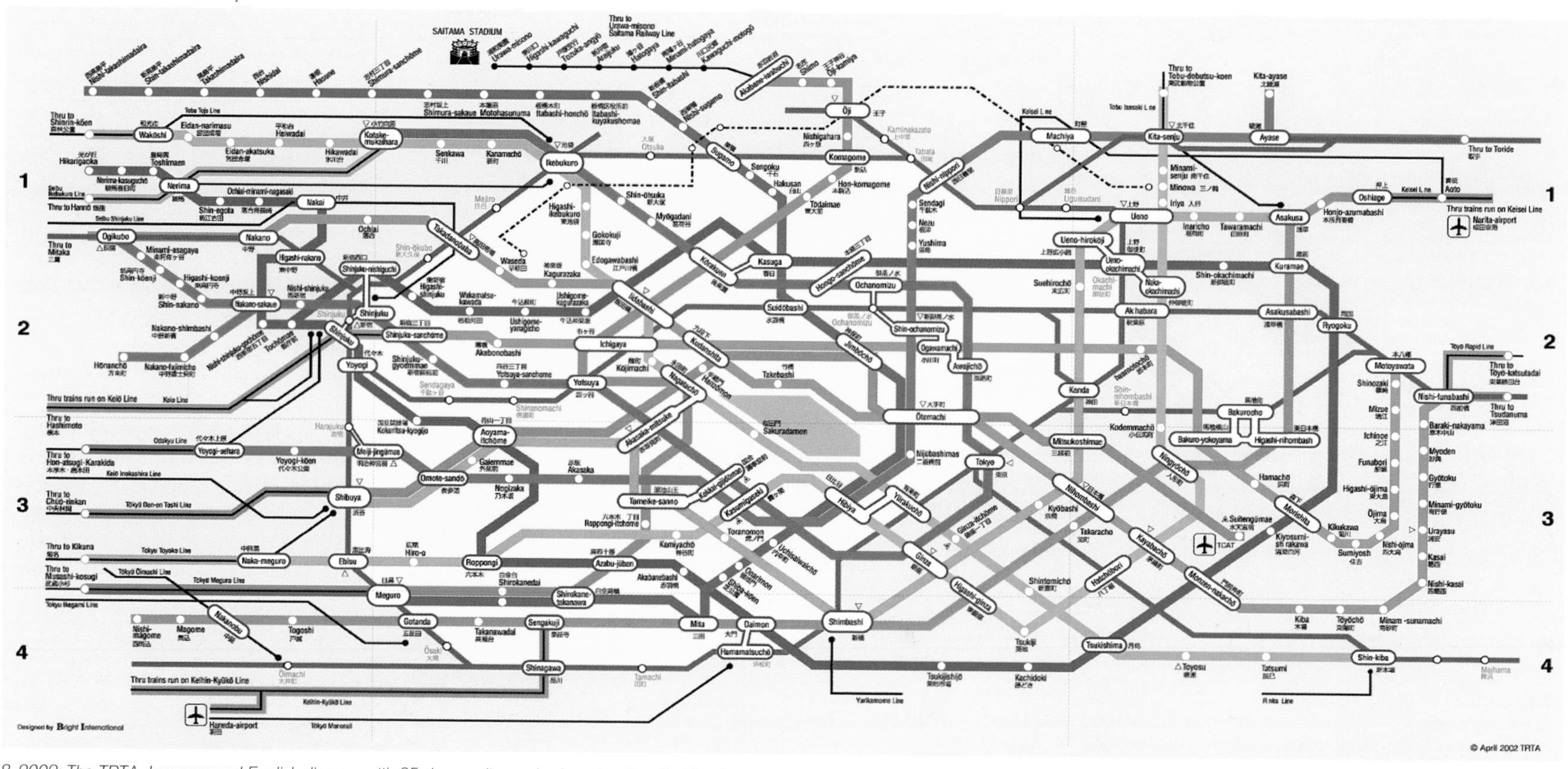

8. 2002: The TRTA Japanese and English diagram with 35-degree diagonals elongates the city. The diagram was made for them by the Bright International company.

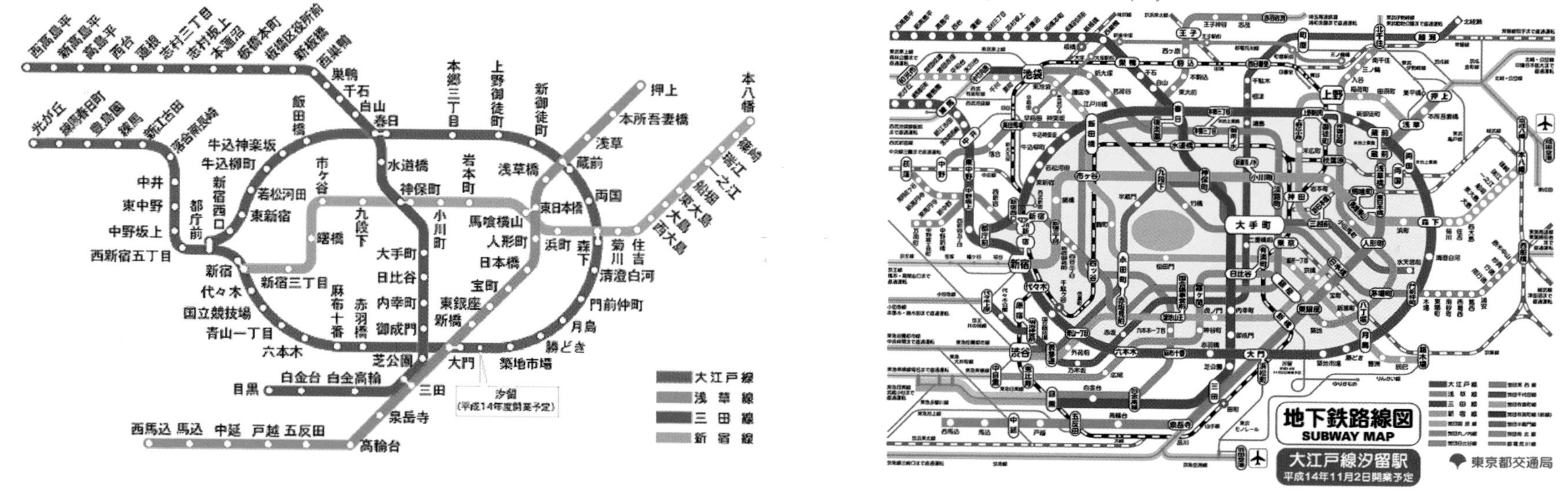

9. 2003: TOEI network on its own and almost resembling a new character in the language!

10. 2004: TOEI-produced joint-system diagram based on 45-degree angles.

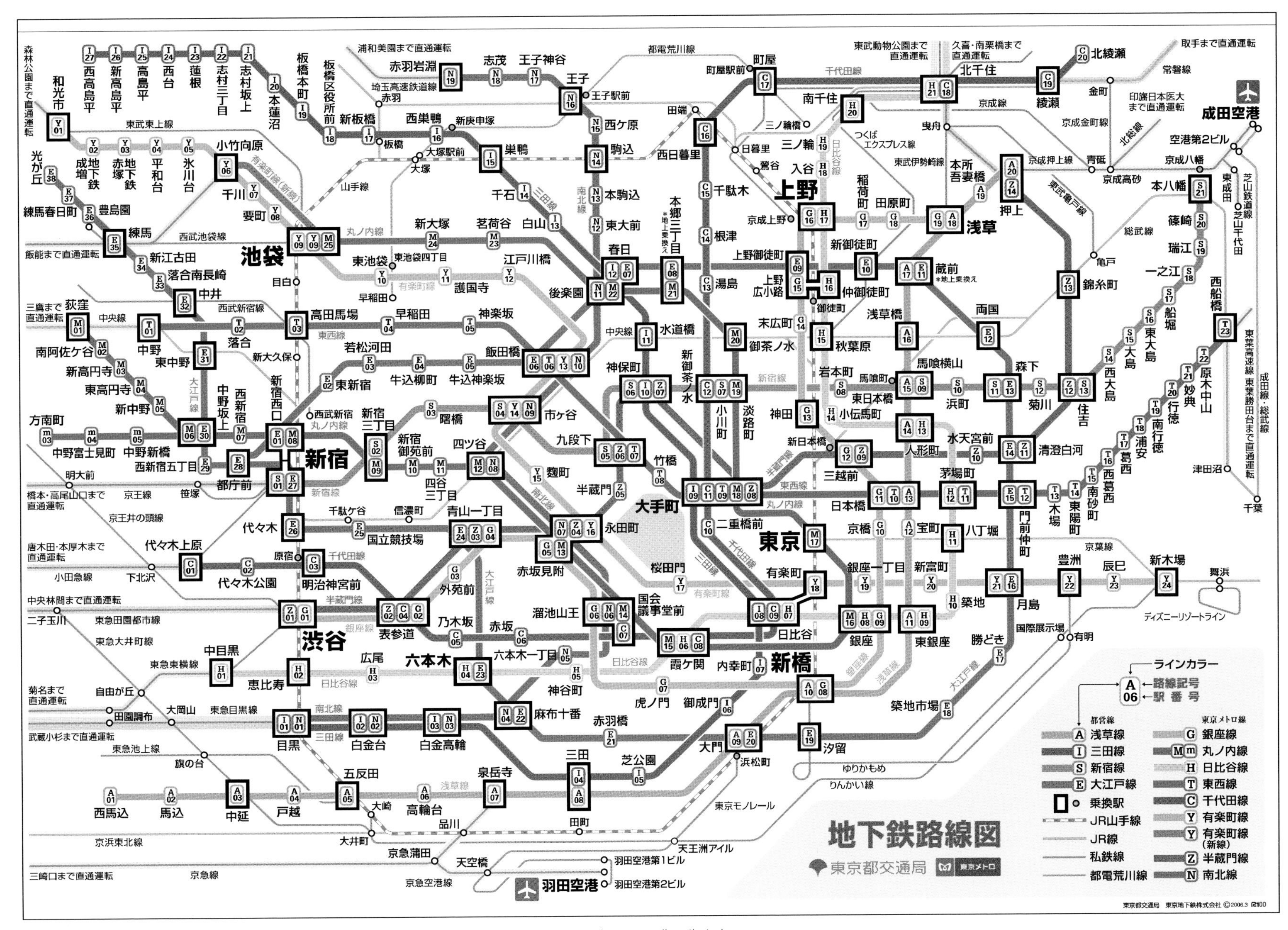

10. 2007: Joint TOEI/Tokyo Metro diagram using Latin letter forms as line identifiers and station numbers as well as their Japanese names.

alphabet letter, generally the first letter of the line's name.

The new joint Tokyo Metro/TOEI map now shows the line letter and station number at every station, as well as the old Japanese line and station name, theoretically making it easier to navigate. However, the resulting diagram (10) does resemble something closer to a computer-chip circuit board—possibly not such a bad idea for such a well-connected modern metropolis.

Subte
Callao
Línea B a F. Lacroze y L. N. Alem
U

M
TMB

Zone 2

Barcelona	046
Boston	048
Budapest	050
Buenos Aires	052
Hamburg	054
Hong Kong	056
Lisbon	058
Mexico City	060
Montreal	062
Munich	064
Osaka	066
San Francisco	068
Seoul	070
St. Petersburg	072
Washington DC	074

On this spread can be seen street "totems" luring passengers onto some of the oldest underground railways, like those in Budapest, Buenos Aires, and Barcelona; these cities, with less-complex networks, have had to make fewer changes in map production. Also in this section are some cities with newer rapid-transit systems, like Washington DC, Mexico City, and Seoul, the latter of which are expanding rapidly. Here can be found, as Diseño Shakespear, creators of the impressive rebranding of Buenos Aires's Subte and map, put it, "design for rational handling of information."

Urban population: 2.7 million. Route length: 63.7 miles. Stations: 124. First section open: 1924. Underground: 98 percent.

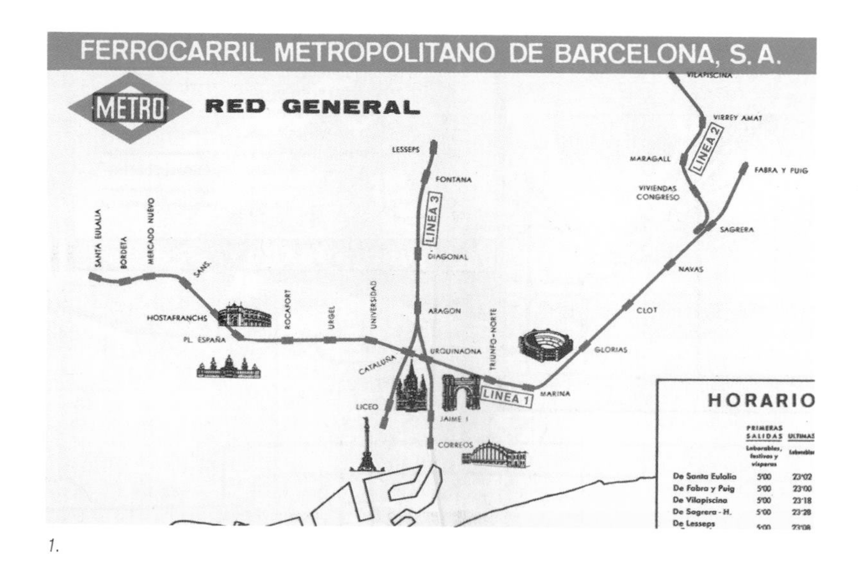

1.

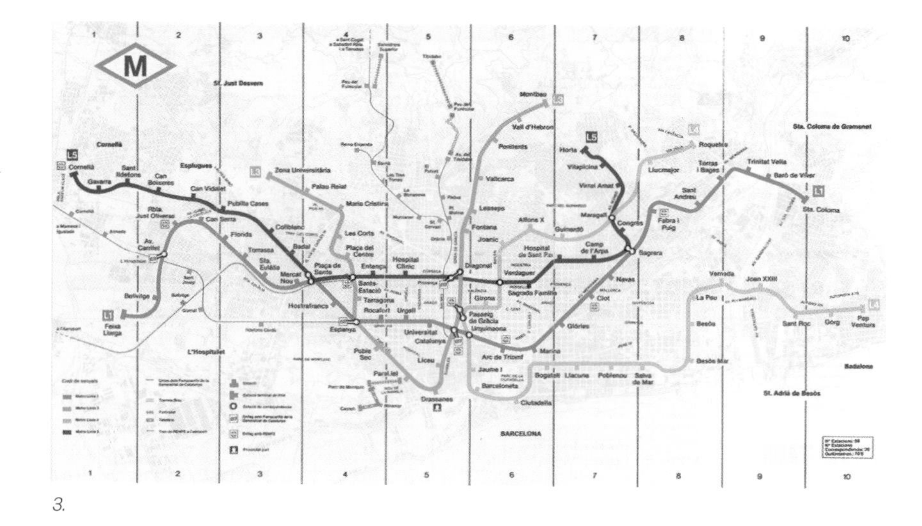

3.

Barcelona's current Metro map (4) is a successful hybrid. While it shows some topographic detail, it manages to retain all the attributes of a schematic. This is partly due to the city's grid-like street layout, under which many of the earlier lines were buried, but also because the designers have employed some familiar cartographic devices that shout "diagram." For example, angles are almost all at the signature 45 degrees. Here are neat ticks for all stations, black circles with white bull's-eyes and white-line connectors marking the interchanges. Curves are all standardized and some unnecessary deviations are ironed out. No station names break the cardinal rule of crossing over a line. The strong line colors are resplendent alongside some key roads. This stems from the map's heritage, when Metro lines were shown as an "overprint" on commercial street plans. Only the Zona Universitària terminus arm of Line 3 and Tram routes to Francesc Macià take an idiosyncratic lean in the 30-degree direction because they follow the physical trajectory of Avinguda Diagonal. However, Avinguda Meridiana, (between Marina and Sagrera stations) as its name suggests, physically runs true north–south, so the sea should be on the right of the map! Most Barcelona maps (as was the case in Chicago, see p. 16) skew the city roughly 40 degrees to put the sea at the bottom!

The first diagram to strip the topography away is likely to have been the pioneering 1966 version (1). Although road-free network plans popped up, like this 1983 example (2), which was the origin of the hybrid plan in use today, most publicity in the last three decades of the twentieth century, like the 1989 pocket map (3), also showed the streets. Both the 1983 and 1989 maps come from the period when the Metro started gaining an identity as a confident, innovative, effective, and growing system. The 1983 diagram was Barcelona's earliest use of the white-line connectors, while the 1989 version introduced the thickening of the lines—both devices still in use on today's schematic.

Barcelona, home to Spain's first railway, the 1848 line to Mataró (now part of R1), opened the Sarria suburban line in June 1863. It became an interurban electric tram in 1916, then a street railway until put belowground in 1929. Any plans to link the city's peripheral termini would

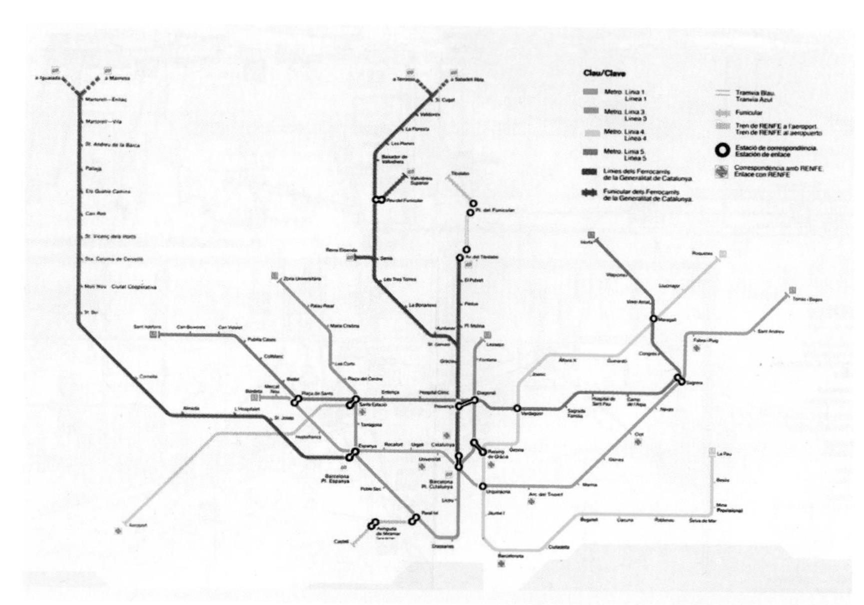

2.

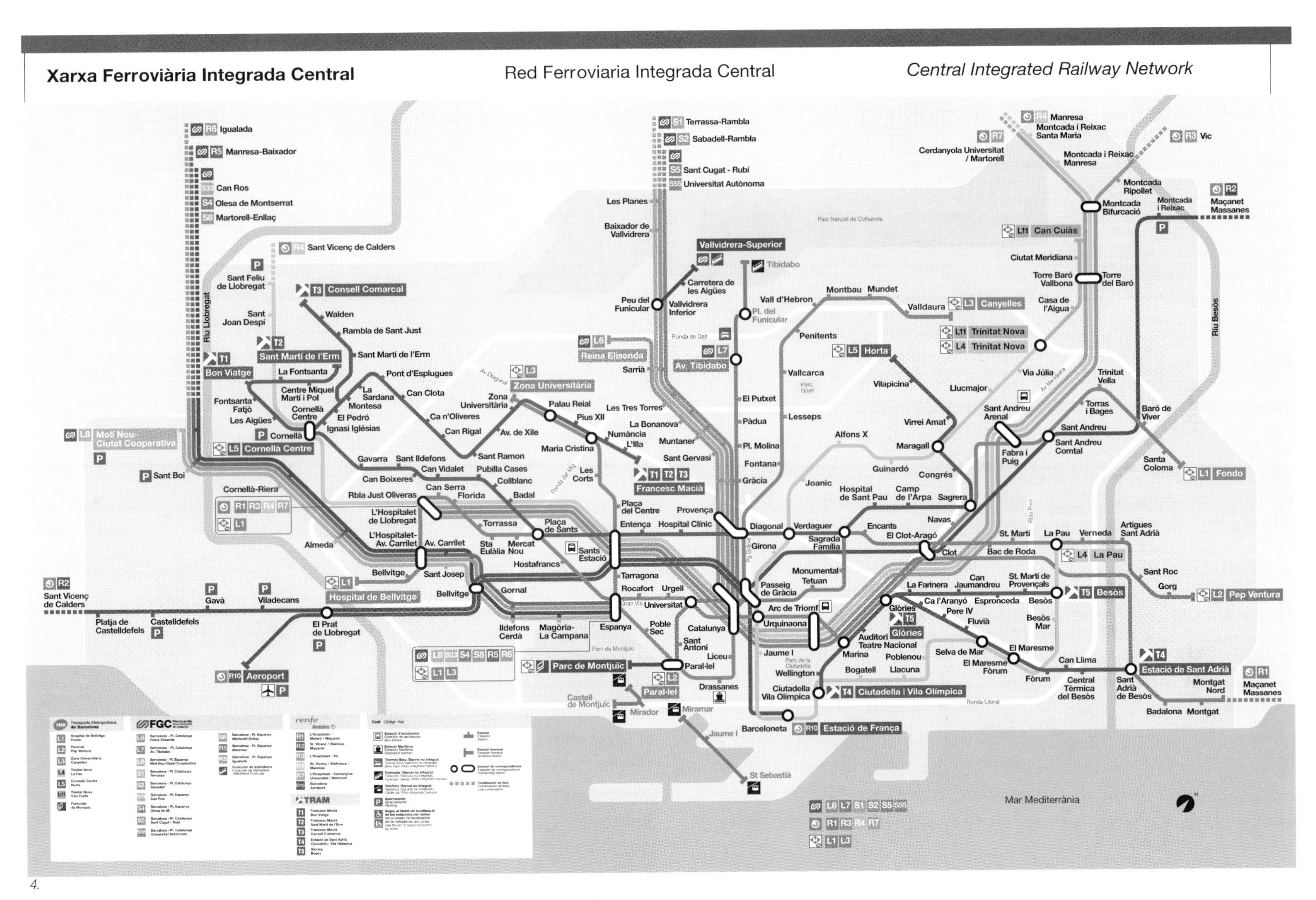

4.

have necessitated unjustifiable damage to the historic city center—one of the great survivors of medieval Europe. So two routes built by separate companies provided the genesis of a true urban-rail network. Just after the First World War, work began on the Gran Metropolitano de Barcelona between Catalunya and Lesseps, which opened in 1924. Intermediate stations followed later, as did a branch to Correos utilizing tunnels excavated as early as 1908 in anticipation of an underground railway. The second line, the Ferrocarril Metropolitano Transversal de Barcelona, opened in 1926 and linked with the Gran at Catalunya. Built with the intention of forming a physical link to the mainline tracks, a feature seldom utilized, it has the wider Iberian gauge allowing carriages to be 10 feet 2 inches across—among the most spacious subway trains in the world.

After the Spanish Civil War, during the growth of the 1950s and '60s, expansion recommenced. Barcelona is notable for numerous and frequent alterations to extensions—such as the modern Line 2, which has veered all over the north of the city since its conception. Major new works, including Line 9, are now under way.

Boston

Urban population: 5.8 million. Route length: 63.7 miles. Stations: 117. First section open: 1897. Underground: 27 percent.

Boston has one of the cleanest-looking urban-transit maps (5), with its bright red, green, yellow, and blue lines, neat 45-degree diagonals, and smooth curves. Given the city's coastal location, the many stylized water features, and the inclusion of several key interstate routes, the illusion is that this is a topographic map, but there is much artistic license here and it is certainly more of a diagram.

The dedication to excellence in graphic design can be traced right back to this very early color diagram issued in 1926 (1 and front cover 2). It shows in orange what had been the first subway outside

2.

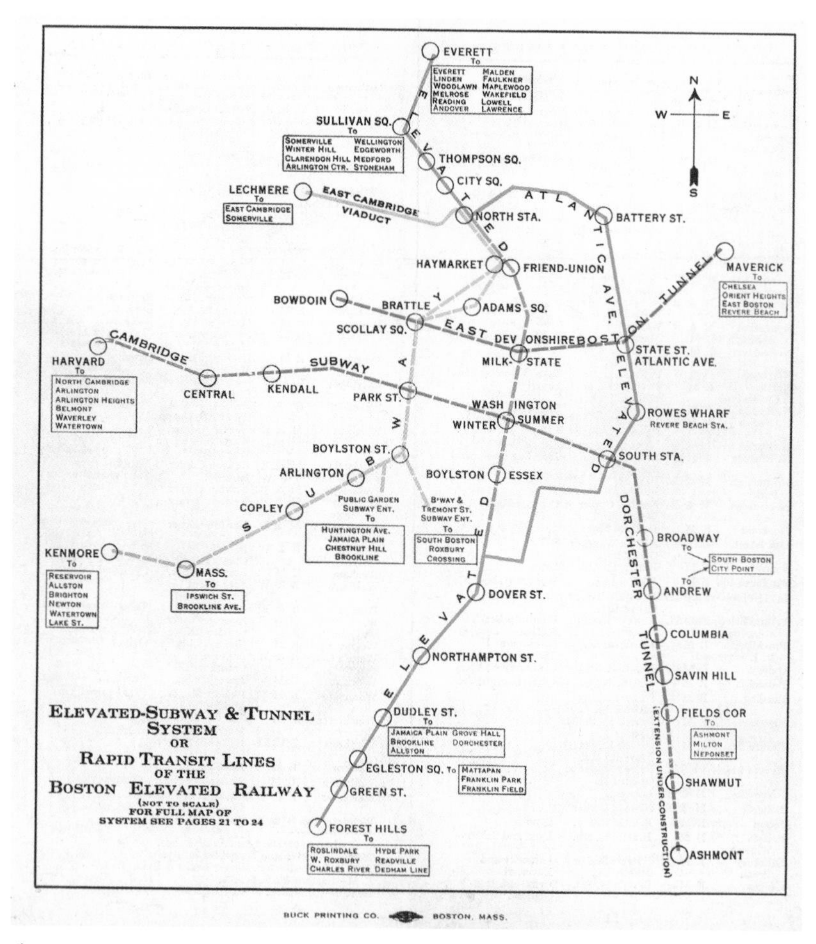

1.

Europe, which now forms part of the Green Line, as well as showing in light blue dashes the Cambridge Subway, the Dorchester Tunnel (later to form the Red Line), and the East Boston Tunnel, now part of the Blue Line.

It is a common misconception in Europe that in North America the car is "king" and always has been. Certainly, when viewing the rapacious growth of U.S. highway construction, it may be hard to imagine anything different. Boston, however, the center of shrewd, intellectual New England, could not have developed into a working city without a major commitment to public transport. This can be spotted partly in the evolution of these stylish and elegant diagrams; the 1967 pocket card version (4) was a classic of its time in the States. Simplifying the system to its most basic perpendicular form, it is as functional as it is graceful.

From a design perspective, the early experimentation in 1974 with an unusual black background (3) is indicative of a pioneering attitude going back to Boston's roots.

Born in the 1600s on a bay-side peninsula, Boston owed its early growth to ferries and bridges. The horse-drawn vehicles that moved the populace

3.

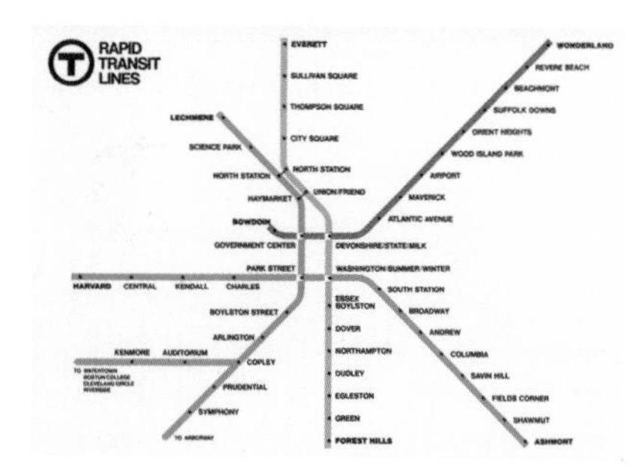

4.

through its early development had by the mid-nineteenth century been converted to early streetcars, drawn over metal rails placed in the road.

As in all major Victorian cities, thousands of horses proved incapable of dealing with increasing traffic demands. In 1889 Boston became the first major city to introduce tramway electrification, following a lead set by the much smaller city of Richmond, Virginia.

Although electric cars proved their worth in speed and capacity, surface congestion increased. The solution to overcrowding in downtown Tremont Street was to submerge the tracks in a subsurface tunnel. Thus in 1897 Boston opened the first subway in the United States, still in use today between Park Street and Boylston.

Innovation continued in the early 1900s with Boston's first elevated railway—or El, as it became known—which began running high above Sullivan Square in 1901. In time the Els were to dive underground, utilizing the Tremont Street tunnel, and eventually became part of today's Orange Line.

The high elevated lines had to swerve around tight radii to dodge buildings as they hugged the street pattern, which necessitated pioneering articulated cars. Another first dates to 1904:

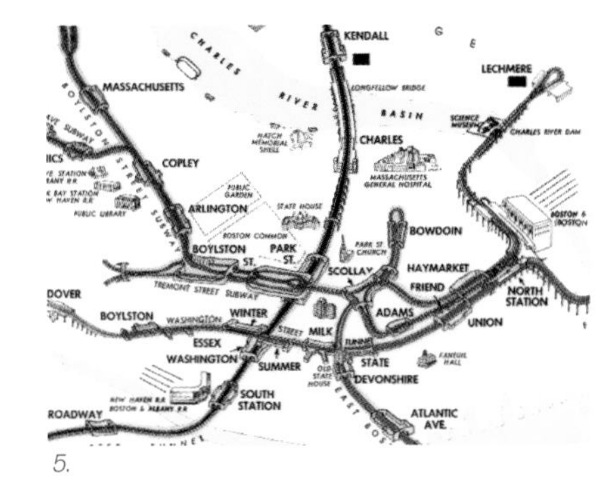

5.

construction of an underwater tunnel beneath Boston Harbor, which now forms part of the Blue Line to Maverick. Regional transport planning began in 1974 with the Massachusetts Bay Transportation Authority (MBTA). The system has been consistently extended throughout the twentieth century, bringing the demolition of many older Els, perceived as streetscape eyesores—hand drawn on a mid-1970s map (5).

The three subway lines marketed as the T, with the light-rail Green Line, have recently been augmented by a new Silver Line service. It is bus rapid transit with pneumatic tires in reserved bus lanes and two separate tunnel sections that could one day be joined together by another new tunnel. That would link the busy Washington Street corridor (route of a former El) to the city's Logan Airport, though the future of the Silver Line is still uncertain.

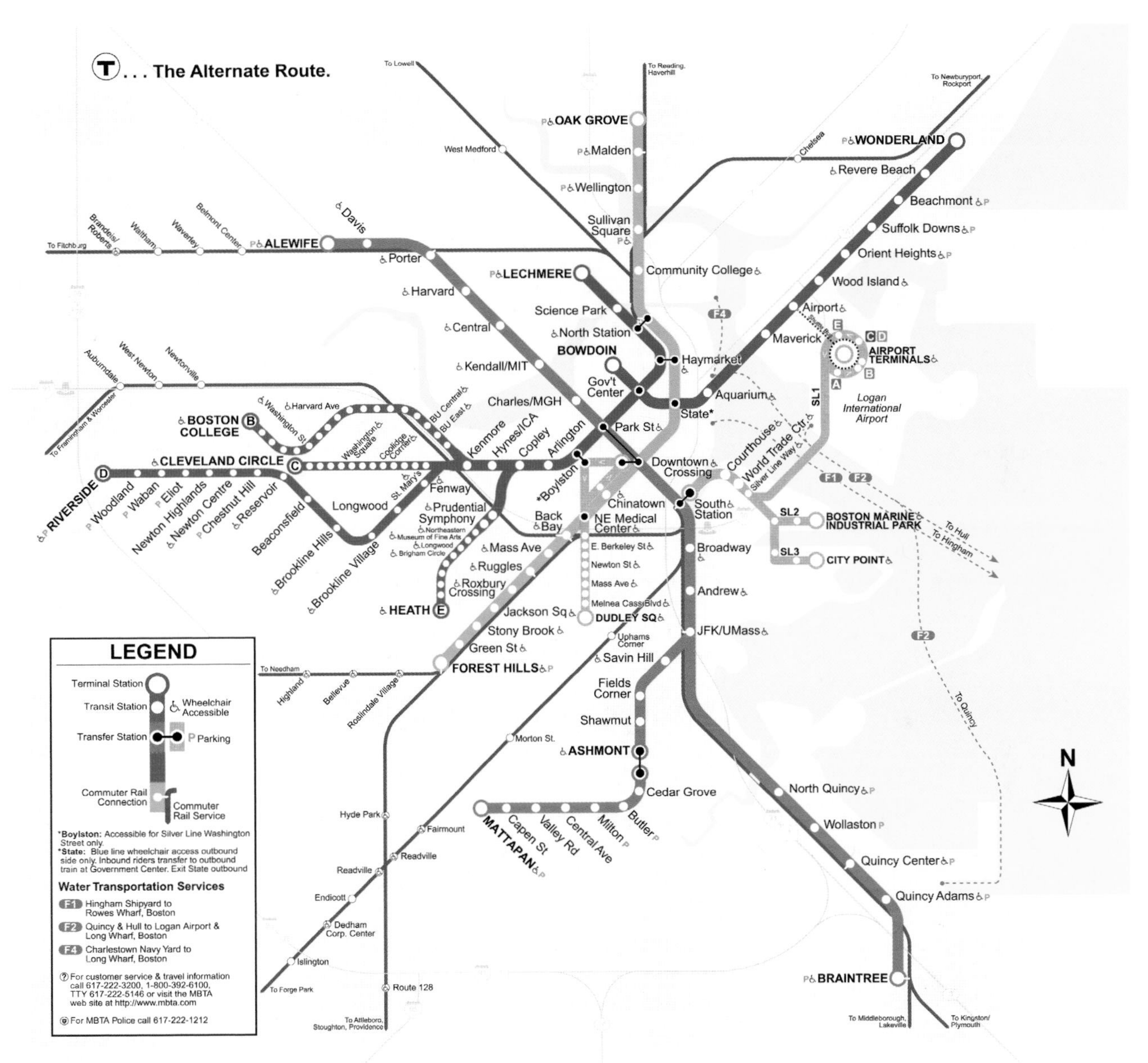

6.

Maps reproduced by kind permission of and copyright MBTA 2007. Antique maps: Mike Ashworth Collection, UITP, author.

Budapest

Urban population: 1.9 million. Route length: 19.9 miles. Stations: 41. First section open: 1896. Underground: 95 percent.

1.

2.

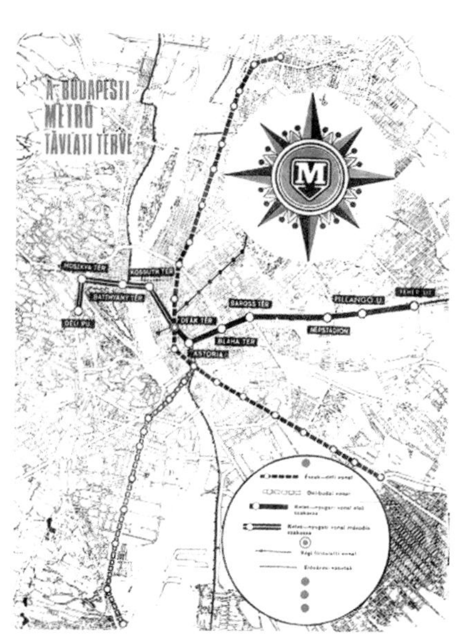

3.

Budapest has a good, simple, clear diagram (9) with bright colors, no topographical features apart from the river Danube, a strongly adhered-to pattern of 45-degree diagonals, and no station names breaking the route lines. In addition, an endearing feature of the 2003 diagram (8) was the use of the M logo as station markers.

Named Milleneum Földalatti Vasút, the Millennium Underground Railway, to celebrate 1,000 years of the Hungarian state, this landmark construction (now the yellow M1 Line) was continental Europe's first subway when it opened in 1896 (6). It was second only as an electric underground railway to the 1890 City & South London line. The architectural splendor (7) of the original Földalatti, with its exquisite tiling, was restored as part of the celebration of its centenary when it gained modern but evocative entrance signs (1).

However, the Hungarian lead was not to be added to for another seven decades when under the planned economy the second line, the red M2 route, opened on an east–west trajectory in 1970, exhibiting a rather harsh, almost austere functionalist design on stations and maps (2).

The third line, a north–south route, opened its doors in 1976 as the blue M3 Line, part of a planned four-line network (3). Since then there have been no more new lines, but a 1987 proposal would have joined up the still-disparate HEV tracks and added a new subway line with several branches (4).

The Municipality of Budapest and the Central Government are both committed to rectifying what they see as a long-overdue "oversight in development," proposed as long ago as the 1960s and included within the unfulfilled 1987 plan, to bring more of the Buda side into the metro system.

Now as the country reorientates itself under European Union membership, the city is regenerating and the fourth metro line is being built (5). Shown here in purple, M4 will link mainline rail stations Kelenföldi pu in the southwest and Keleti pu in the northeast; it will cross M3 at Kálvin tér, go under the Danube to the popular tourist spa and castle areas around Szt Gellért tér, and then veer through Buda's more comfortable suburbs.

It will be 4.5 miles long, with ten stations, and will benefit half a million Budapest residents and tourists. Construction, including an additional four stations toward the northeast, was due to start in 1998, with the line expected to open by 2003, but though work has now started it's not yet clear when the first trains will run.

A major refit of the Blue and Red lines, which started in 2004, and a rebranding of the system have led to a much improved ambience, which has not only boosted traffic but also encouraged film crews, who have made several commercials on the system!

6.

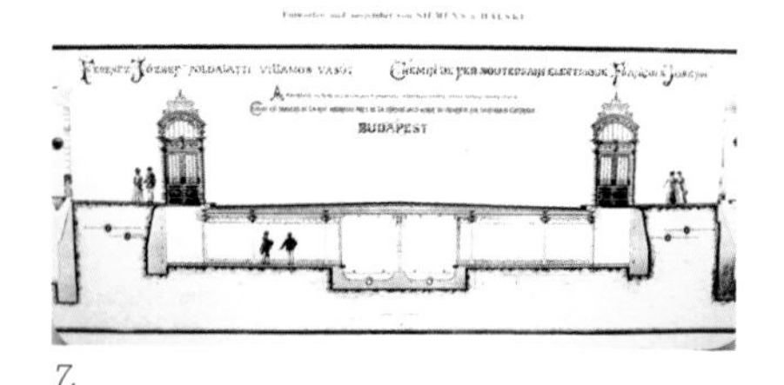

7.

4.

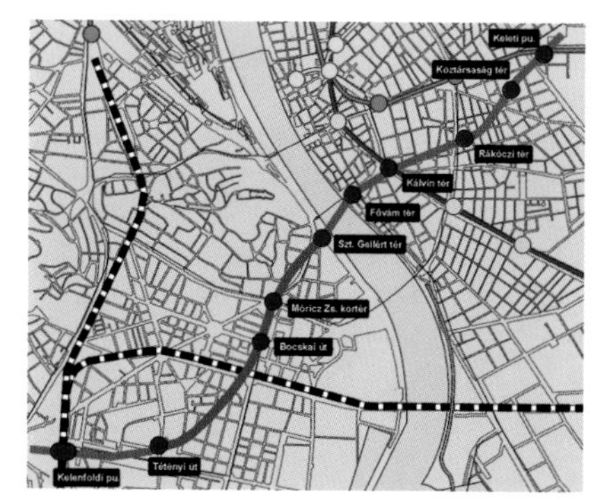

5.

Stations have recently been renovated.

Much old ex-Soviet-style rolling stock remains.

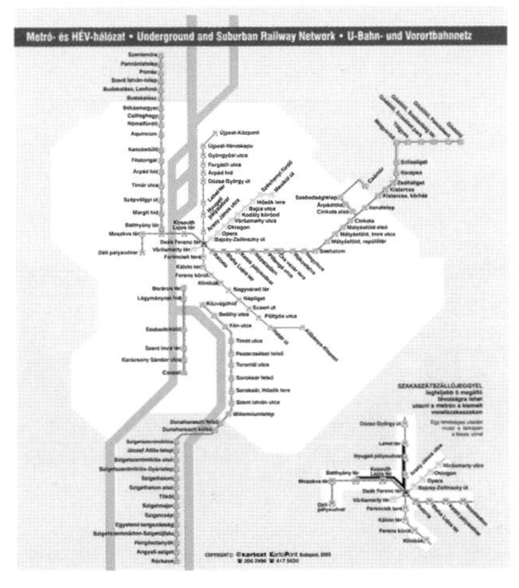

8.

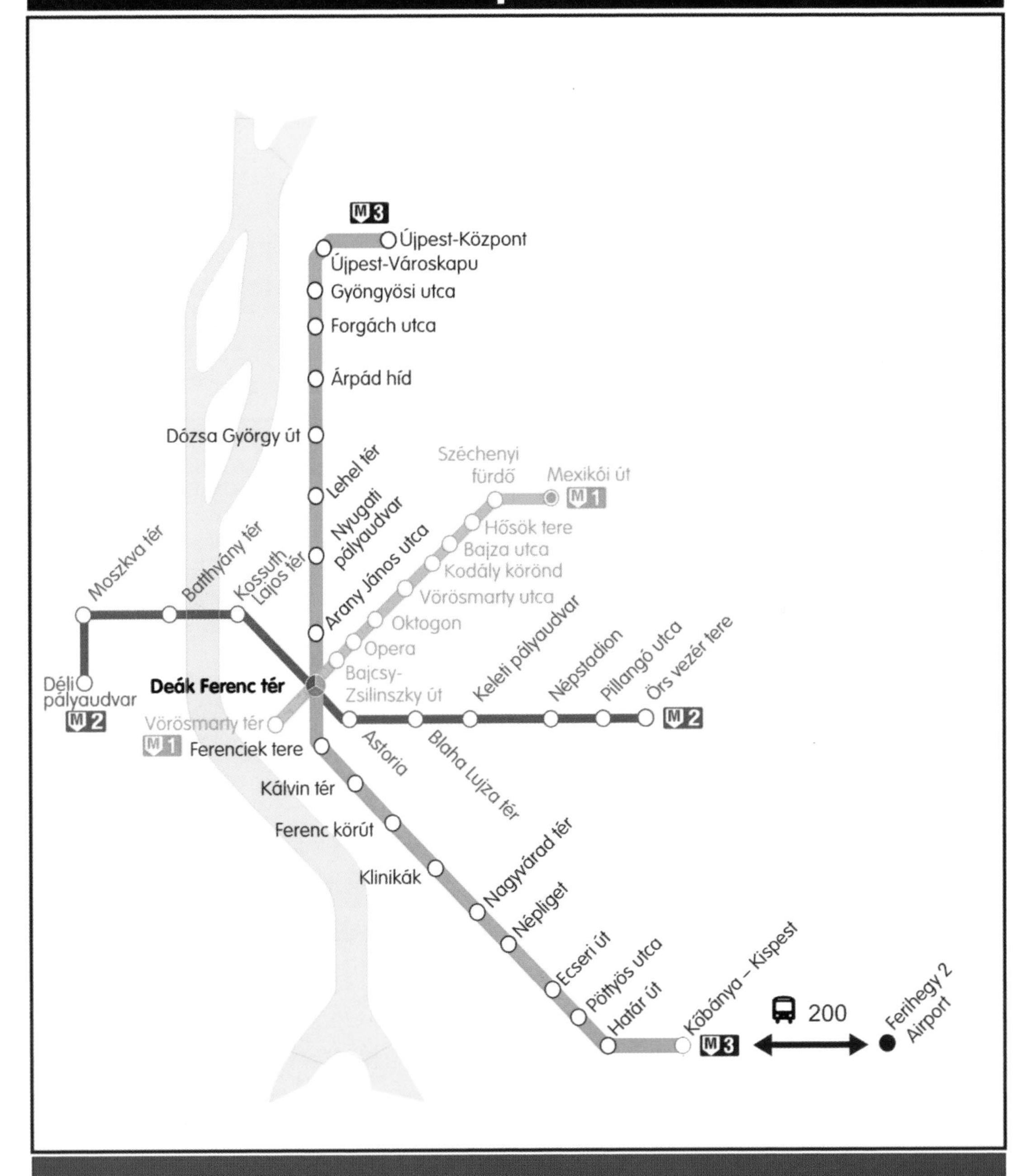

9.

Maps and photos reproduced by kind permission of and copyright BKV 2007. Antique maps: UITP and author.

Buenos Aires

Urban population: 12.4 million. Route length: 30.4 miles. Stations: 63. First Section open: 1913. Underground: 100 percent.

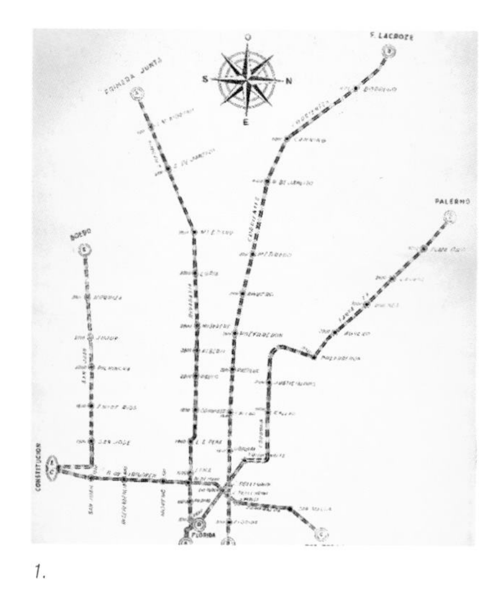

1.

The present Buenos Aires map is one of only a small number of diagrams to regularly be set against such a profoundly dark or black background (6), the others being Montreal, Oslo, Toronto, Naples, Kobe, and the planned system of Chelyabinsk.

Printing with such a large amount of black not only requires accurate registration, or alignment of the design elements in the printing process, but is often felt by some to be more difficult to read because the human eye can have problems with smaller letters reversed out of such dark backgrounds. But sometimes going against the grain pays, and this wonderful official diagram by Lorenzo Shakespear of Diseño Shakespear, along with a complete rebranding of the system (4, 5), seems to have pulled it off beautifully and is now a distinctive part of life in the Argentine capital.

It hasn't always been this way. A 1955 map (1) exhibits a total lack of color, though it forms an effective diagram. The 1972 in-car map (2) depicted the street pattern above in incredible detail. Even up until 1995, the diagram in popular use was a rather messy affair (3).

The current diagram and new corporate identity for the Subte, from the Spanish *subterraneo*, (underground), has brought about a complete sea change in the public perception of what had been a fairly forlorn network.

3.

4.

For many smaller systems, especially in earlier years, line designation was done by name, usually taking that of the terminus station. The trend toward color tagging appears to have been a marketing development brought about by expansions of the more complicated systems.

Buenos Aires was the first city in South America to build a metro, much of which was thanks to American and British design and capital, as with the development of the country's mainline railways. The first line opened in 1913, running from Plaza de Mayo to Plaza Miserere. Quickly expanded to fourteen stations in about a 3-mile run, this formed the backbone of the system. It also included the extraordinary sight of underground trains rising from their subterranean lair and running down the middle of streets to access the depot!

It was not until the early 1930s that a second line was built from L. N. Alem to Federico Lacroze. The same decade also saw work commence on the third and fourth lines. There was a hiatus in development until 1966, when Line E was extended to Bolívar. It reached the current terminus at José María Moreno in 1973.

The latter years of the twentieth century saw much stagnation in the system, characterized by declining passenger numbers and lack of investment, neither of which was conducive to expansion. Design and corporate identity were last on the agenda, as the earlier maps show. However, there are now huge plans for extensions, including four new lines that will lead to a network of 103 stations on 43.5 miles of route.

Work started on the construction of the first new north–south line, Line H, in 2000, and it is due to open at the end of 2007.

2.

5.

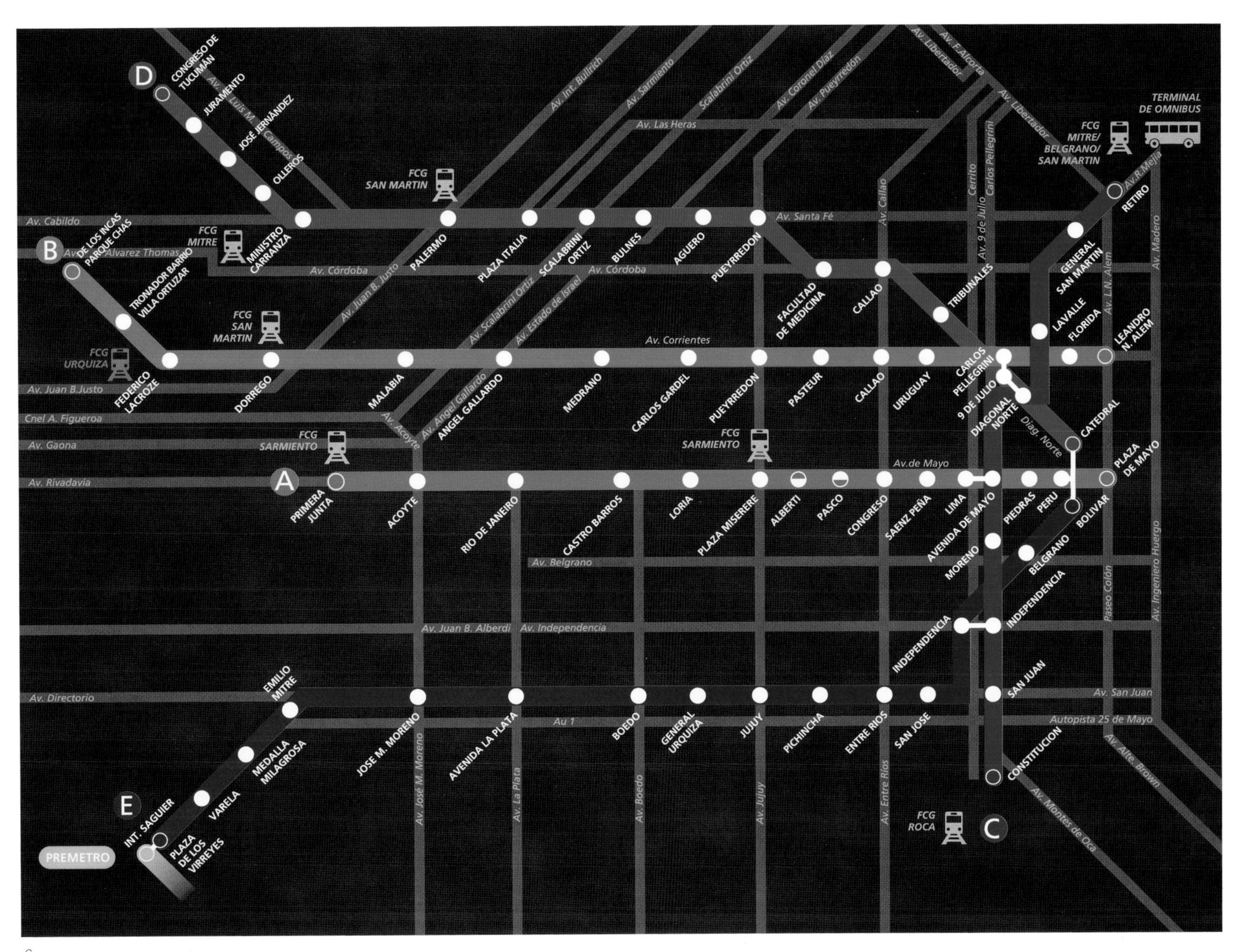
D
CONGRESO DE TUCUMÁN
JURAMENTO
JOSÉ HERNÁNDEZ
OLLEROS
MINISTRO CARRANZA
PALERMO
PLAZA ITALIA
SCALABRINI ORTIZ
BULNES
AGUERO
PUEYRREDON
FACULTAD DE MEDICINA
CALLAO
TRIBUNALES
9 DE JULIO
DIAGONAL NORTE
CATEDRAL
GENERAL SAN MARTIN
LAVALLE
RETIRO
TERMINAL DE OMNIBUS
FCG MITRE/ BELGRANO/ SAN MARTIN
FCG SAN MARTIN
FCG MITRE
FCG URQUIZA
FCG SARMIENTO
FCG ROCA
B
DE LOS INCAS PARQUE CHAS
TRONADOR BARRIO VILLA ORTUZAR
FEDERICO LACROZE
DORREGO
MALABIA
ANGEL GALLARDO
MEDRANO
CARLOS GARDEL
PUEYRREDON
PASTEUR
CALLAO
URUGUAY
CARLOS PELLEGRINI
FLORIDA
LEANDRO N. ALEM
A
PRIMERA JUNTA
ACOYTE
RIO DE JANEIRO
CASTRO BARROS
LORIA
PLAZA MISERERE
ALBERTI
PASCO
CONGRESO
SAENZ PEÑA
LIMA
AVENIDA DE MAYO
PIEDRAS
PERU
BOLIVAR
PLAZA DE MAYO
MORENO
BELGRANO
INDEPENDENCIA
INDEPENDENCIA
SAN JUAN
CONSTITUCION
C
E
EMILIO MITRE
JOSE M. MORENO
AVENIDA LA PLATA
BOEDO
GENERAL URQUIZA
JUJUY
PICHINCHA
ENTRE RIOS
SAN JOSE
MEDALLA MILAGROSA
VARELA
INT. SAGUIER
PLAZA DE LOS VIRREYES
PREMETRO
Av. Cabildo
Av. Alvarez Thomas
Av. Córdoba
Av. Juan B. Justo
Cnel A. Figueroa
Av. Gaona
Av. Rivadavia
Av. Directorio
Av. Santa Fé
Av. Las Heras
Av. Corrientes
Av.de Mayo
Av. Belgrano
Av. Juan B. Alberdi
Av. Independencia
Av. San Juan
Au 1
Autopista 25 de Mayo
Av. Luis M. Campos
Av. Int. Bullrich
Av. Sarmiento
Scalabrini Ortiz
Av. Coronel Díaz
Av. Pueyrredón
Av. Libertador
Av. F. Alcorta
Av. R. Mejía
Av. Callao
Av. 9 de Julio
Cerrito
Carlos Pellegrini
Av. L.N. Alem
Av. Madero
Av. Ingeniero Huergo
Paseo Colón
Diag. Norte
Av. Scalabrini Ortiz
Av. Estado de Israel
Av. Angel Gallardo
Av. Acoyte
Av. Jose M. Moreno
Av. La Plata
Av. Boedo
Av. Jujuy
Av. Entre Rios
Av. Alte. Brown
Av. Montes de Oca

6.

Hamburg

Urban population: 2.6 million. Route Length: 62.5 miles. Stations: 89. First section open: 1912. Underground: 70 percent.

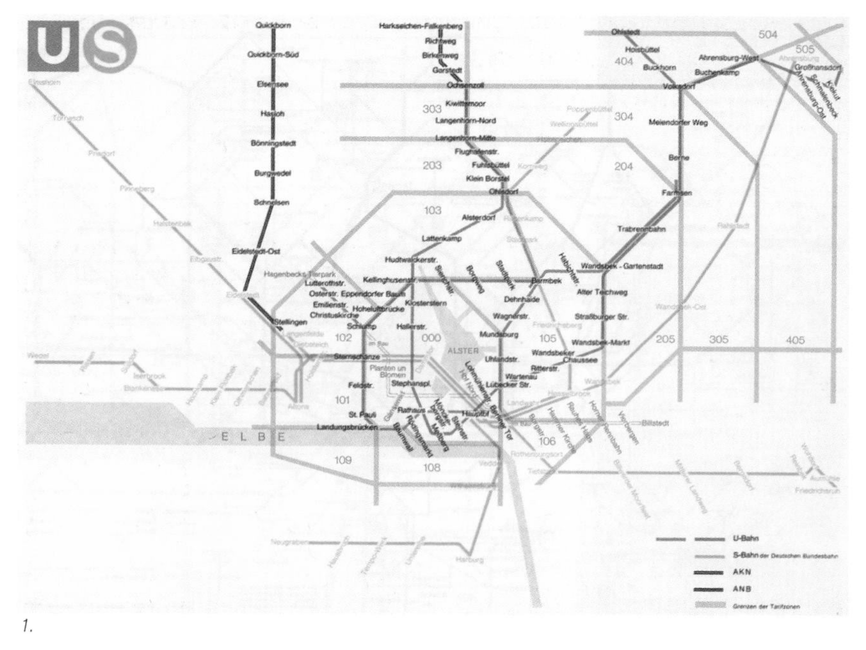

1.

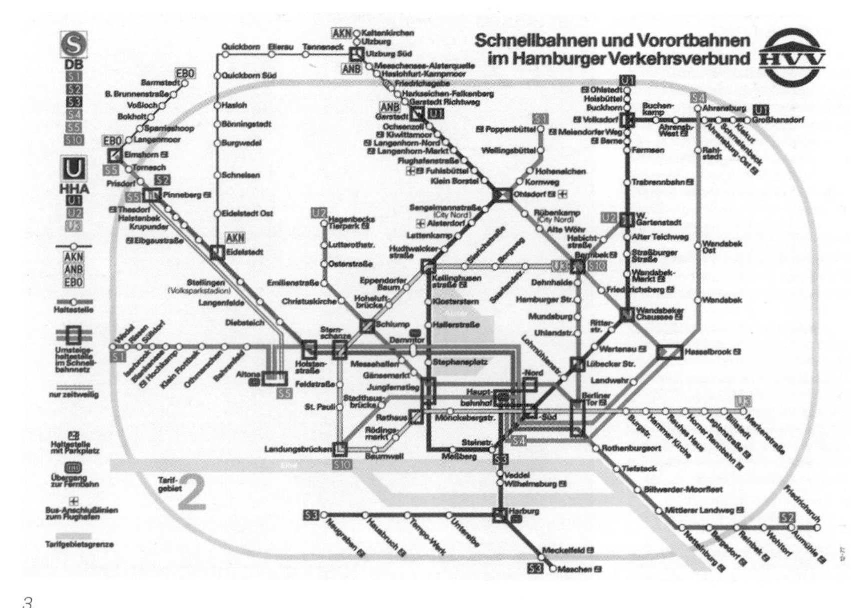

3.

The most noticeable thing about Hamburg transit maps is their lack of curves—except for the elongated oval of the Greater Hamburg boundary and at the corners of Lake Alster. The current style of diagram (4) began life just after the initial ring was added to in the mid-1960s (1), with a new line from Ochsenzoll to Ohlstedt & Großhansdorf (now on the U1). On this there are a few slight curves and the diagonals are not yet standardized. There is a bold but unsuccessful attempt to remove ticks or circles for station names and rely simply on the text of the station name itself crossing the line, but even with just two routes this gets pretty confusing in the Berliner Tor–Hauptbahnhof region.

2.

By 1972, on the eve of the U2 Line's completion, this detail from a large station wall poster (2) shows the true ancestry of the current diagram was clear, with bold colors and standardized 45-degree diagonals—but again this map also completely lacks curves.

The 1977 pocket diagram (3) included some equally rectilinear devices for interchange stations, consisting of black outlines with the route crossovers showing underneath—check Ohlsdorf, Hauptbahnhof, and Jungfernstieg!

There are now so many S-Bahn lines converging at the Hauptbahnhof that it is beginning to resemble the Frankfurt situation (see p. 119), where it might be difficult to tell them apart at a quick glance. Finding these quirky anomalies on the diagram of the country's second city is all the more surprising given normally high levels of urban-rail diagram standards in Germany.

The current U-Bahn diagram integrates S-Bahn and regional rail services, which could be extended, and whose frequent and efficient services complement those of the U-Bahn in the city center.

The Hamburg "Hochbahn," as it is affectionately known, started life in 1912 as a partly elevated, partly subterranean circle-line service. There

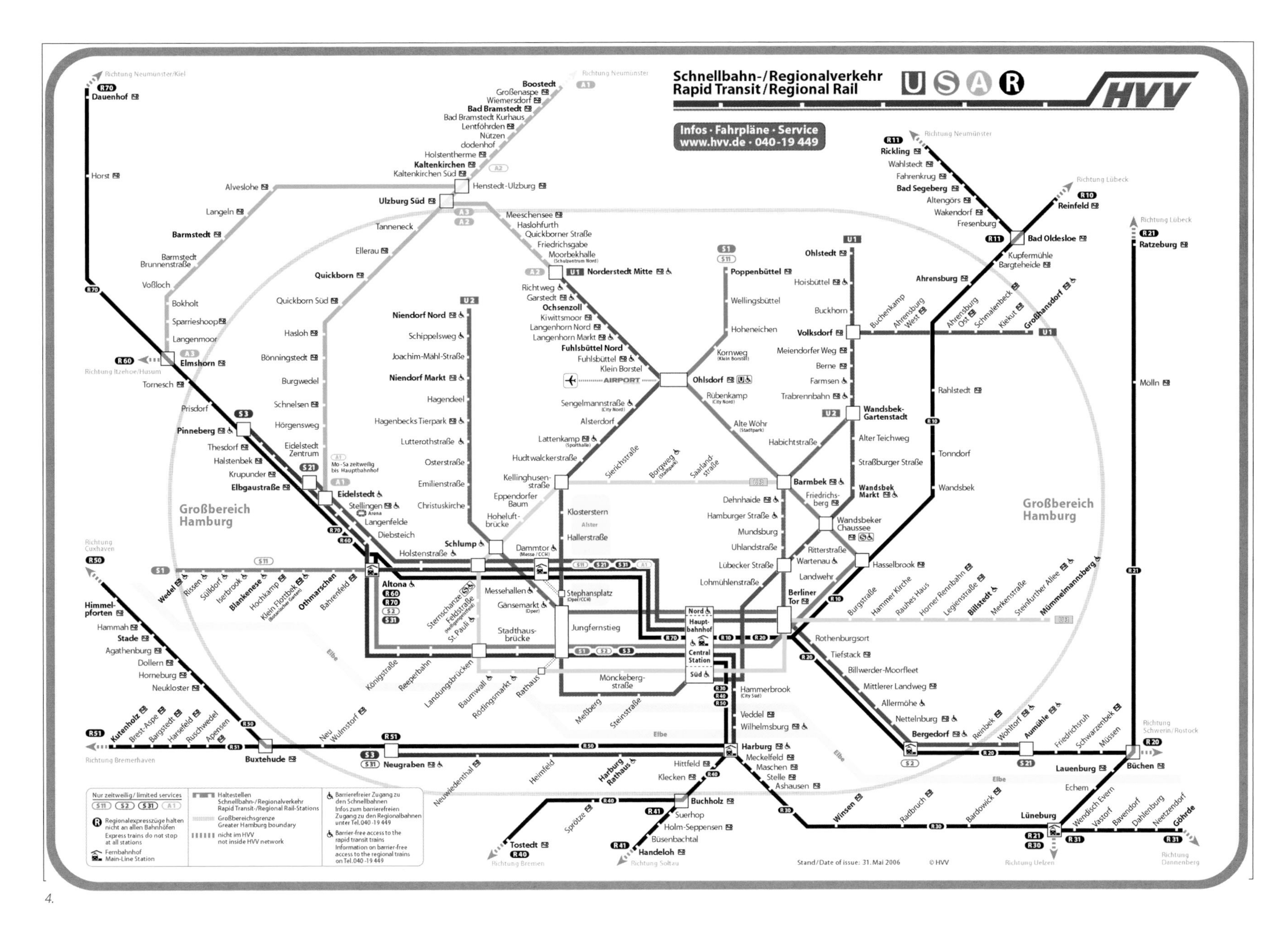

4.

was a massive reorganization in 1973, in which the circle was split up to become parts of the new lines U2 and U3. But it was not until 1985 that these sections were joined by new branches stretching to the outlying areas of the city.

It took until 1996 to complete the three-line network currently in service. A new U4 line is now being added as a branch off the U2 from Jungfernstieg to the HafenCity, a huge docklands development. An S-Bahn extension will reach the airport in 2008, and while the U3 will re-create a ring line with a handle, the future U2 will run from Niendorf Nord to Mümmelmannsberg.

Hong Kong

Urban population: 6.7 million. Route length: 56.5 miles. Stations: 53. First section open: 1979. Underground: 60 percent.

Designed and mostly built when the province was still a British colony, Hong Kong's sleek, seven-line mass-transit system has a diagram (5) that leans in an almost affectionate way toward that of London. Notice the straightening of the wiggles, equalization of station spacing, and the 45-degree diagonals, all in clear contrast to the circuitous route some of the lines run geographically (1).

The bilingual diagram, interestingly called a "system map" by the operator, Mass Transit Railway (MTR), uses black circles with white centers for all stations but elongates them into a more rounded oblong shape on interchanges (5). Both the linear and the more topographic varieties have been around since the system's inception, although the less diagrammatic were slightly more commonplace, as in the 1984 version (2).

One gets the feeling that virtually everything the MTR attempts is infused with a strong sense of purpose. This rigid approach to good design has often been the result of building a rapid-transit system through the world's most crowded urban space—take for example the enclosed elevated section at Tsuen Wan, erected to reduce noise pollution to surrounding housing.

MTR also issues smart plastic ticket passes where the map is often depicted on the reverse (3), which have become collector's items (4). Even the in-car line diagrams light up to show the trains' position on the route and flash the next station.

2.

1.

Hong Kong metro cars are some of the broadest in the world, and the trains are among the fastest. The system has been extended steadily since opening in 1979, with bold engineering solutions, like the Tsing Ma and Kap Shui Mun bridges, the longest of their type, to cope with the many water crossings of this archipelagic city.

Since the transfer of control to China in 1997, the MTR has continued to expand, with the long line out to the new Chek Lap Kok Airport on Lantau Island, and the renovation of existing facilities. There are numerous extension projects, which, along with the Kowloon–Canton Railway extensions, could double the size of the existing urban-rail network.

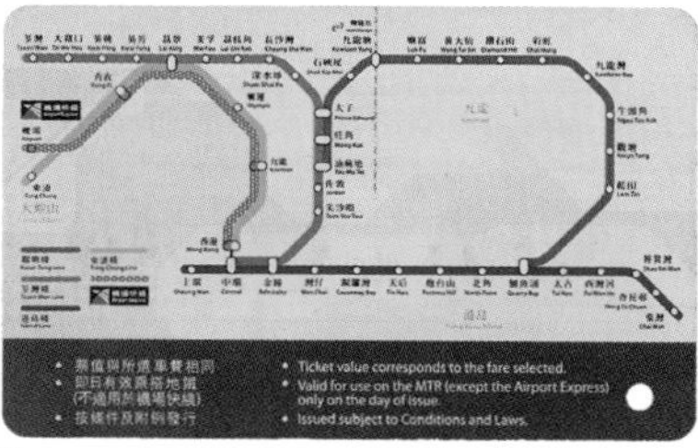

3.

4.

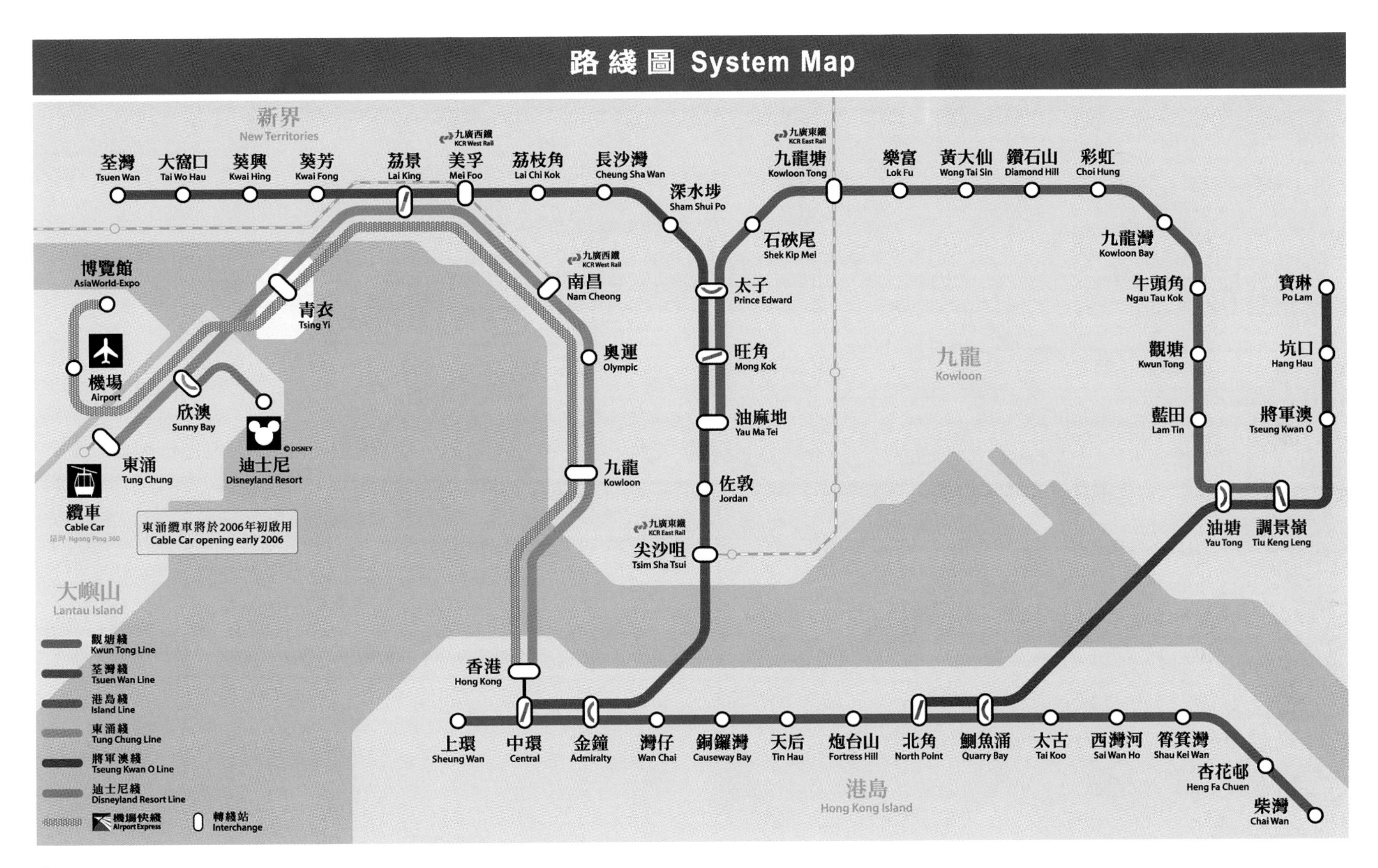
路綫圖 System Map
新界
New Territories
九廣西鐵
KCR West Rail
九廣東鐵
KCR East Rail
荃灣
Tsuen Wan
大窩口
Tai Wo Hau
葵興
Kwai Hing
葵芳
Kwai Fong
荔景
Lai King
美孚
Mei Foo
荔枝角
Lai Chi Kok
長沙灣
Cheung Sha Wan
深水埗
Sham Shui Po
九龍塘
Kowloon Tong
石硤尾
Shek Kip Mei
樂富
Lok Fu
黃大仙
Wong Tai Sin
鑽石山
Diamond Hill
彩虹
Choi Hung
九龍灣
Kowloon Bay
牛頭角
Ngau Tau Kok
觀塘
Kwun Tong
藍田
Lam Tin
寶琳
Po Lam
坑口
Hang Hau
將軍澳
Tseung Kwan O
油塘
Yau Tong
調景嶺
Tiu Keng Leng
博覽館
AsiaWorld-Expo
機場
Airport
青衣
Tsing Yi
欣澳
Sunny Bay
迪士尼
Disneyland Resort
© DISNEY
東涌
Tung Chung
纜車
Cable Car
昂坪 Ngong Ping 360
東涌纜車將於2006年初啟用
Cable Car opening early 2006
南昌
Nam Cheong
奧運
Olympic
九龍
Kowloon
太子
Prince Edward
旺角
Mong Kok
油麻地
Yau Ma Tei
佐敦
Jordan
尖沙咀
Tsim Sha Tsui
九龍
Kowloon
大嶼山
Lantau Island
香港
Hong Kong
上環
Sheung Wan
中環
Central
金鐘
Admiralty
灣仔
Wan Chai
銅鑼灣
Causeway Bay
天后
Tin Hau
炮台山
Fortress Hill
北角
North Point
鰂魚涌
Quarry Bay
太古
Tai Koo
西灣河
Sai Wan Ho
筲箕灣
Shau Kei Wan
杏花邨
Heng Fa Chuen
柴灣
Chai Wan
港島
Hong Kong Island
觀塘綫
Kwun Tong Line
荃灣綫
Tsuen Wan Line
港島綫
Island Line
東涌綫
Tung Chung Line
將軍澳綫
Tseung Kwan O Line
迪士尼綫
Disneyland Resort Line
機場快綫
Airport Express
轉綫站
Interchange

5.

Lisbon

Urban population: 2.5 million. Route length: 23 miles. Stations: 44. First section open: 1959. Underground: 72 percent.

The present Lisbon diagram gives a wonderful sense of spaciousness, light, and clarity (6). This must at least in part come from the unique idea of naming the lines so poetically and with such whimsically stylized pictograms.

The Blue Line "Gaivota" is the Seagull Line; Yellow Line "Girassol" is the Sunflower Line; Green Line "Caravela" is the Sail Ship Line; and Red Line is the line of the "Oriente" or, in other words, the Far East!

2.

The main map opposite fulfills all the essential criteria for an excellent urban-rail diagram, with diagonals at 45 degrees, standardized curves, stations equally spaced, and their names never crossing over lines.

The typography (in a font designed specially for the system) and line symbolism are used as the basis of all publicity material. This includes station signage, in-car line diagrams (3), and a whole rail-system map, all of which use this clean style and font. As can be seen from a comparison to the more geographic and topographic map (4), the lines on the main diagram have been heavily distorted to straighten them out. The in-car whole-route diagram (4) does this even more obviously by squeezing the entire system into a landscape rectangle for mounting inside the trains.

The red *M* logo is found across the city and is easy to spot among the street furniture. A strong emphasis has been placed on station design and finishes (5), mostly based on evocatively Portuguese ceramic tiling. Many schemes are the product of one artist, Maria Keil, who has been working on the system since its inception in 1959.

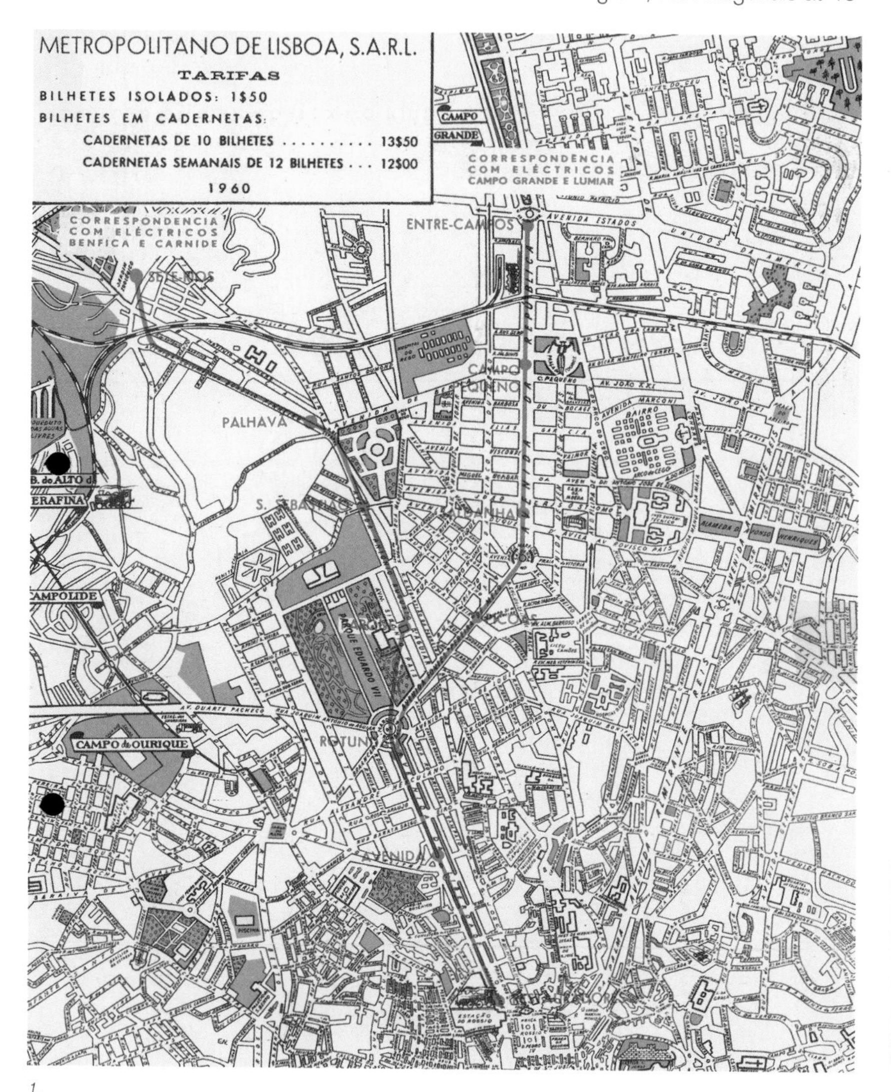

1.

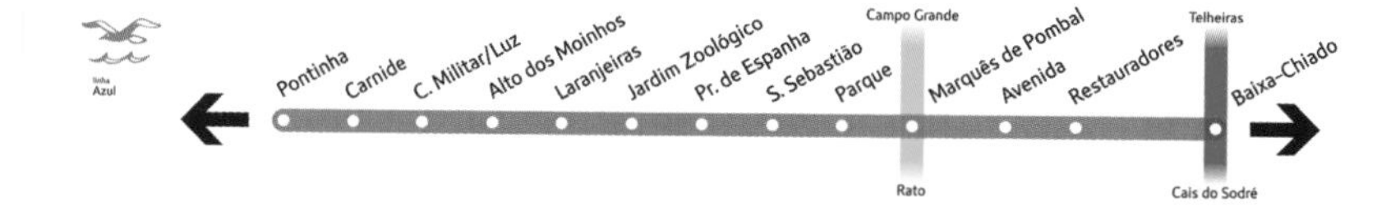

3.

4.

5.

Lisbon decided to join the ranks of subway cities after the end of World War II. Utilizing funds allocated from the Marshall Plan, construction started in 1955. The first part was a Y shape formed from a common section of line from Restauradores to Rotunda (now Marquês de Pombal), where it split to form two branches running to Sete Rios (now called Jardim Zoológico) and Entrecampos, as a 1960 map shows (1).

This initial route grew in a somewhat organic fashion, with continual extensions and improvements until 1998, when the Red Line opened with a distinctly new trajectory. The system has plans for many extensions, several of which are under construction and due to open soon. Existing lines and stations are also being reconstructed to improve services.

The airport is not yet served, and one could be excused for thinking the system is avoiding it—all the lines heading in that direction having taken sharp turns away from it. However, a branch of the Red Line to Aeroporto is now planned, despite the fact that the airport itself is planning to relocate.

6.

Maps reproduced by kind permission of and copyright Metropolitano de Lisboa 2007.

Mexico City

Urban population: 22 million. Route length: 110 miles. Stations: 151. First section open: 1969. Underground: 65 percent.

1.

Mexico City's Metro aspirations go back to 1917, but even after further planning in the 1950s it was not until the city won the right to host the 1968 Olympics that work began in earnest. Unfortunately, the formidable problems of building an underground railway in an earthquake zone and under the old lake-bottom silt that much of the city sits on meant that the work was not completed until after the event. The network now totals over 100 miles, but plans for a full fifteen lines by 2015 are now on hold.

Most of the diagrams reflect a geographic approach. Lines tend to meander rather unnecessarily, depicting corners that riders barely notice. The tiny 1972 pocket map (1) shows just the first three lines complete, with pictograms for each station.

The 1985 pocket version (2) was one of the few that straightened the lines somewhat and thereby simplified the network. By 1989 a fairly detailed background road layout had appeared in faint blue or gray (3).

The current official version (5) retains the geographic feel—minus the roads—with very wiggly lines, but it does show the scale of expansion. In addition to color coding all eleven lines, Mexico City has embraced an idea only hinted at elsewhere: it has given every station its own unique emblem in order to help the illiterate (4). Only the much smaller city Fukuoka, in Japan, has tried this approach outside South America. The pictograms represent images of each station's locality. Talisman station on Linea 4, where the 10,000-year-old remains of a woolly mammoth were found during construction, has an elephant! Occasionally, such individuality is taken a stage further. Bellas Artes station has representations of architectural details from other world systems, such as the Guimard-style entrance of a Paris Métro station.

RED GENERAL
SISTEMA DE TRANSPORTE
COLECTIVO

HORARIO DE SERVICIO

DIAS	LINEAS 1-2-3	LINEAS 4-5-6-7
DIAS LABORABLES	5:00-0:30 HRS	6:00-0:30 HRS
SABADOS	6:00-1:30 HRS	6:00-1:30 HRS
DOMINGO Y DIAS FESTIVOS	7:00-0:30 HRS	7:00-0:30 HRS

LINEA 6— INSTITUTO DEL PETROLEO—MARTIN CARRERA INAUGURACION 1986

LINEA 7— AUDITORIO—BARRANCA DEL MUERTO INAUGURACION 1985

2.

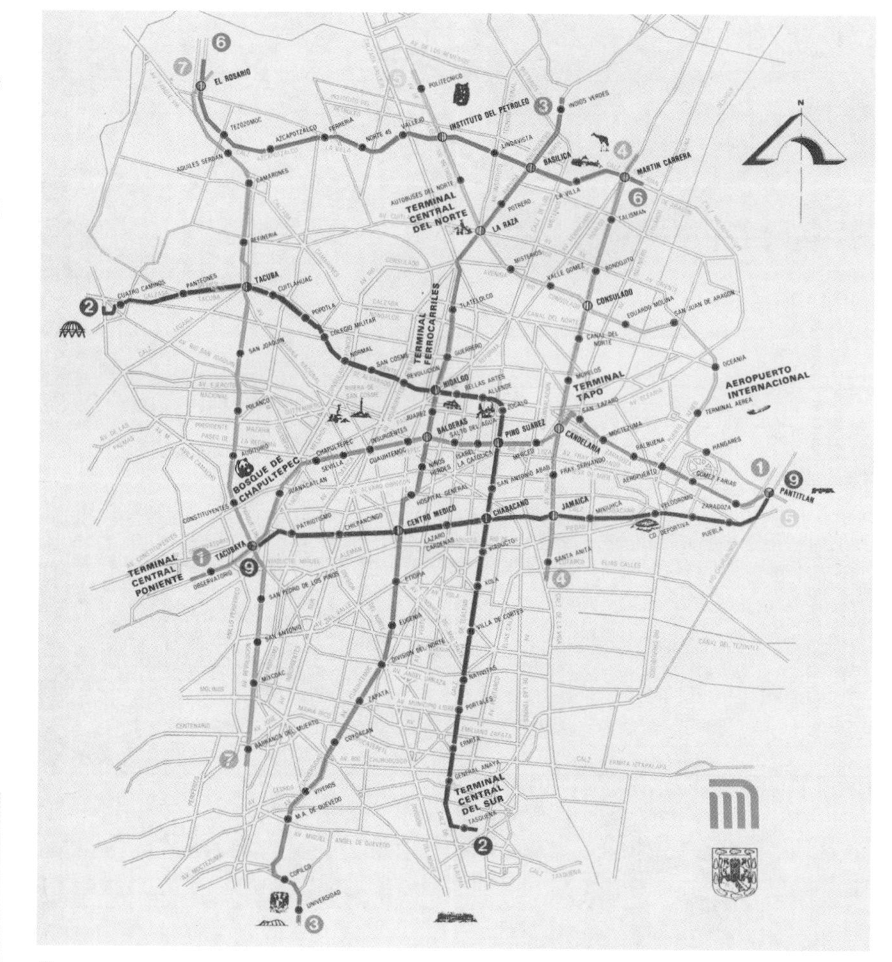

3.

4.

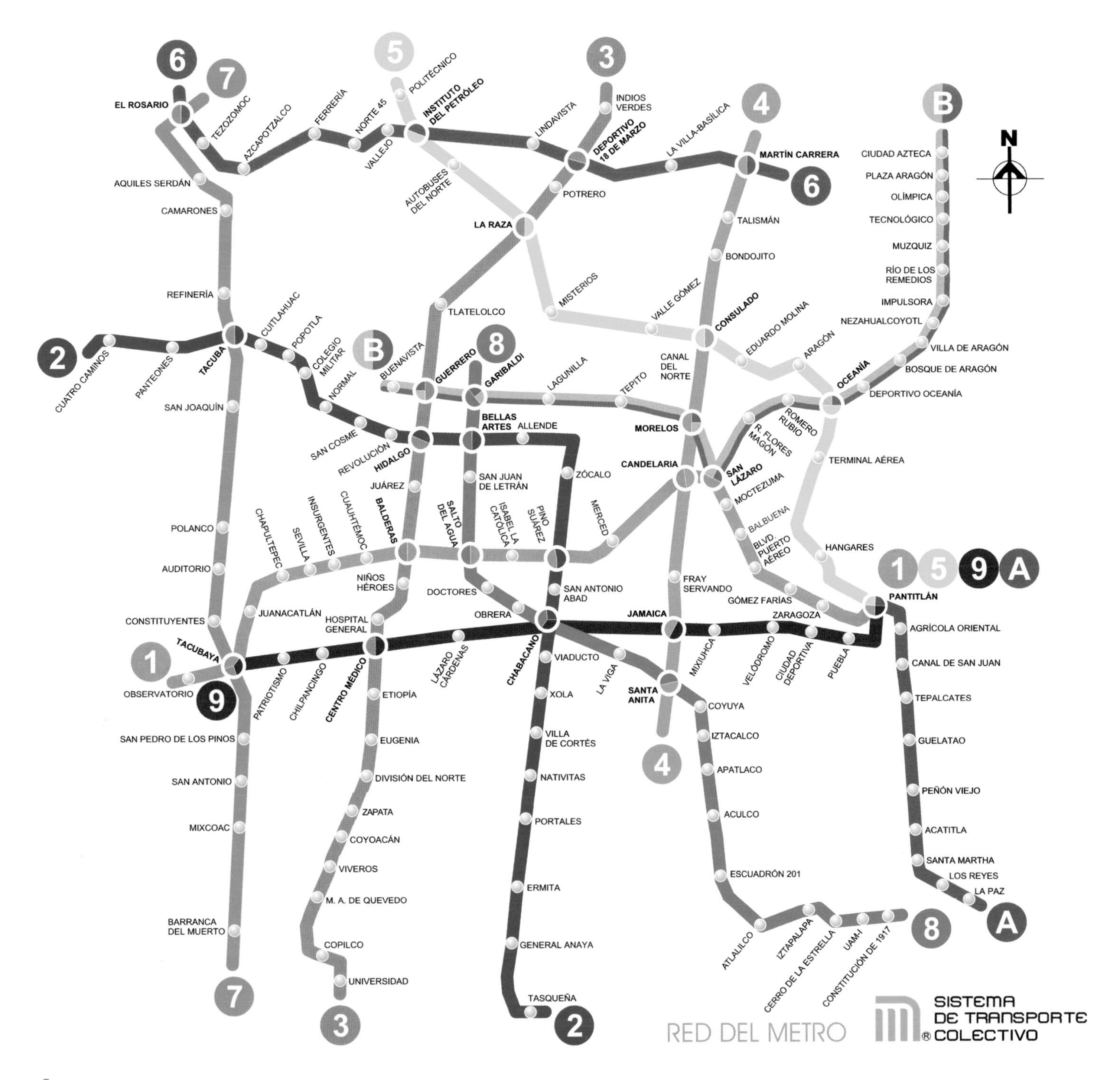

EL ROSARIO
TEZOZOMOC
AZCAPOTZALCO
FERRERÍA
NORTE 45
VALLEJO
POLITÉCNICO
INSTITUTO DEL PETRÓLEO
AUTOBUSES DEL NORTE
LINDAVISTA
INDIOS VERDES
DEPORTIVO 18 DE MARZO
POTRERO
LA RAZA
LA VILLA-BASÍLICA
MARTÍN CARRERA
TALISMÁN
BONDOJITO
AQUILES SERDÁN
CAMARONES
REFINERÍA
TACUBA
CUITLAHUAC
POPOTLA
COLEGIO MILITAR
NORMAL
SAN COSME
REVOLUCIÓN
HIDALGO
CUATRO CAMINOS
PANTEONES
SAN JOAQUÍN
TLATELOLCO
BUENAVISTA
GUERRERO
GARIBALDI
LAGUNILLA
TEPITO
MISTERIOS
VALLE GÓMEZ
CONSULADO
EDUARDO MOLINA
ARAGÓN
CANAL DEL NORTE
MORELOS
BELLAS ARTES
ALLENDE
SAN JUAN DE LETRÁN
ZÓCALO
JUÁREZ
CANDELARIA
SAN LÁZARO
R. FLORES MAGÓN
ROMERO RUBIO
OCEANÍA
CIUDAD AZTECA
PLAZA ARAGÓN
OLÍMPICA
TECNOLÓGICO
MUZQUIZ
RÍO DE LOS REMEDIOS
IMPULSORA
NEZAHUALCOYOTL
VILLA DE ARAGÓN
BOSQUE DE ARAGÓN
DEPORTIVO OCEANÍA
TERMINAL AÉREA
MOCTEZUMA
BALBUENA
BLVD. PUERTO AÉREO
HANGARES
POLANCO
AUDITORIO
CONSTITUYENTES
CHAPULTEPEC
SEVILLA
INSURGENTES
CUAUHTÉMOC
BALDERAS
SALTO DEL AGUA
ISABEL LA CATÓLICA
PINO SUÁREZ
MERCED
NIÑOS HÉROES
DOCTORES
SAN ANTONIO ABAD
FRAY SERVANDO
GÓMEZ FARÍAS
PANTITLÁN
JUANACATLÁN
HOSPITAL GENERAL
OBRERA
JAMAICA
ZARAGOZA
AGRÍCOLA ORIENTAL
TACUBAYA
OBSERVATORIO
PATRIOTISMO
CHILPANCINGO
CENTRO MÉDICO
LÁZARO CÁRDENAS
CHABACANO
VIADUCTO
LA VIGA
SANTA ANITA
MIXIUHCA
VELÓDROMO
CIUDAD DEPORTIVA
PUEBLA
CANAL DE SAN JUAN
TEPALCATES
SAN PEDRO DE LOS PINOS
ETIOPÍA
XOLA
COYUYA
IZTACALCO
GUELATAO
EUGENIA
VILLA DE CORTÉS
SAN ANTONIO
DIVISIÓN DEL NORTE
NATIVITAS
APATLACO
PEÑÓN VIEJO
ZAPATA
ACULCO
MIXCOAC
PORTALES
ACATITLA
COYOACÁN
SANTA MARTHA
VIVEROS
ESCUADRÓN 201
LOS REYES
ERMITA
LA PAZ
M. A. DE QUEVEDO
BARRANCA DEL MUERTO
ATLALILCO
IZTAPALAPA
CERRO DE LA ESTRELLA
UAM-I
CONSTITUCIÓN DE 1917
COPILCO
GENERAL ANAYA
UNIVERSIDAD
TASQUEÑA
RED DEL METRO
SISTEMA DE TRANSPORTE COLECTIVO
N

5.

Montreal

Urban population: 3.5 million. Route length: 37.8 miles. Stations: 65. First section open: 1966. Underground: 100 percent.

Beautifully combining a gentle Francophile style with the more mechanical force of contemporary North American design, diagrams of the Montreal Métro have always maintained a distinctive feel.

The identity of Montreal itself draws heavily upon the elegance of its European motherland. As the center of staunchly independent Quebec, the hybrid Anglo-French nature of the city is reflected in virtually every detail of the system. It was initiated using the rubber-tired trains pioneered by RATP in Paris in the 1950s. Secondly, given that French is the official language, the majority of station names conjure up Parisian life with such evocative echoes as Champ-de-Mars and Vendôme.

Thirdly, the lines when first built were numbered, although generally the color shown on the diagram now acts as the line indicator.

The system opened with architecturally striking stations and a keen awareness of clear design on signage using a neat, condensed sans-serif face.

The earliest diagram shown here (1) dates from 1976, a decade after the system opened, when Montreal was center stage for the Olympic Games. Métro received a makeover around this time, producing an in-car diagram with the first use of black as a background (2). In 1978 the bold shift to black was introduced on the system diagram itself (3), which includes an as yet unfulfilled extension of the Blue Line west of Snowdon.

Pocket maps with a white background (4) are also common, but a wonderful black version is still available and is seen in trains (5). As on most French system maps, the terminal points of lines, such as at Longueuil or Saint-Michel, are reversed out in blocks (4 and 5). Juxtaposed against this are the North Americanisms, such as stations like Peel, McGill, or Monk, named simply after streets they are on.

Black is relatively rarely used by cartographers, probably because of the difficulty with legibility under certain lighting conditions, but it has to be said that it makes Montreal's diagram all the more striking.

Rather than using the usual graphic "ticks," the city has employed other methods of indicating stations, ranging from large domino-style white dots to the operator's cubelike logo. The current open white circles are perfectly evenly spaced, giving a surrealistic equality between all stops.

At the same time the black was introduced, the orientation was tilted, giving a definitive Parisian leaning!

Plans for a subway in the biggest French-speaking city in the Americas go back to the early twentieth century, and were revived during almost every decade. The system has grown steadily, although for several years there was much talk but few new extensions delivered.

More recently, two new routes were announced. A 3-mile elongation

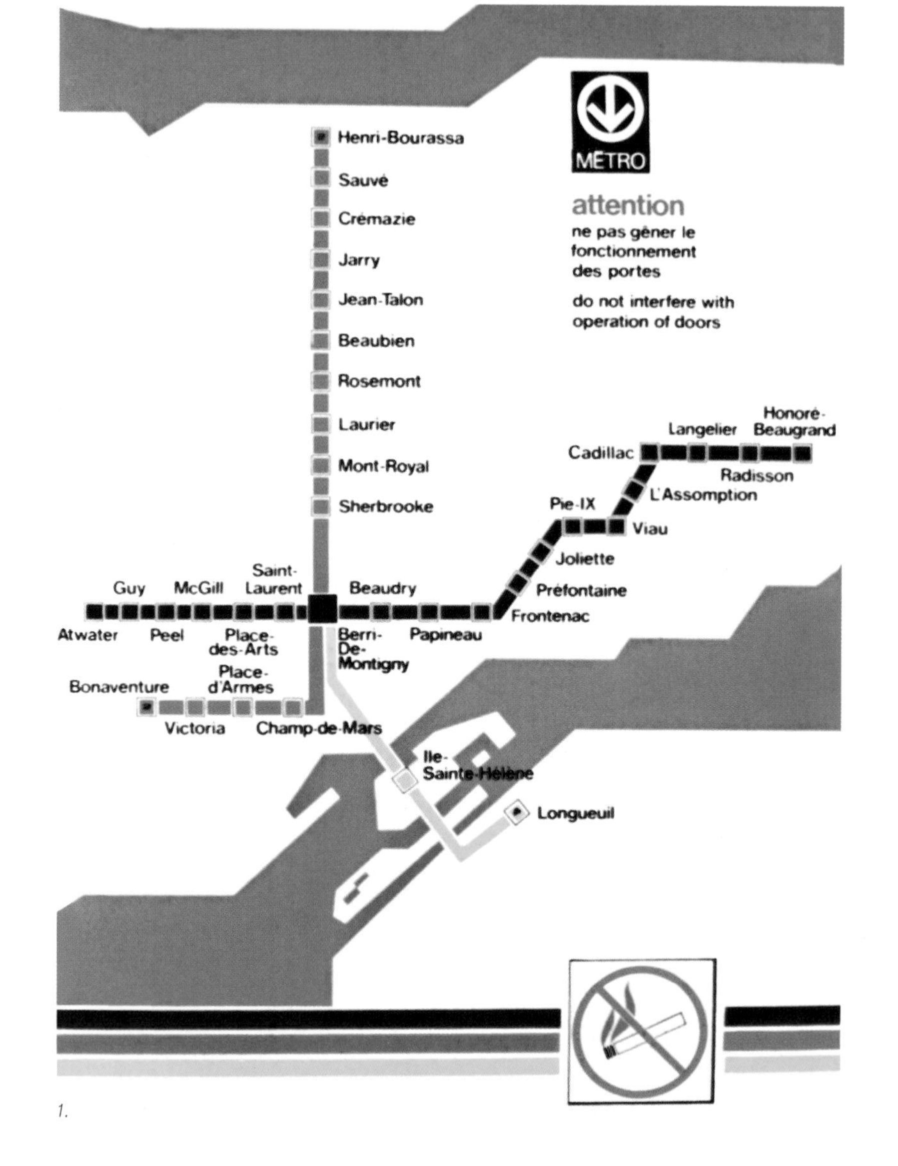

1.

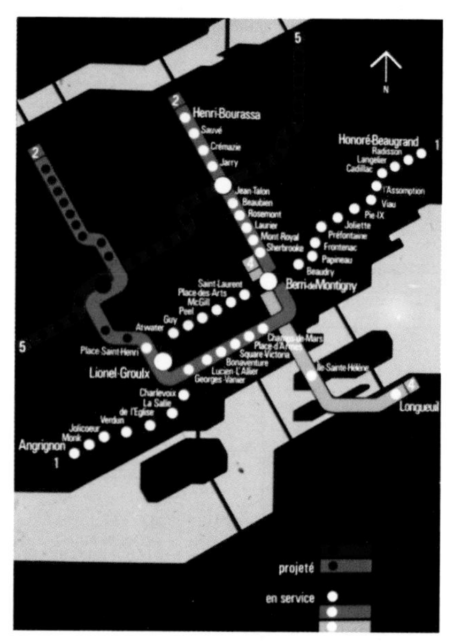

3.

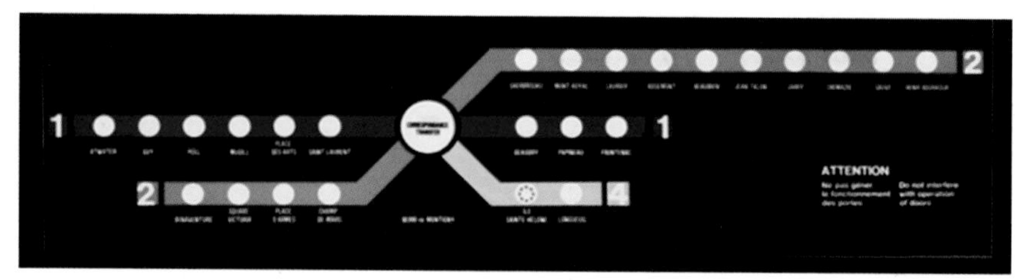

2.

Beri-UQAM station interior.

of the Orange Line, already under construction, will bring Métro to the suburban town of Laval, and it is intended that the Blue Line will gain at least five more stations as it heads toward Anjou.

There is also talk of extending the smaller Yellow Line at both ends—by four stops toward Roland-Therrien—and creating a useful new downtown link from the busy Berri-UQAM station to bustling McGill.

Montreal station interiors are generally vast, light, spacious, and full of art and sculpture. Some are so impressive that they could be compared to modern equivalents of some grandiose Moscow Metro stations.

4.

5.

Maps and photo reproduced by kind permission of and copyright STCUM 2007. Antique maps from Benoît Clairoux. Photo by Jean René Archambault.

Munich

Urban population: 1.3 million. Route length: 56 miles. Stations: 96. First section open: 1971. Underground: 90 percent.

The Munich diagram is reminiscent of most other German transit maps, with quite thin lines, cubes or oblongs for interchanges, and ticks as station markers (3). Diagonal lines are standardized at 45 degrees, the graphics include clean, simple curves, and no station names break over the lines.

The earliest map of the system, dating from 1972 (2), uses 45-, 50-, and 55-degree diagonals and includes S-Bahn lines as well. The 1975 version, in the same graphic style, shows the U3/U6 tunnel complete to Harras (1).

The Bavarian capital had subterranean aspirations as long ago as 1941, when a 2,000-foot-long tunnel at Goetheplatz was constructed for a proposed rail link between the north and south of the city. This was finally utilized as the backbone of Munich's plans to sink many of the downtown tram routes below the surface, and in 1964 construction on the first route between Goetheplatz and Kieferngarten was started.

The city won the right to stage the 1972 Olympic Games, so work was quickly reprioritized to convert most of the tram subways to full-scale heavy urban rail. This was the genesis of an initial two-line system that included a branch to the Olympic Stadium served by the U3.

The system opened in 1971 and was augmented by the commencement of the new S-Bahn tunnel right through the city center.

Munich now has six underground lines—although essentially the system comprises three main tunnels, each with branches. MVG, the U-Bahn operator, has some major plans for expansion. The U3 is being extended to meet U1 at the OEZ, U4 could be taken a little farther east, and U5 might also thrust a few stops more to the west.

Most central-area stations, built in the 1960s and '70s, with their flat concrete surfaces and contemporary "municipal" colors, have a rather bland air of functionality. Because they appear very "period" now, it is easy to forget these stations were greeted with fanfare as the height of modernity and chic when they were first brought into service.

Georg-Brauchle-Ring station on U1

Starting with U2's Königsplatz, the stations opened in the past twenty years show much more individuality. Many have found ways to bring light down to platform level and utilize a more varied palette of colors. Königsplatz itself reflects the treasures stored in the museum above: passengers can see into parts of exhibitions while waiting for trains. Candidplatz has made a virtue out of needing more than the average number of supporting pillars by covering them all with a painter's swatch book of different colors.

The creatures from Hellabrunn Zoo adorn U3's Thalkirchen station. Another beauty is Prinzregentenplatz

Schnellbahnen im Münchner Verkehrs-und Tarifverbund

MVV

1.

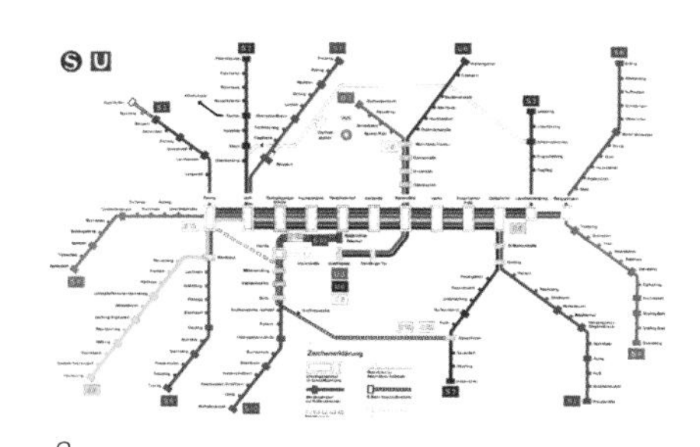

2.

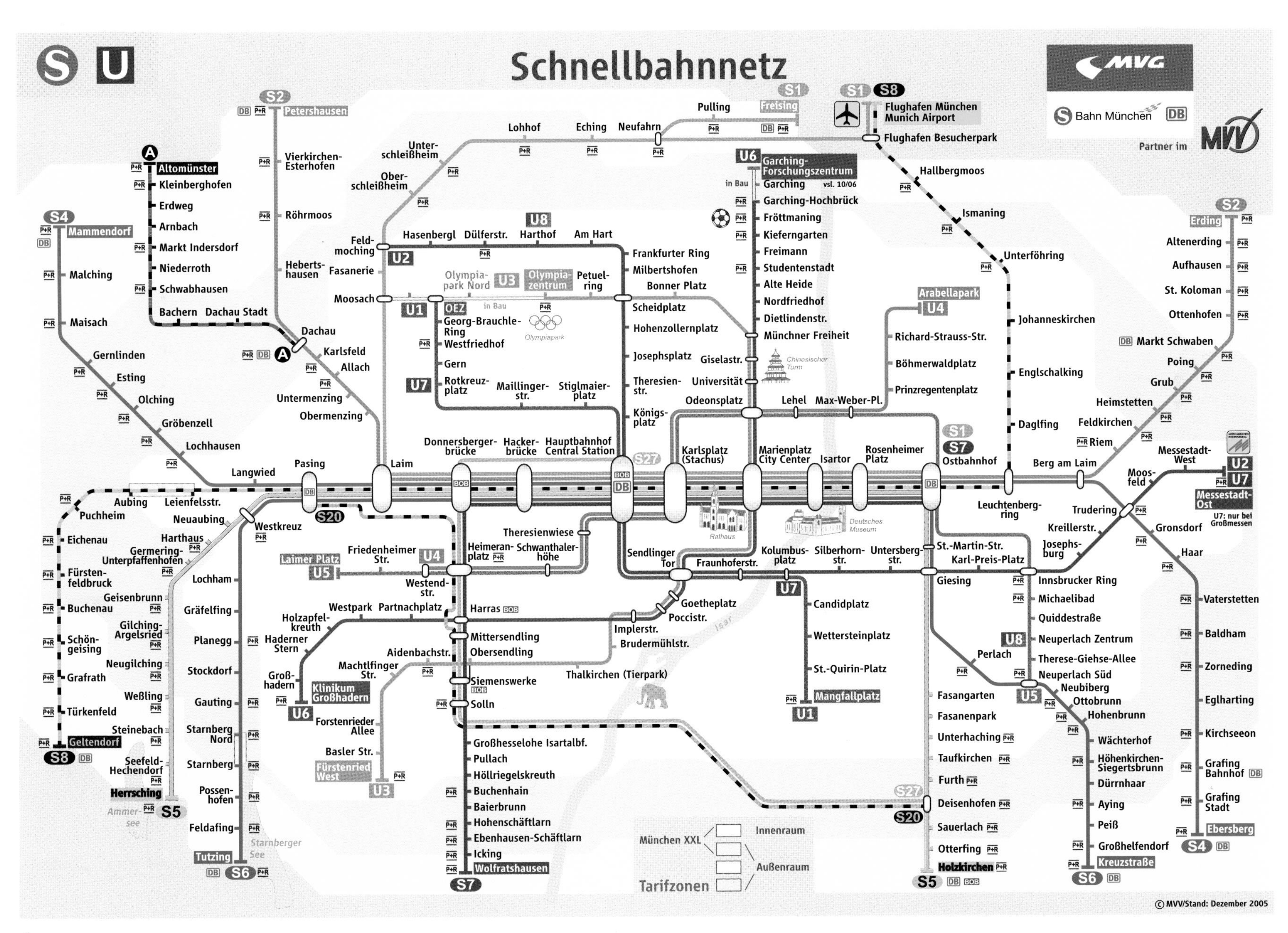

3.

on U4, which uses marble and granite to entice passengers to the nearby historic Prinzregententheater. The walls of Westfriedhof have been left uncovered, revealing just the bare concrete lit by massive domes of multiple colors. All the U2 stations from Innsbrucker Ring to Messestadt Ost are linked with red colors. All of this gives the system a strong corporate identity. No wonder Munich has a mass rapid transit system to be proud of!

Osaka

Urban population: 8.8 million. Route length: 77.9 miles. Stations: 121. First section open: 1933. Underground: 90 percent.

Greater Osaka, Japan's second-largest conurbation, is a gigantic urban expanse of 730 square miles. Notable among other achievements for being the first Japanese city to build a linear induction subway, it is now one of the busiest mass-transit systems in the world.

In the main diagram (4), the eight subway lines are broad and in striking colors. The downtown area is so criss-crossed with lines that virtually every station is an interchange to another line.

Despite the regimented, latticelike perpendicular nature of the central area, the curves and diagonals are less disciplined. Like many East Asian cities, Osaka now numbers its stations, with the first letter of the line name followed by the individual station number. These are shown inside large open spheres, which are essentially wide black circles with white centers. This device can be traced back at least to the English-language diagram produced for Expo 70 (1), where an elongated oval with a black edge was used, with the station name inside. This diagram was surprisingly well spaced compared to other versions of the period, like one issued just two years later showing a proposed extension in a more geographic setting (2).

A beautiful cover to a brochure about the modernization of the first line features a diagram of the system so stylized that it clearly resembles a new character in the vast Japanese logogram alphabet (3).

The current diagram (4) also shows

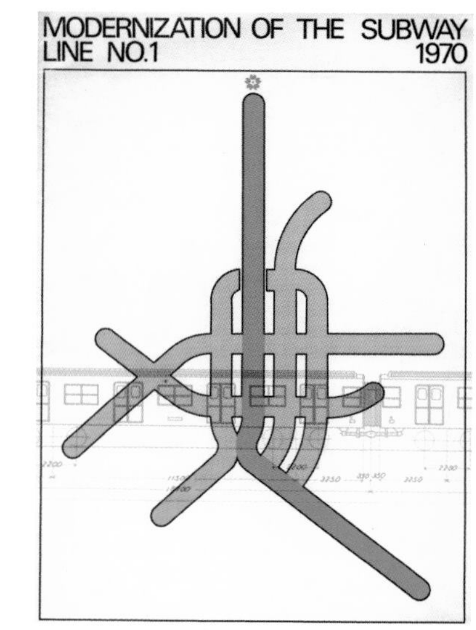

3.

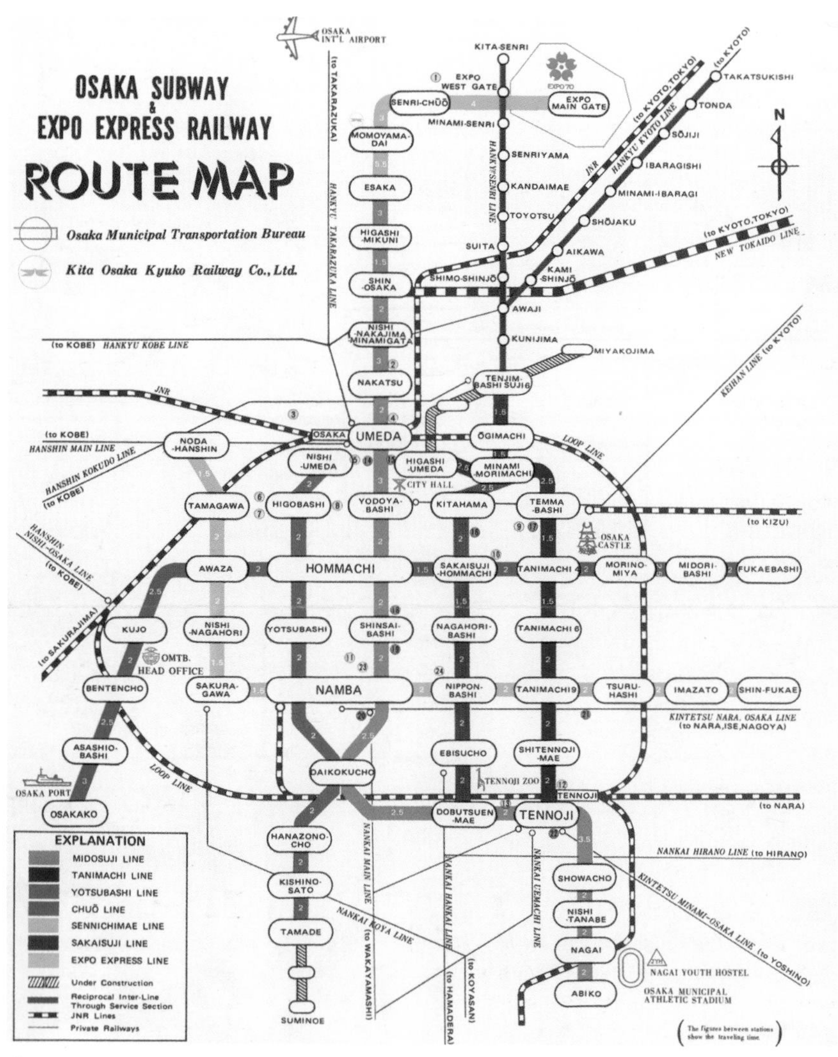

1.

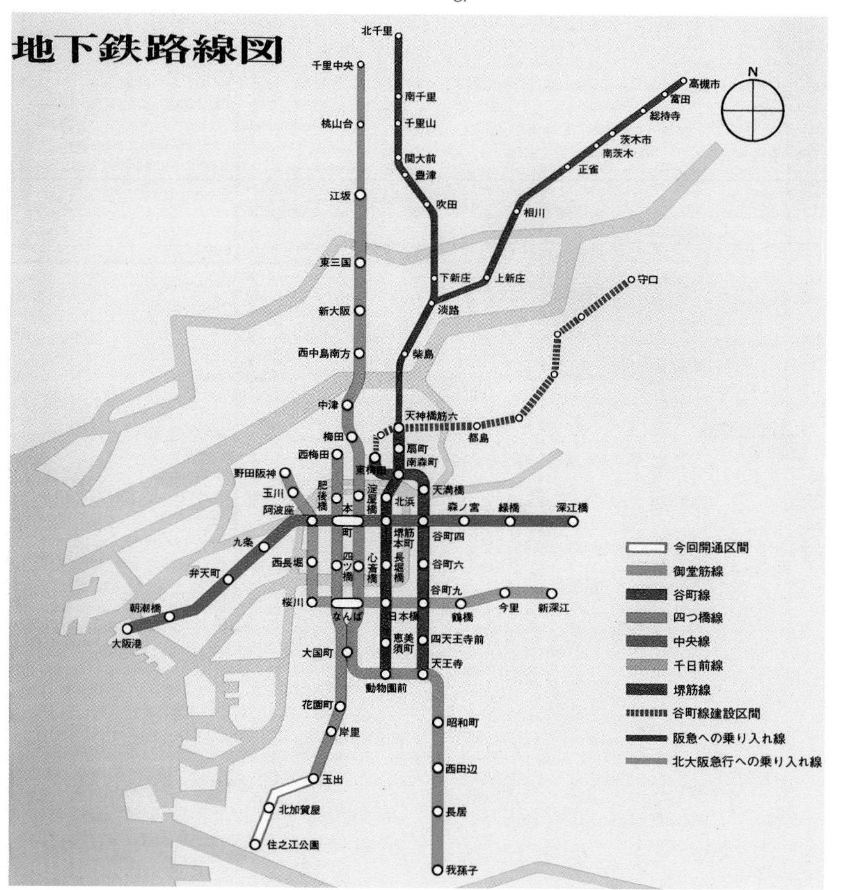

2.

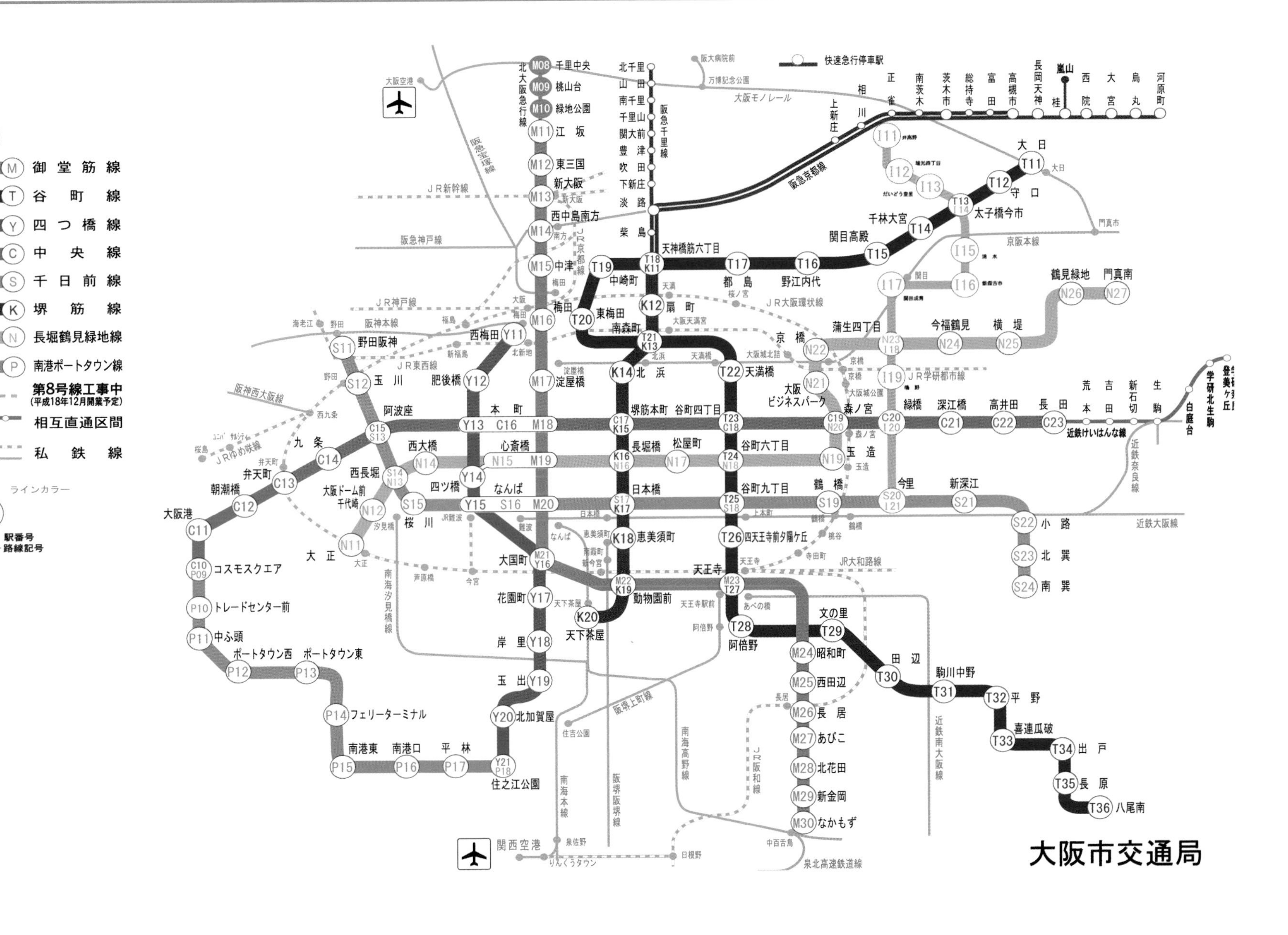

4.

the highly successful port line and the "new tram" in light blue, plus connecting mainline and suburban railways in gray.

Osaka was the first subway to experiment with a radical new form of traction power, the linear induction motor, which uses magnetism to propel the trains along Line 7, opened just in time for the 1990 Expo, and on the new Line 8, opened in 2006.

As in the capital, signage and maps are increasingly appearing with approximate English translations to aid travelers' journeys.

Maps and photo reproduced by kind permission of and copyright Osaka Municipal Transportation Bureau 2007. Antique maps and photo from UITP.

San Francisco

Urban population: 7.1 million. Route length: 113.2 miles. Stations: 43. First section open: 1972. Underground: 20 percent.

San Francisco has an outstanding commitment to public transportation. Few places on the North American continent boast such a diversity of services. From the clanging of the renovated cable cars—now an incomparable landmark—sliding gracefully up and down the dizzy gradients of Powell, Hyde, California, and Mason streets to its sleek, efficient underground system, BART, brazenly traversing the infamously unstable San Andreas Fault, this is undoubtedly a public transport lover's paradise.

BART was first approved by state-wide public vote and constructed at enormous cost in the mid-1960s and early '70s, during the very height of the automobile's dominance. As the system has expanded, throwing tentacles into northern California, BART maps have now dispensed with almost all diagrammatic elements in favor of a purely geographic representation (6) arguably going even further than New York's (see p. 32).

Winding track routes are shown with incredible detail, leading to what can appear as a fairly wobbly line—not perhaps intended but certainly appropriate for an earthquake-prone zone!

This may be perfectly reasonable on a system with so many long surface sections, but perhaps a core ingredient of simplicity is slightly lost. Take for example the "bunching" of names around MacArthur, 19th St./Oakland, and Oakland City Center/12th St. stations, or the meandering Dublin/Pleasanton–Daly City line—is it really so necessary to show on a small map exactly how the line negotiates every bend in the Castro Valley?

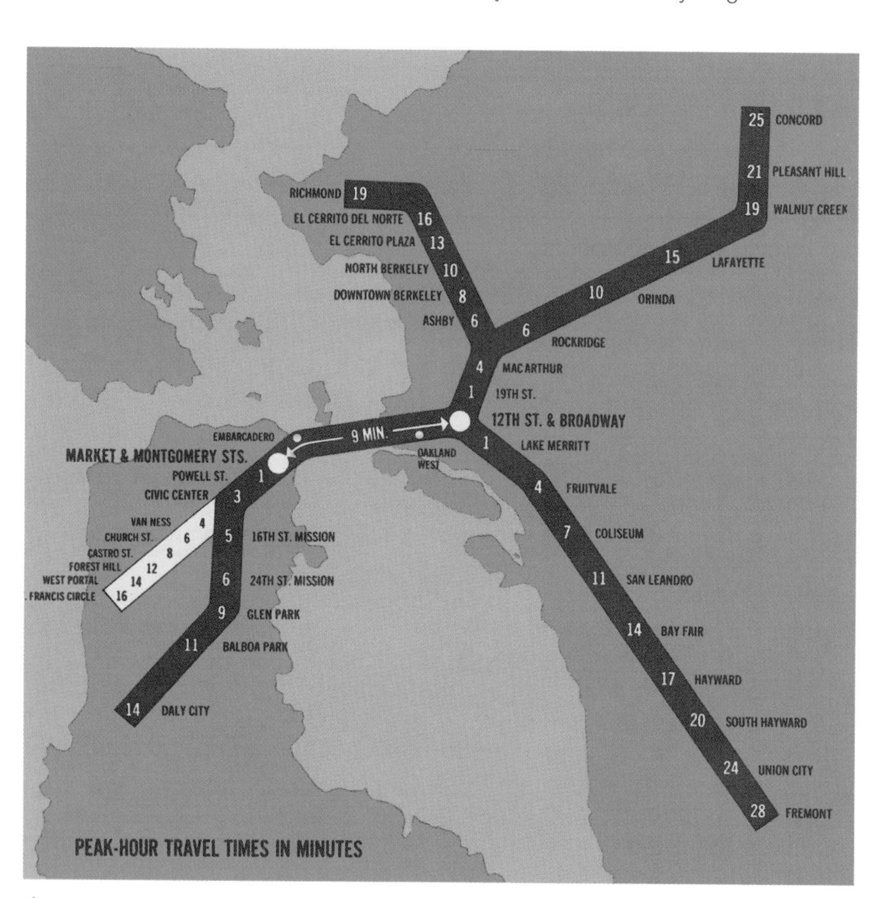

1.

In the future, BART may yet reach farther-flung cities like Byron or Tracy; if this happens, surely retaining such slavish topographic accuracy will either make the map very large or force the central area to become so cramped that a magnifying glass might be needed with every copy!

In 2007 some experimentation is under way to gauge passengers' reactions to several new designs, some of which bear all the hallmarks of more traditional urban-rail diagrams based on standard 45-degree angles.

Early diagrams, however, like the late 1960s version (1), were patently highly stylized and even showed travel times from Embarcadero or 12th & Broadway; here the number of minutes became the station marker on the line.

By the mid-1980s, this had simplified into a four-legged *X* (2). Metallic Regional Connections diagrams from 1981, with 45-degree angles, can still be found fixed to walls in some outlying stations (3).

In addition to BART, San Francisco also has a unique combination of streetcar/light rail/underground called MUNI Metro, which runs on a long section of subterranean track downtown, above BART and below a renovated streetcar line, along the superserved Market Street thoroughfare. MUNI then surfaces for street operation after Van Ness station (J Church and N Judah lines) and West Portal station (K Ingleside, L Taravel, and M Oceanview lines). MUNI use a schematic form for their system diagram (4).

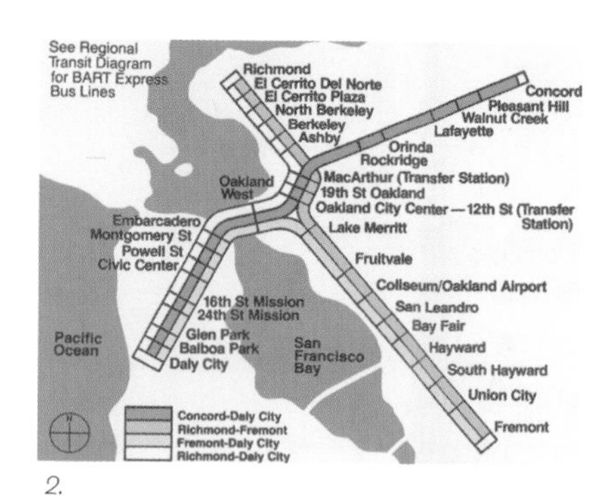

2.

3.

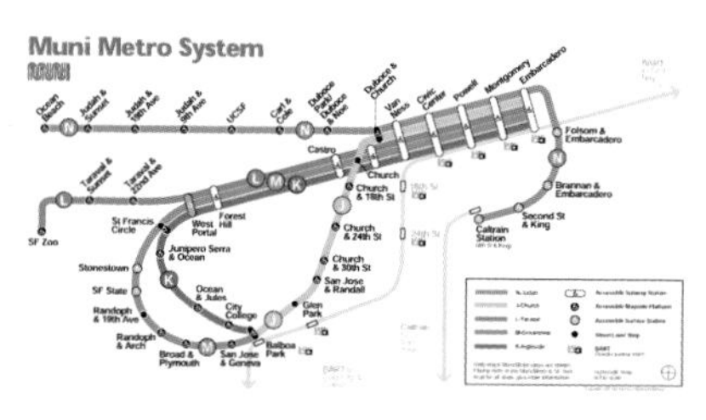

4.

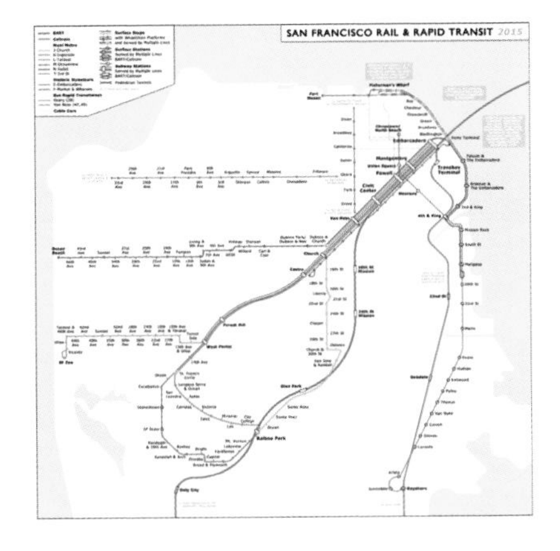

5.

Complementing all these rail-based systems, there are trolleybuses that take power from overhead wires, feeding the sprawling residential areas fringing Golden Gate Park.

BART recently opened a much-needed extension to the international airport and Millbrae. The new light-rail service between Caltrain Bayshore and Caltrain Terminal stations, then onward into a subway section below Third Street, with a terminal in Chinatown, is now well under construction.

All possible extensions in the city are shown on Steve Boland's Rail & Rapid Transit 2015 diagram (5). These multiple modes have often been featured in popular culture, appearing in everything from *The Simpsons* to the cult movie *Koyaanisqatsi,* and they help make the city a compulsory destination for anyone from the wildly dedicated transport enthusiast to the mildly interested tourist!

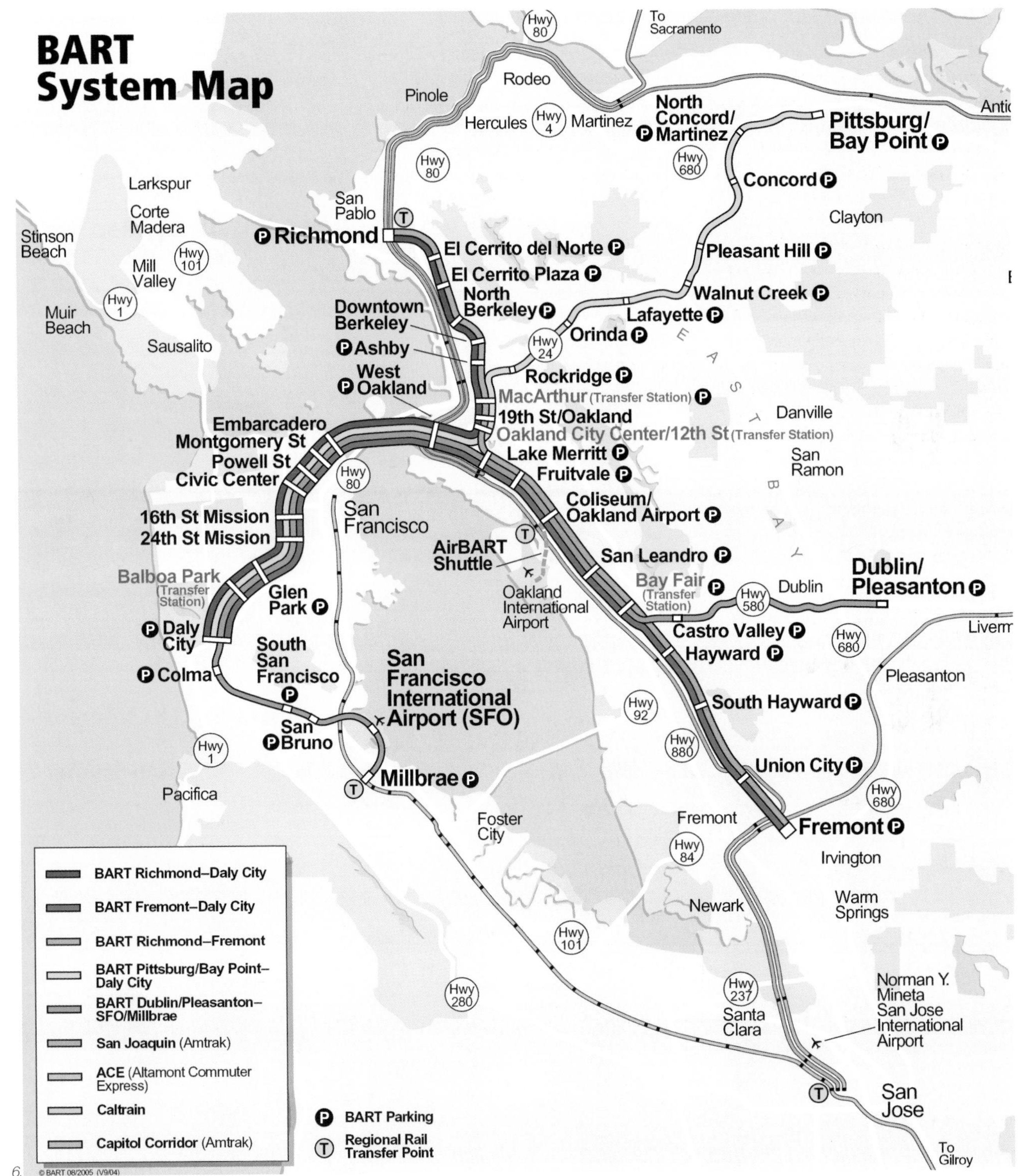

6.

Maps and photo reproduced by kind permission of and copyright BART 2007. Muni map: San Francisco Municipal Railway. Antique maps and photo from UITP. 2015 map: Steve Boland.

Seoul

Urban population: 22.8 million. Route length: 177.7 miles. Stations: 286. First section open: 1974. Underground: 70 percent.

If subway construction has become a bit of a fascinating habit in Southeast Asia, then South Korea has become an addict! From nothing in the early 1970s, the Seoul urban-rail system is now one of the most impressive modern systems in the world. By size alone it should already be in this book's Zone 1, but owing to its newness it has not yet produced enough historical material to warrant four full pages.

The vast urban spread of the Seoul National Capital Area (which claims to be the second most populous in the world after greater Tokyo) has gained 170 miles of new lines in the past thirty years. Two companies operate them, with Korean National Rail (Korail) trains using some of the subway tracks, too.

Seoul Metro (formerly known as Seoul Metropolitan Subway Corporation) runs lines 1–4, and its diagram is shown here (7). The Seoul Metropolitan Rapid Transit Corporation, SMRT, runs the newer lines 5–8. The earliest map here is from a 1985 brochure about the proposed network extension (1).

Both the 1996 Korean (2) and English (3) schematics are so distorted that they almost resemble a character of the Korean alphabet! Most recent offerings show the regional tendency to number individual stations, which is quite a handy trend for the visitor. Seoul uses a three-digit code starting with the line number, then two digits for the station, though Incheon Subway stations are prefixed with an *I* and Korail-line stations with a *K*.

A photo of a diagram above the ticket machines shows how dense the system has become (4). In fact, this complex network has been made a lot more accessible by these impressive diagrams, which in keeping with most world urban-rail maps make use of 45-degree diagonals (though not exclusively) and are quite heavily distorted from the geography the lines follow belowground. One SMRT example (5) is extreme.

Most Seoul diagrams also employ a most distinctive regional feature: the fabulous interchange symbol called a "Taefuk," which looks more like a stylized cyclone or typhoon—an often apt description of the way commuters rush between lines.

Line 2 has some of the closest-

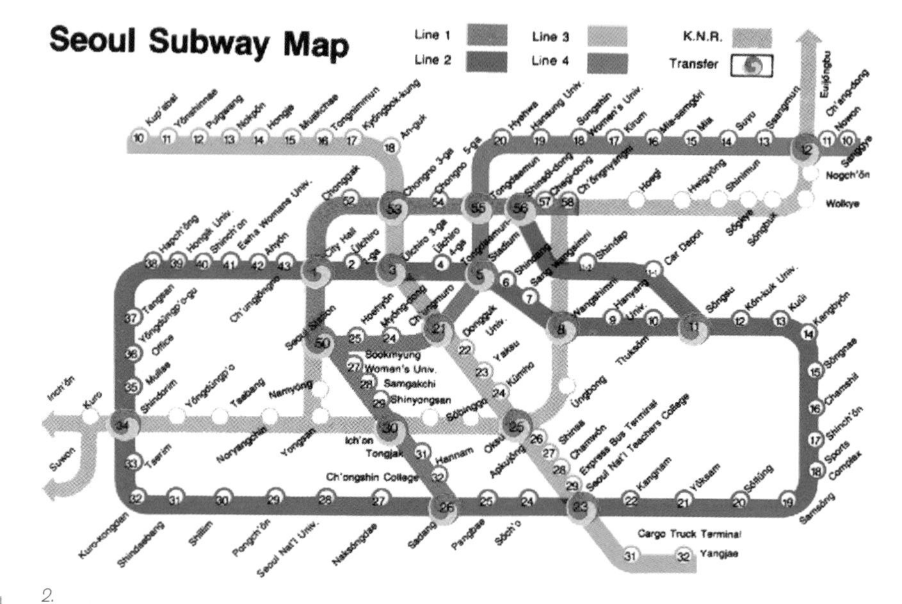

2.

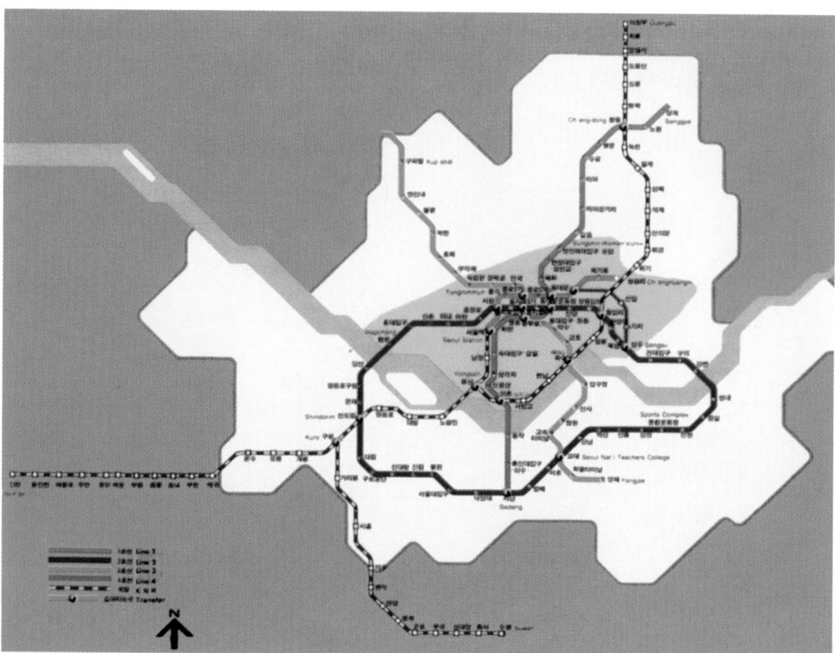

1.

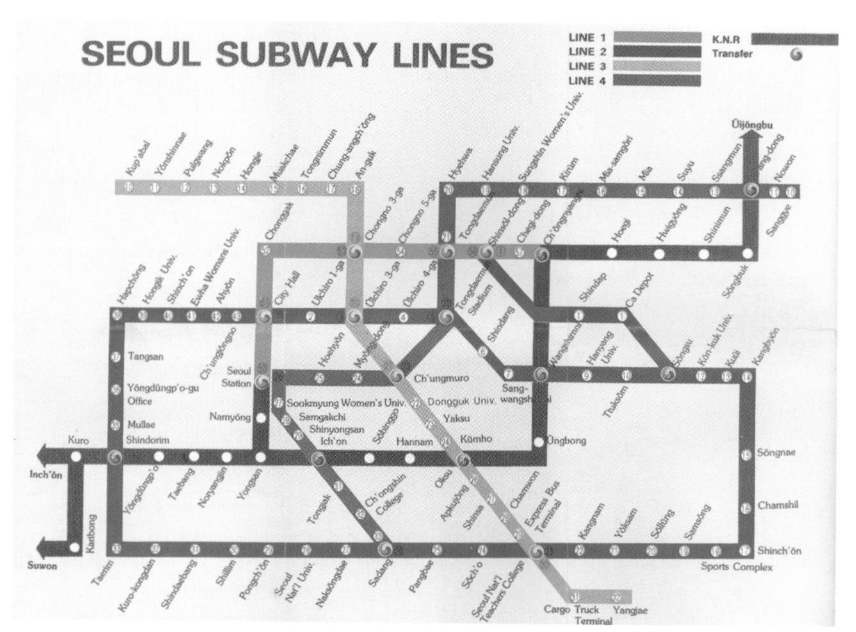

3.

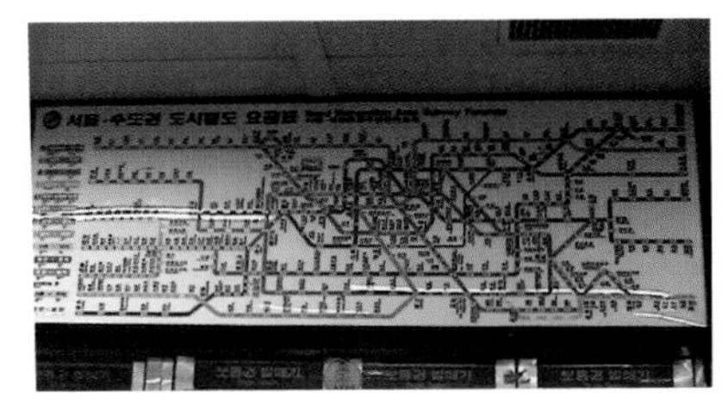

4.

spaced trains on any world subway; the headway (time between each train) can be as little as thirty seconds! Line 1 now links directly to the neighboring city of Incheon (see also p. 133), and the Yellow Line runs into the nearby town of Bundang. A further line, 9, is also under construction (6). Closely following the river, it should be open in 2009.

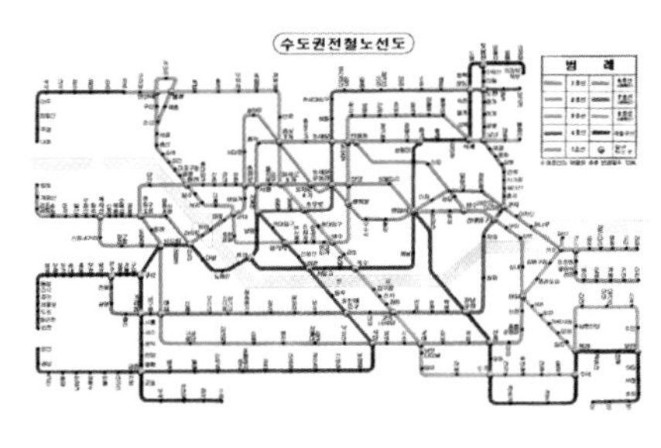

5.

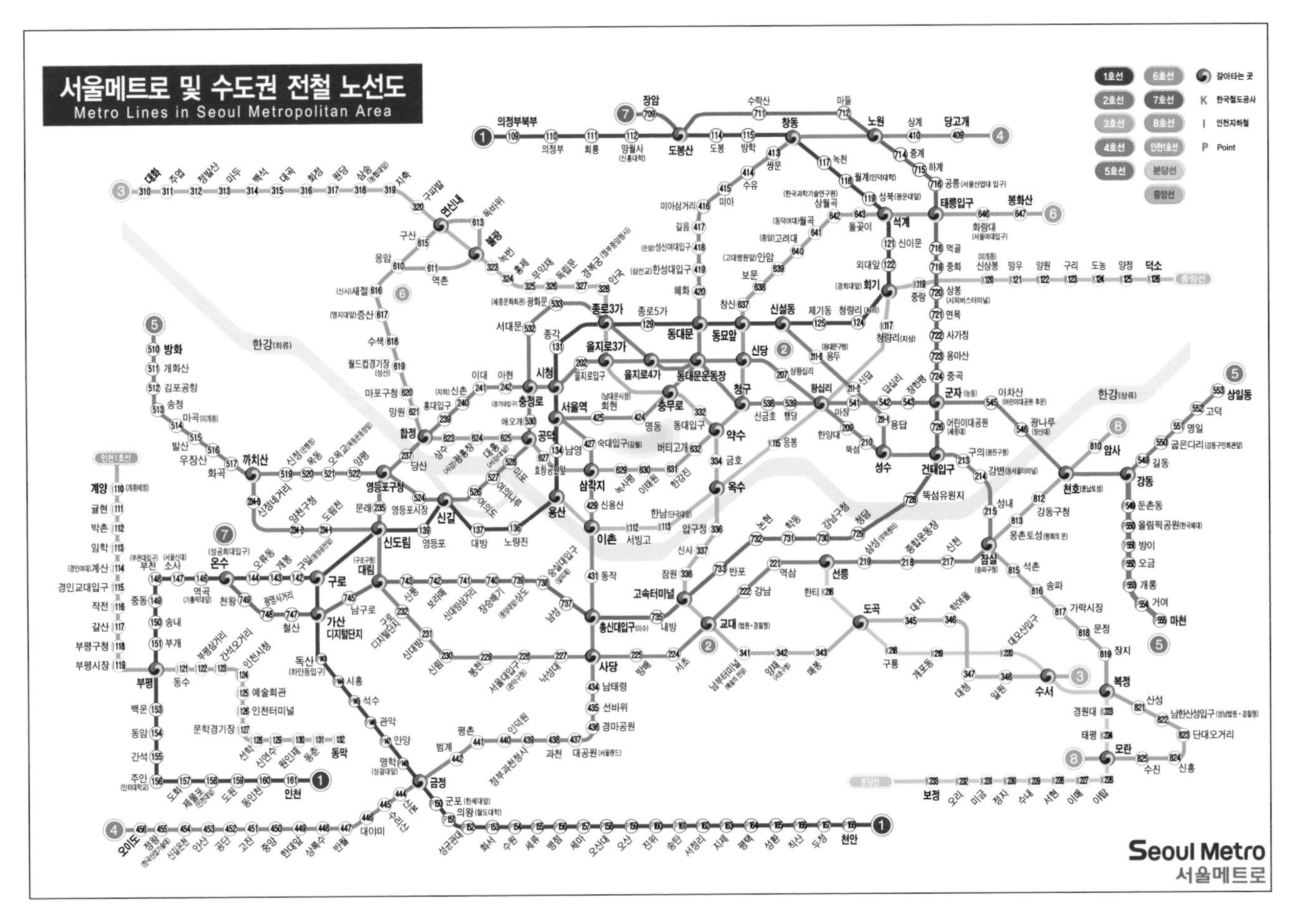

7.

6.

Maps reproduced by kind permission of and copyright SMS/SMRT 2007.

St. Petersburg

Urban population: 4.7 million. Route length: 64.7 miles. Stations: 54. First section open: 1955. Underground: 58 percent.

No half measures here; this is metro construction on the grandest scale, second only to the subterranean splendor of Moscow. Some even claim Avtovo is the world's most beautiful subway station. The name, the huge *M* logos all over the city, the stations, and even the current diagram (5) resonate with Russia's capital. St. Petersburg's strident blue Line 2 column corrals the other three lines into order, as Moscow's brown circle line anchors its comrades into harmony.

The diagram is supremely "Russian Metro" in style. Heavily distorted from geographic reality, the simplicity is enhanced by the use of bold colors for each line and airbrushed 3-D-style interchange symbols. Unlike Moscow's diagram, though, no text breaks over the lines and there is a complete absence of any diagonals. An advertising agency, Kommet, has assumed control of selling space on the system, recently covering an entire car in candy ads!

Kommet is also responsible both for producing the current diagram and coming up with ideas for network improvements, which are numerous, although such extravagant extensions are unlikely to be fulfilled.

The twice-renamed city, which celebrated its 300th anniversary in 2003, was finally granted a subway by the Soviet state in the late 1940s. Line 1 opened in all its ostentatious glory in 1955 between Ploshchad Vosstaniya and Avtovo, linking four (and later all) of the city's five mainline railway terminals.

Almost outdoing the grandiose cathedralesque stations on some of the finer sections of the Moscow Metro, Line 1 was as much an art and architecture exhibition as a public-transit system. Construction was a protracted affair due to the marshy soil conditions below the city, which

3.

1.

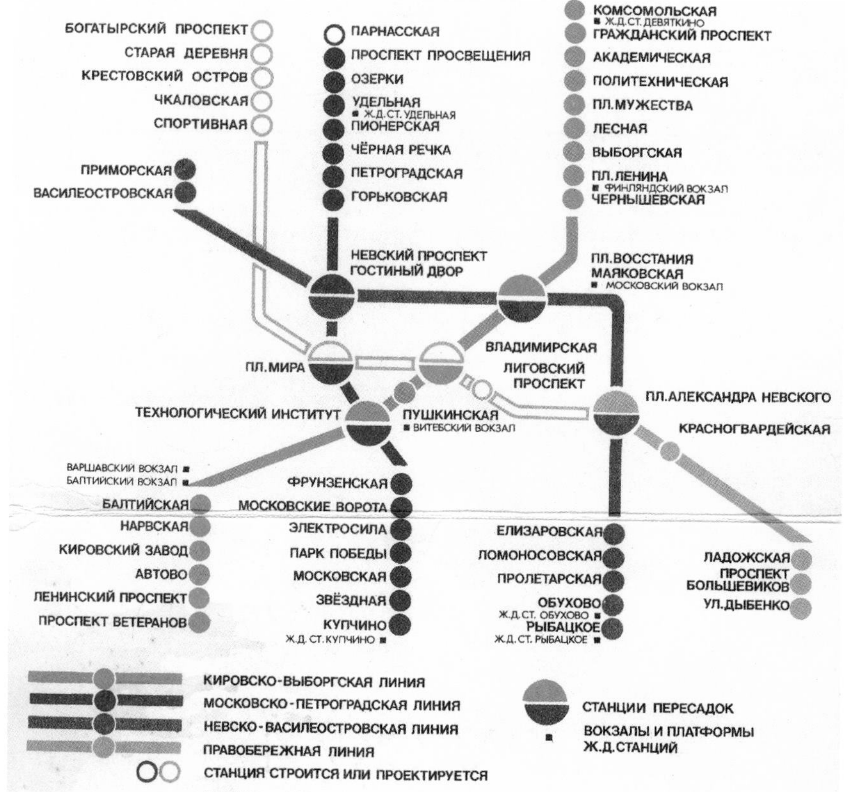

2.

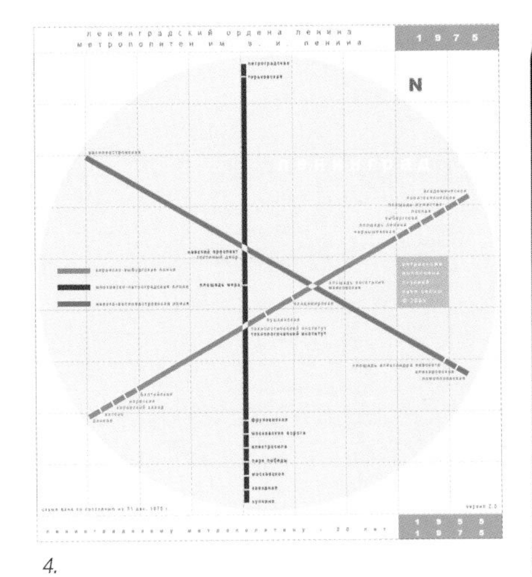

4.

resulted in the tunnels needing to be on average 200 feet deep and supersealed to stop water leakage.

Lines 1 to 3 were mostly completed by the early 1970s, as seen on a commemorative poster from 1975 (4). There were a few extensions afterward (1), but the first part of the orange Line 4 only opened in 1985 (2) and was extended at either end in the 1990s (3), when signs of the current diagram can be detected.

St. Petersburg was first to fit platform-edge doors nearly fifty years ago and boasts an amazing 3-D metal wall-mounted map on one of the stations, and due to the depths of the tunnels it also claims some of the world's longest escalators. Construction progress is slow but has begun on the new Line 5.

5.

Washington DC

Urban population: 3.9 million. Route length: 106.3 miles. Stations: 86. First section open: 1976. Underground: 48 percent.

The Washington DC Metrorail System Map has become one of the world's most familiar and distinctive—due in part to the system's being North America's second busiest, but also thanks to its idiosyncratic design (3). It is a highly stylized representation of the system, which plays a number of tricks with the true geographic layout, but despite including several topographic features (rivers, parks, monuments, and the Beltway) it still feels very much like a diagram, not a map.

Displaying some of the most enduring urban-rail diagram features, it adds its own unique elements. Notice the bright, strong, bold colors; virtually all angles at 45 degrees; standardized curves; and even the black circle with white center as station markers, but here it's completely enclosed by the thickness of the colored lines and doubled up at interchanges. The fact that four station names break over the route lines and that the text appears both horizontally and diagonally (usually a recipe for disaster) does not necessarily detract from the map's overall effectiveness.

The Washington style of thicker route lines has inspired at least one facsimile, that of neighboring Baltimore (see p. 108). There's distortion not only in the

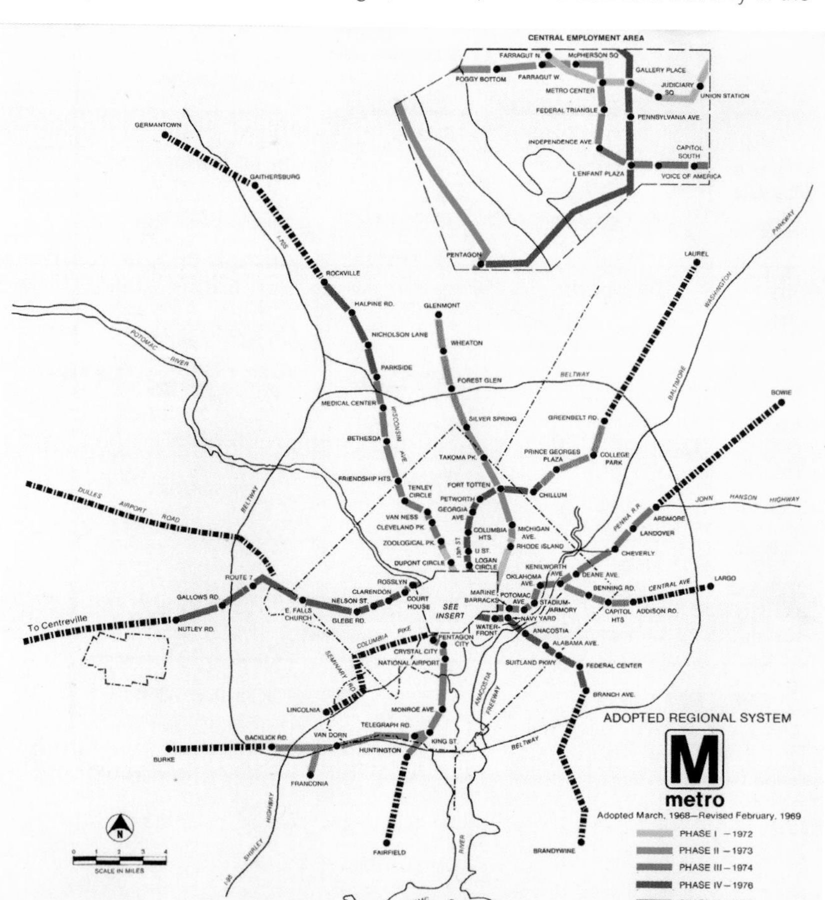

1.

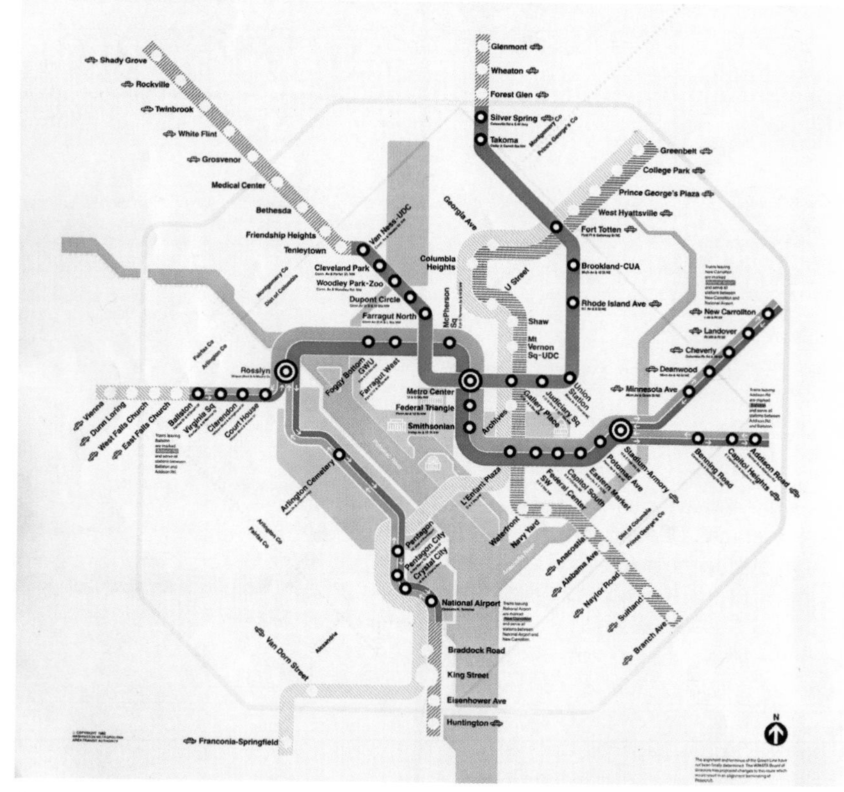

2.

trajectory of the lines but also in the equal spacing between stations, as seen on the western end of the Orange Line for example.

The Metro, which straddles not only DC but adjoining areas of Virginia and Maryland, is still being expanded. Some other, more far-flung sections talked about in 1969 (1) have not yet come to fruition, but the entire "100-mile" network projected on this 1982 diagram (2) is now open.

The Silver Line, which will link Dulles Airport to Stadium-Armory, has started construction, though what the service pattern will be and whether the current diagram design can accommodate another new line through the center without appearing too cramped is yet to be seen!

The U.S. capital lagged behind others in getting mass rapid transit, but the wait was worth it. Lessons of earlier subways, with their often dim and narrow passages that some find claustrophobic or just plain scary, had been well learned by the time the first section of line from Rhode Island Avenue to Farragut North opened in 1976, after forty years of planning, consultation, and changes, and seven years building at a cost of $10 billion.

No surprise that it is considered the most impressive system in the States. It was the first major network with air-conditioning in both stations and trains.

There are platform-edge lights that illuminate as the train approaches. The stations are bright and airy spaces with aesthetically appealing lighting, making a virtue out of the exposed concrete tunnel linings. At street level, stations are mostly set in public open spaces, such as plazas or squares.

The Metro has a strong corporate identity, with a clarity of signage that was probably the clearest and boldest of its time—huge brown pillars support most of the directional information. The system's logo, the capital *M* in white sitting astride the pillar top, has become an icon for Washington DC itself.

There is another, much smaller electric tube railway in Washington, too. The Congressional Subway first ran in 1909 and now has three separate underground lines that connect the U.S. Capitol building with the offices of the Senate and the House.

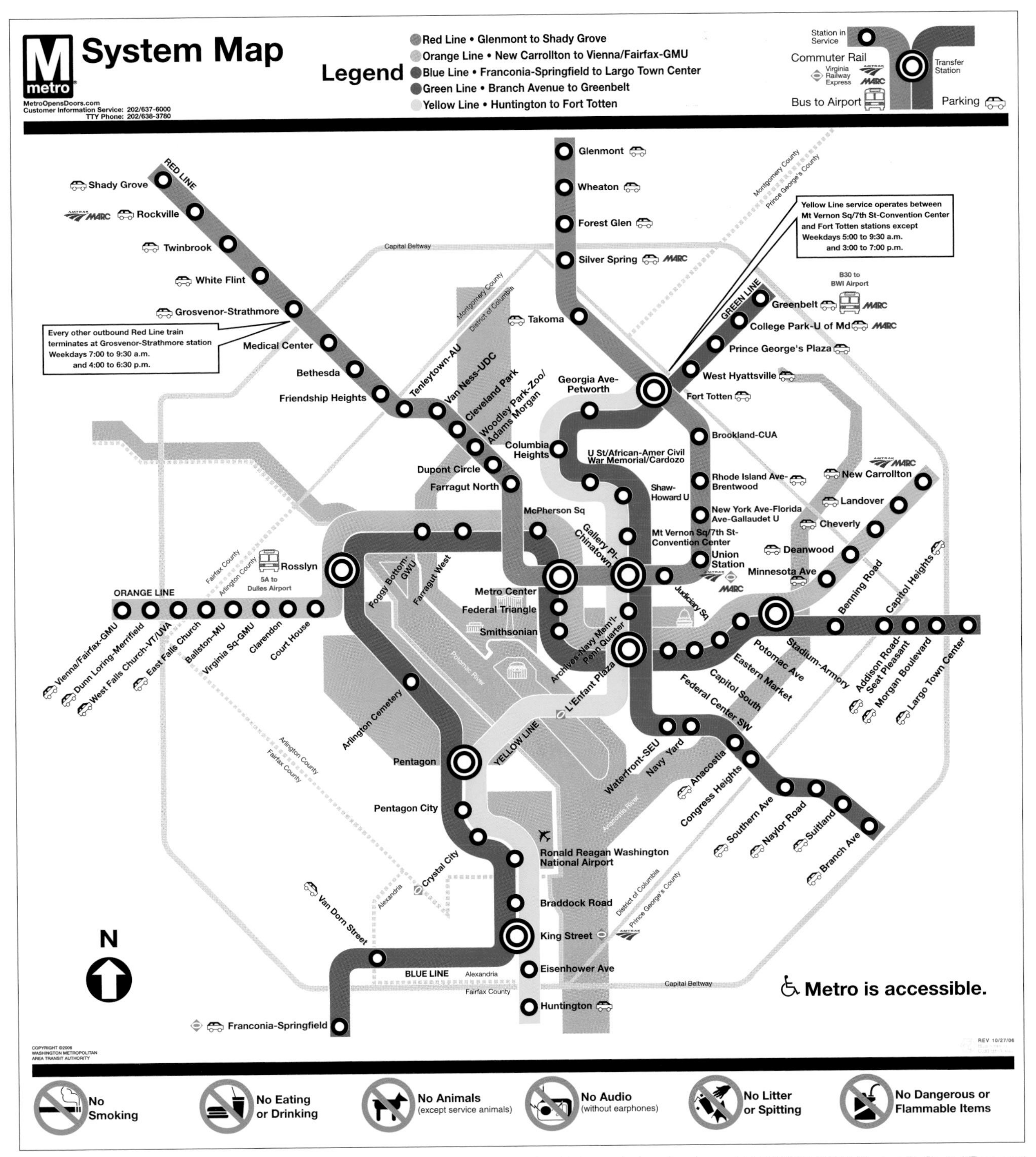

3.

Maps and top photo reproduced by kind permission of and copyright WMATA 2007. Photo left: Capital Transport.

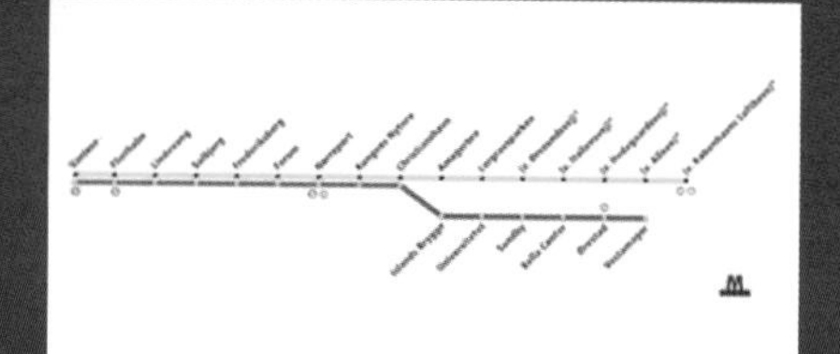

Mapa de Embarque
Linha 1-Azul
Tucuruvi
Parada Inglesa
Jardim São Paulo
Santana
Carandiru
Tietê
Armênia
Tiradentes
Embarque
Luz
São Bento
Sé
Liberdade
São Joaquim
Vergueiro
Paraíso
Ana Rosa
Vila Mariana
Santa Cruz
Praça da Árvore
Saúde
São Judas
Conceição
Jabaquara
EMTU

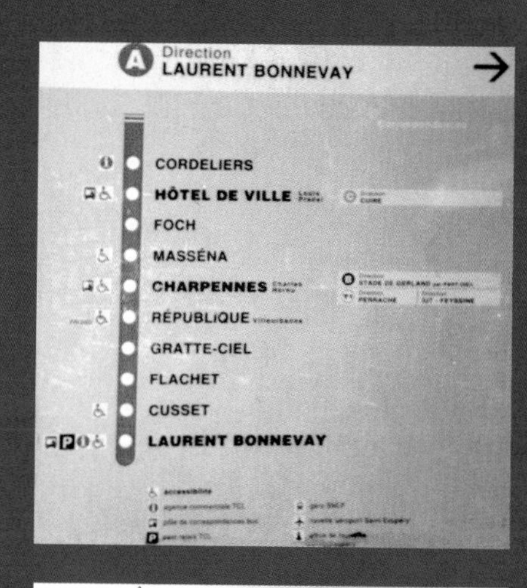

A
Direction
LAURENT BONNEVAY
CORDELIERS
HÔTEL DE VILLE
FOCH
MASSÉNA
CHARPENNES
RÉPUBLIQUE
GRATTE-CIEL
FLACHET
CUSSET
LAURENT BONNEVAY

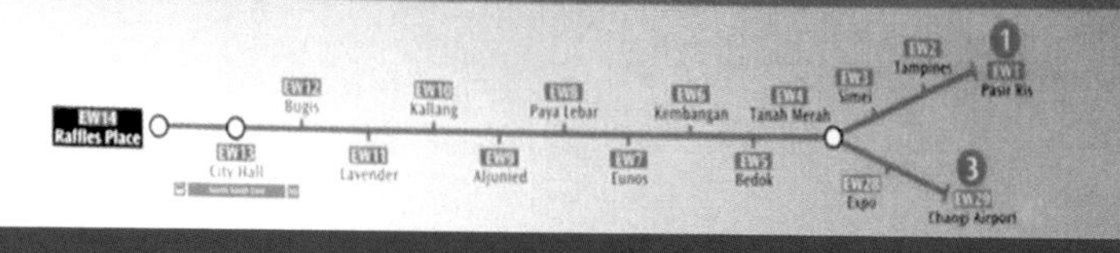

East West Line
EW14
Raffles Place
EW13
City Hall
EW12
Bugis
EW11
Lavender
EW10
Kallang
EW9
Aljunied
EW8
Paya Lebar
EW7
Eunos
EW6
Kembangan
EW5
Bedok
EW4
Tanah Merah
EW3
Simei
EW2
Tampines
EW1
Pasir Ris
CG1
Expo
CG2
Changi Airport

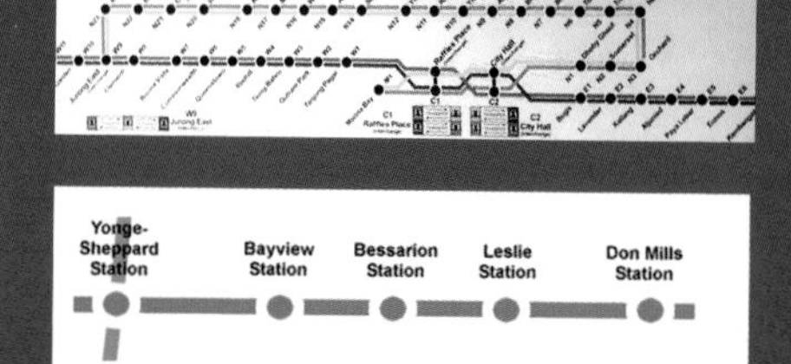

Yonge-Sheppard Station
Bayview Station
Bessarion Station
Leslie Station
Don Mills Station

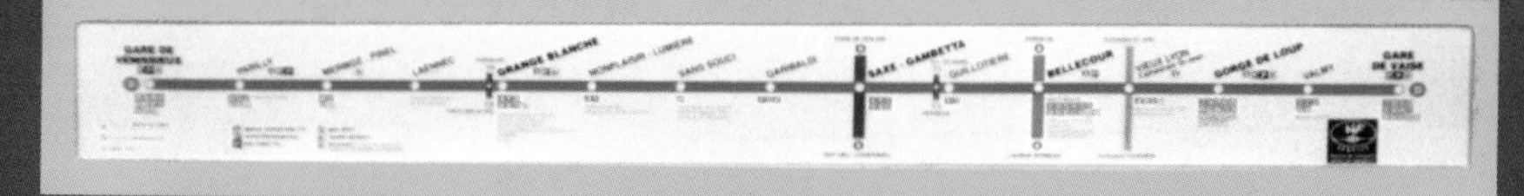

METRÔ RIO
1
Pavuna
Eng. Rubens Paiva
Acari/Fazenda Botafogo
Coelho Neto
Colégio
Irajá
Vicente de Carvalho
Thomaz Coelho
Engenho da Rainha
Inhaúma
Del Castilho
Maria da Graça
Triagem
Maracanã
São Cristovão
Saens Pena
São Francisco Xavier
Afonso Pena
Estácio
Praça Onze
Central
Presidente Vargas
Uruguaiana
Carioca
Cinelândia
Glória
Catete
Largo do Machado
Flamengo
Botafogo
Cardeal Arcoverde
Siqueira Campos
2

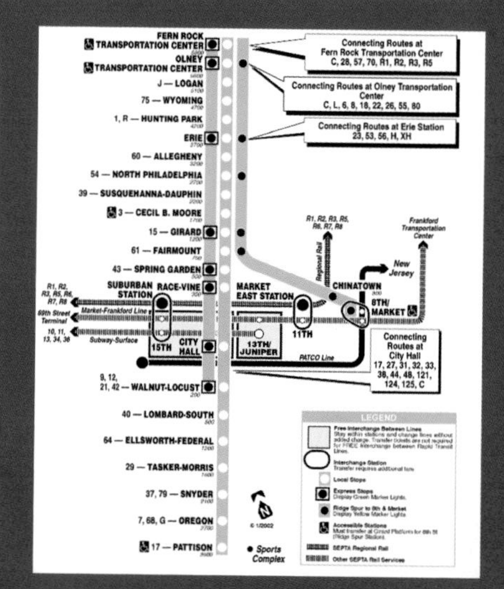

FERN ROCK TRANSPORTATION CENTER
OLNEY TRANSPORTATION CENTER
LOGAN
WYOMING
HUNTING PARK
ERIE
ALLEGHENY
NORTH PHILADELPHIA
SUSQUEHANNA-DAUPHIN
CECIL B. MOORE
GIRARD
FAIRMOUNT
SPRING GARDEN
RACE-VINE
CITY HALL
WALNUT-LOCUST
LOMBARD-SOUTH
ELLSWORTH-FEDERAL
TASKER-MORRIS
SNYDER
OREGON
PATTISON
Connecting Routes at Fern Rock Transportation Center
Connecting Routes at Olney Transportation Center
Connecting Routes at Erie Station
SUBURBAN STATION
15TH
MARKET EAST STATION
13TH/JUNIPER
11TH
CHINATOWN
8TH/MARKET
New Jersey
Frankford Transportation Center
Connecting Routes at City Hall
PATCO Line
Sports Complex
LEGEND

Zone 3

On this spread are some in-car route diagrams and platform maps of systems within Zone 3, which includes many of the smaller or newer urban rapid transit systems that have only had the chance to produce a more limited variety of cartographic material. Systems such as those of São Paulo, Los Angeles, Athens, Beijing, and Singapore are, however, expanding so rapidly that they will soon have their own extensive archive of older maps.

Amsterdam	078
Athens	079
Beijing	080
Bilbao	081
Brussels	082
Bucharest	083
Copenhagen	084
Delhi	085
Glasgow	086
Kiev	087
Kuala Lumpur	088
Los Angeles	089
Lyon	090
Nagoya	091
Newcastle	092
Oslo	093
Philadelphia	094
Prague	095
Rio de Janeiro	096
Rome	097
Rotterdam	098
São Paulo	099
Singapore	100
Stockholm	101
Taipei	102
Toronto	103
Valencia	104
Vienna	105

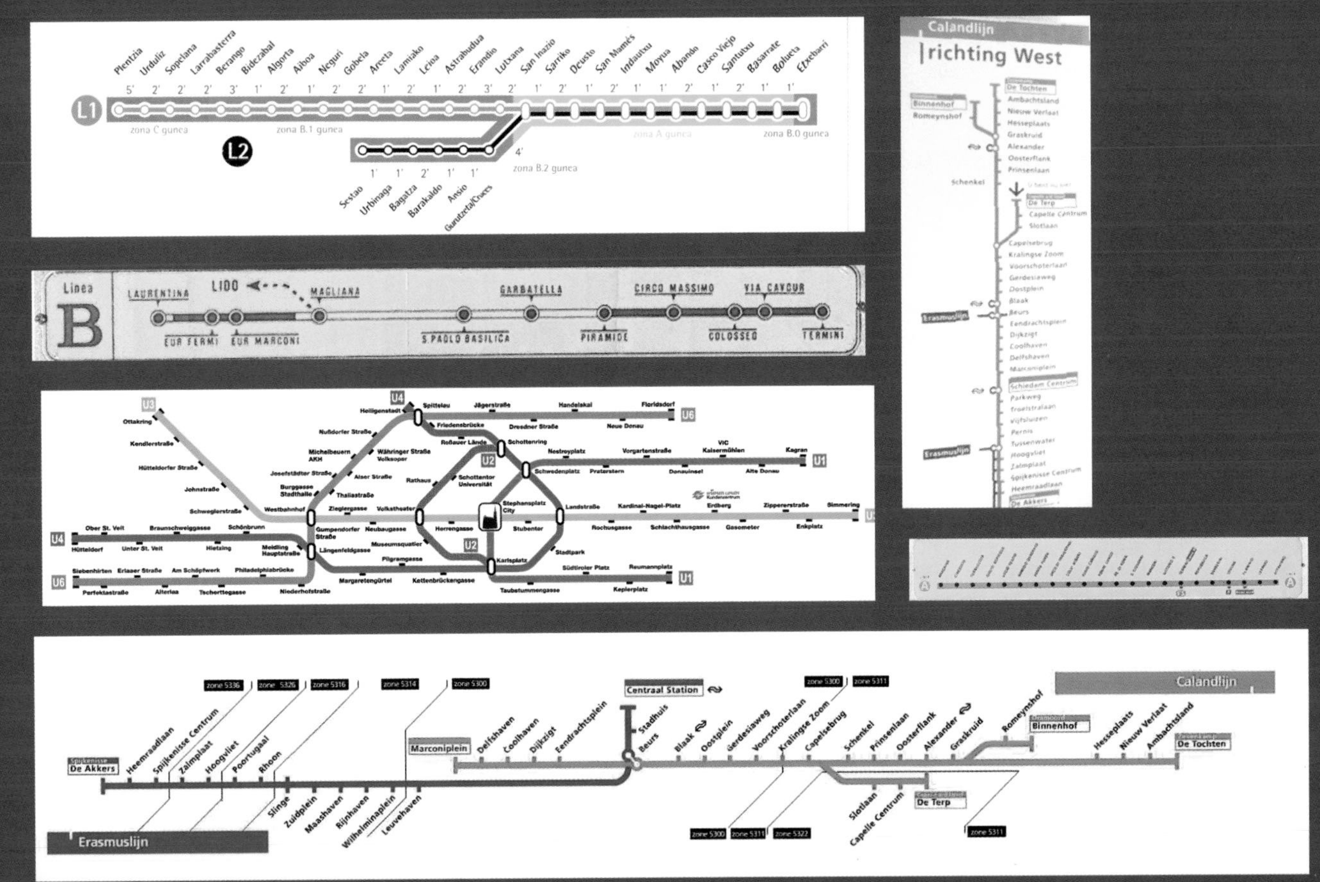

Amsterdam

Urban population: 1.3 million. Route length: 31.7 miles. Stations: 49. First section open: 1977. Underground: 15 percent.

When in the mid-1990s GVB's map designers visited London Underground to discuss using elements of the Harry Beck–inspired design classic, the Brits were only too happy to oblige, leading to an Amsterdam Metro map produced by influential Dutch design firm Bureau Mijksenaar with a loving nod toward the United Kingdom (1). Stations are shown as ticks or double ticks at terminals, and interchanges are marked by circles with white-eye centers linked by white-line connectors. Although Amsterdam has chosen a 65-degree angle for the main rail spine as opposed to the more common 45 degrees, its diagram contains all the hallmarks of good practice for urban-rail diagram design.

Interestingly, with the inclusion of topographic features such as the detailed blue waterways, gray urban areas, and key parkland in green, along with the main commuter lines being portrayed at angles closer to their scenic meanders, the diagram has many maplike qualities. However, features such as the simplification of Metro routes and even spacing of stations help retain a diagrammatic feel.

Prior to the current version, a 60/30-degree-based design was used, as the 1986 version shows (3), though the sketch on its cover (4) seems to show a 45-degree-based map that never existed!

While many cities were mindlessly digging up tram tracks as if the rails were polluting roads, Amsterdam stubbornly invested in its substantial streetcar system, which, alongside its picturesque canals, is now a significant tourist attraction in its own right.

In the mid-1960s there was a proposal to replace the trams with a four-line underground railway system (2), the hub of which would have been Centraal Station. But the idea gained bad press and mass protest when the scale of proposed demolition was revealed, along with the realization that the beloved trams might go, too!

The first section of the Metro, which required the least tunneling by justification of having a substantial elevated section, did not open until 1977. This, now known as lines 53 and 54, had its original southern terminus at Bijlmer, a major new town on the edge of the existing urban area.

GVB progressively opened more extensions to establish a network not dissimilar to that originally envisaged. Of several important sections still missing, the most crucial is the North–South Line, construction of which has finally begun. By 2013 it will link Zuid/WTC to Centraal Station and then cross beneath the river to the poorly served area at Buikslotermeerplein. It is likely to be shown in light blue on future diagrams (5).

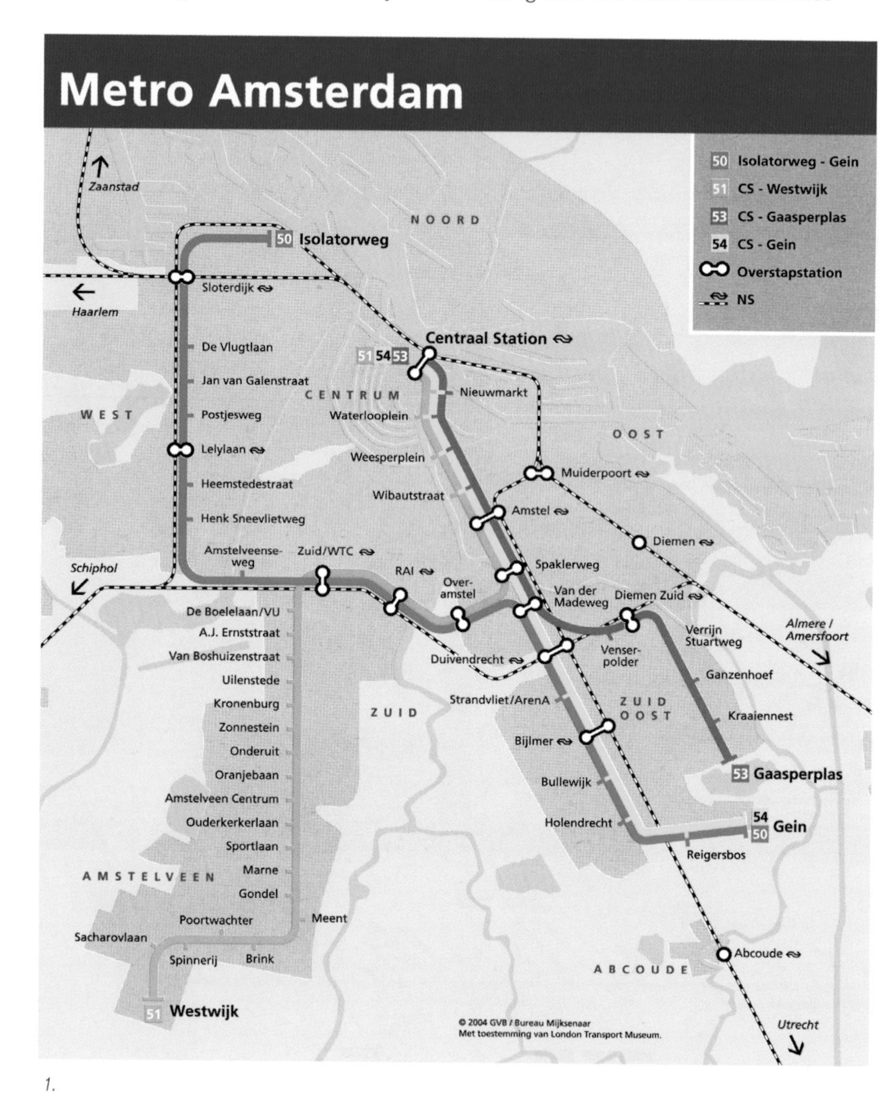

1.

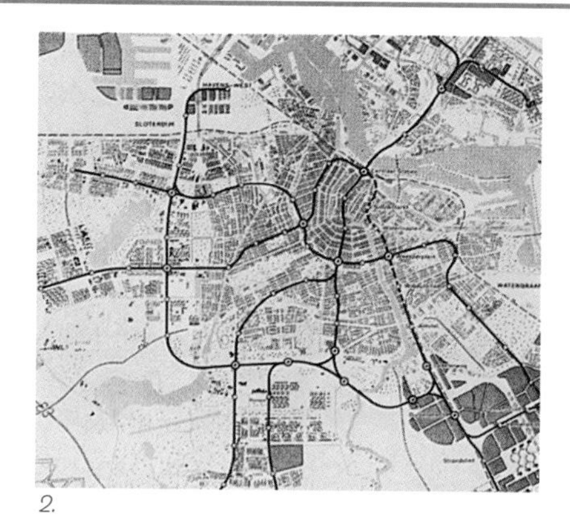

2.

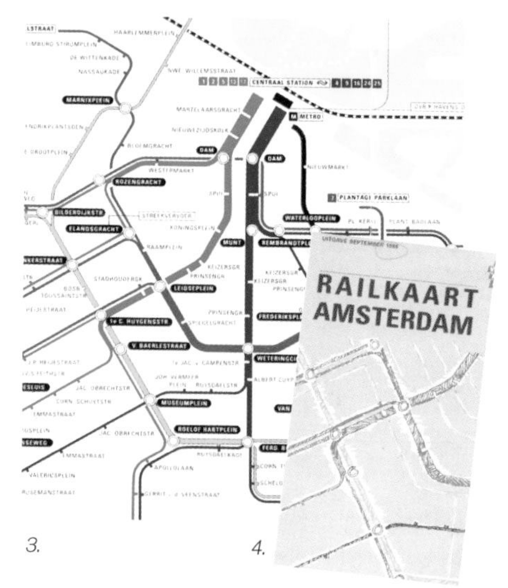

3. 4.

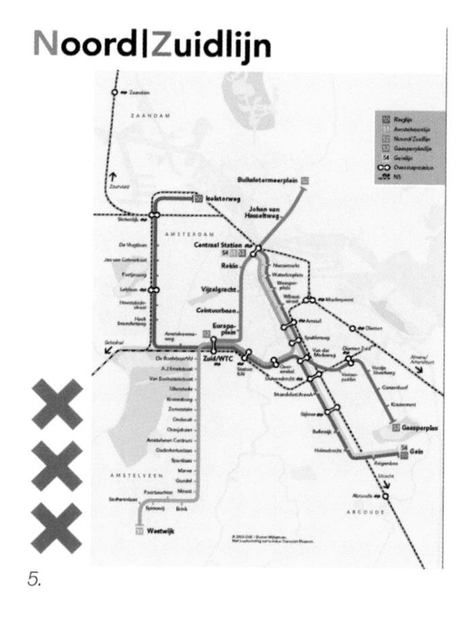

5.

Maps reproduced by kind permission of and copyright GVB 2007, with thanks to Paul Mijksenaar.

Athens

The main Athens diagram (1), as seen on stations and in foldout maps, with its red, blue, and green primaries, exudes confidence. The simplification of diagonals to the standard 45-degree angle along with equalization of distances between stations and effective ironing out of twisting geographical bends make this a truly classic urban-rail diagram.

At first sight the inclusion of such topographic features as the mountain ranges, major roads, and green spaces might lead to its classification as a map rather than a diagram, until one notices that even these features are stylized by horizontal and vertical lines and 45-degree diagonals! The black circle with white-eye center works well as a station marker, and there are effective double ticks for terminal stations.

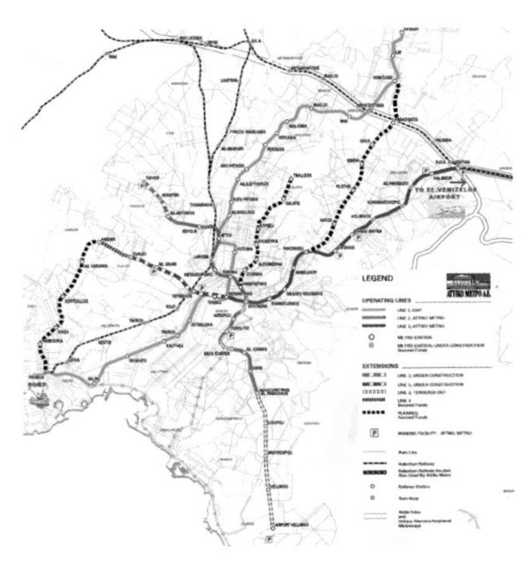

3.

For the 2004 Olympic Games, the blue line was given a massive long extension to the airport; however, this is generally shown only on more geographic maps.

These more geographic versions of the system are quite prevalent in the city and are seen on the Web site (2), and this one does show the new long run out to the airport. There is also a geographic map of the plans for the central-area extensions (3).

The Attiko Metro logo, comprising a blue stylized *M* in a green circle, works perfectly either in small print or as a vast station portal. It is becoming shorthand for the city, as the emblem of any half-decent urban-rail system should.

Now known as Line 1, Athens's first mass-transit route was the Athens–Piraeus Electric Railway (ISAP), a title gained following electrification in 1904, though it has its origins in a line opened in 1869, which ran from the elegant harbor station toward the center of the city at Thissio. Extended in 1895 to Omonia Square, it was effectively converted to metro status in 1957 with the construction of a mile-long tunnel north and a surface route toward Kifissia.

As the Greeks choked their way into the automobile age of the 1970s and '80s, it was clear that Athens needed help. Plans were laid to augment the existing ISAP line with two new underground lines to alleviate the horrendous pollution caused by a million and a half cars.

The city won the largest single contribution of European money for the country's biggest-ever construction project, to build the two stunning full-scale heavy metro lines that were of great help to the city when it staged the 2004 Olympics.

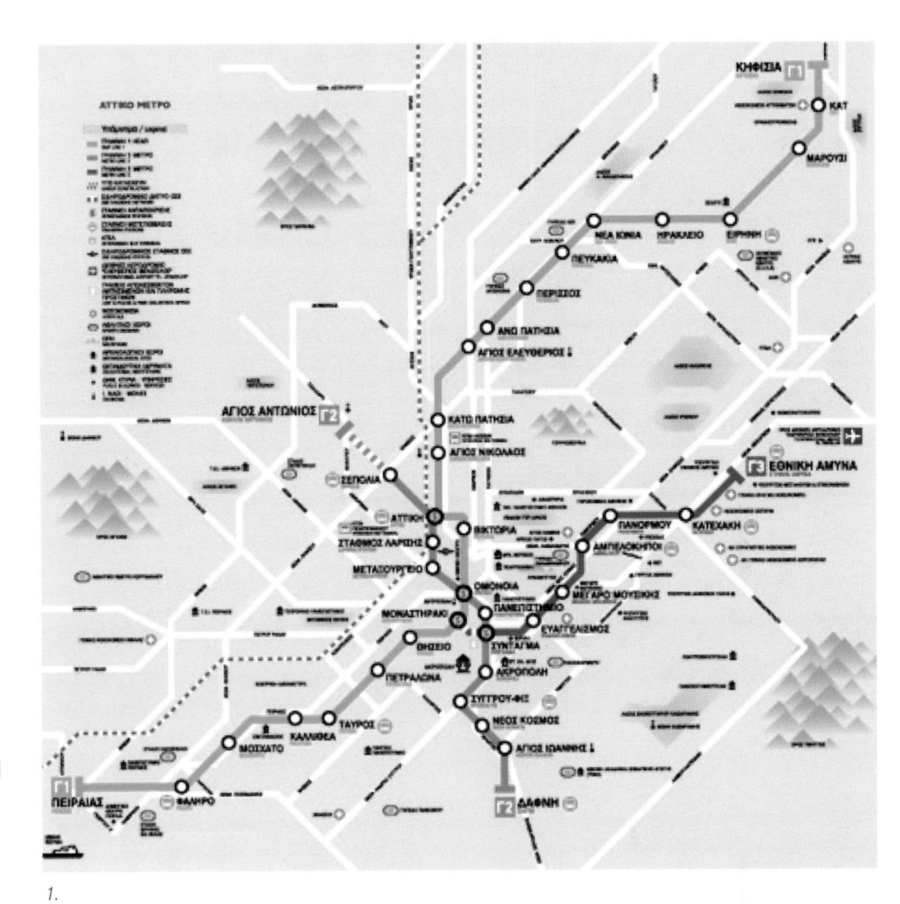

1.

The excavations, not surprisingly, unearthed literally thousands of archaeological artifacts and tombs from the long history of this great civilization, and many of these were used to enhance the stations, which scream "culture" from every mosaic floor and wall mural. Opened with the rebranded Line 1 in 2000 as Attiko Metro, the new network immediately eased communications, cut congestion, reduced pollution, and proved the city's expensive investment was justified.

So if you thought all the majesty and grandeur of the ancient Greek capital was only at street level—try going underground!

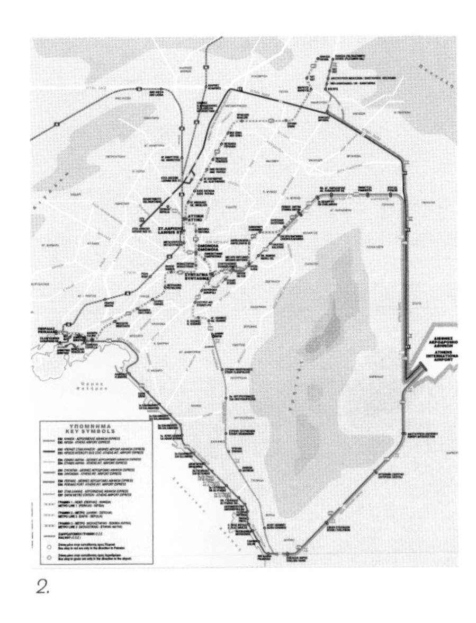

2.

Beijing

Urban population: 13.8 million. Route length: 70.8 miles. Stations: 64. First section open: 1977 (1969). Underground: 100 percent.

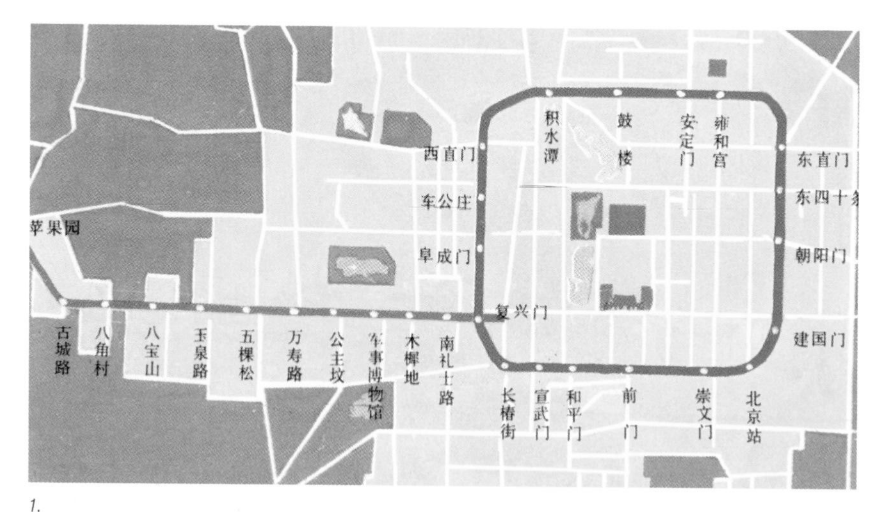

1.

For a city of its size, Beijing has been slow in developing mass transit, forcing the public to rely heavily on street transportation, including the ubiquitous bicycle! Faced with much-increased car ownership following China's rapid technological development, and the environmental consequences of that trend, public transport is now high on the agenda. The country's first subway originally opened in 1969 and now forms the southern section of circular Line 2 and the western arm of what has become the lengthy east–west Line 1.

Due to the simplicity of the system and the secretive nature of the regime (only officials were allowed to use it until 1977 and no foreigners were permitted until 1980), early system maps are rare, but a 1988 one (1) dates from just after the opening of the full circle. The bright colors of districts are typical of East Asian city maps.

The tunnels themselves follow uncomplicated trajectories. Diagrams further simplify the network with some straightening-out of the line and a little distortion on the circular route. Such alterations, however, are inconsistent; for example, the offset bend in the top left-hand corner of the circle reproduces a geographical deviation taken by the tunnel itself. A corresponding deviation taken by the actual tunnels on the opposite side of the circle is, for some reason, smoothed over on the early diagram!

In 2004, Beijing Top Result Metro, an agency charged with selling ad space on the system, introduced a new diagram (2), which nicely smoothes out all the curves; the basic concept of this was also used for station wall plans. The contemporary Top Result diagram shows the recently opened Line 13 (3) on a version with even less clutter, and Line 5 should be open at the end of 2007 or early the following year.

There are now numerous extensions planned for the 2008 Olympic Games (4). At the current rate, it seems likely most will come to fruition. China is one of the world's powerhouse economies, and if all the proposals do come to life, Beijing could potentially soon have one of the world's densest subway systems.

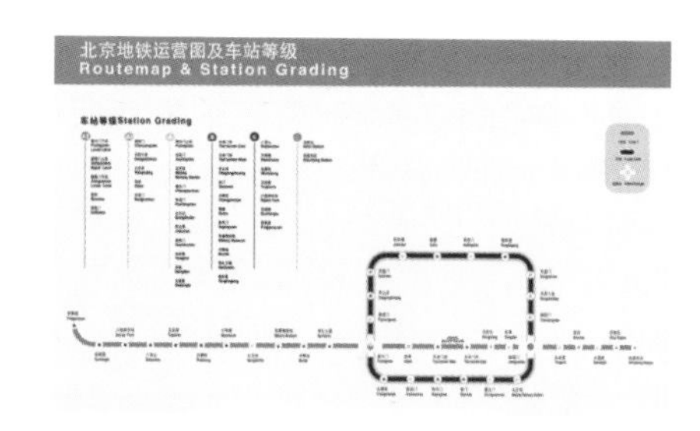

2.

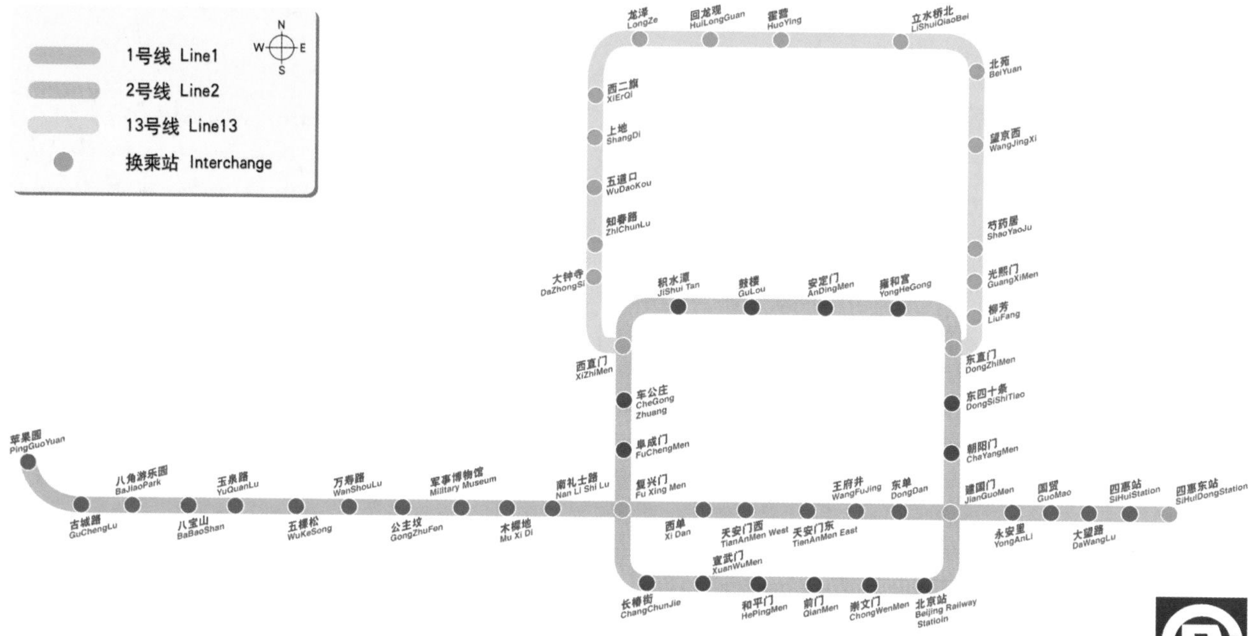

3.

北京市区轨道交通线网近期建设方案（2008年）

4.

Maps reproduced by kind permission of and copyright Beijing Map Publishing House, BSC, and Beijing Top Result Metro 2007.

Bilbao

Urban population: 925,000. Route length: 23.8 miles. Stations: 36. First section open: 1995. Underground: 35 percent.

Yet more proof, if it were needed, that Spain takes mass rapid transit very seriously, the Bilbao system is a most impressive modern construction with all the central-area underground stations entered by amazing glass armadillo-esque entrances designed, like the entire system, by Sir Norman Foster and affectionately dubbed Fosteritos (1). This helps explain Foster and Partners' ethos for the project: "The building of tunnels for trains is usually seen in isolation from the provision of spaces for people—even though they are part of a continuous experience for the traveler, starting and ending at street level. The Bilbao Metro is unusual in that it was conceived as a totality: architectural, engineering, and construction skills were integrated within a shared vision."

The three-circle emblem, station nameplates, and corporate identity are in Foster's favorite font Rotis (aka Libre Sans Serif), and enable easy identification as you walk throughout this beautiful old city. The logo marks an interesting departure from the traditional Spanish Metro symbol of a red diamond.

1.

The schematic itself (2) sticks to 45-degree diagonals but attempts to stay true to the geography of the urban area. This means Line 1's spiderlike arm, which genuinely winds out many miles to the coast, has a tendency to keep the map in portrait format, although a useful foldout plan of downtown was available in the late 1990s (3) and current pocket maps have a neat line diagram on the reverse (see p. 76).

Line 2 is currently being extended farther, which will give the Basque city a 27.5-mile network with forty-one stations by 2011, and Line 3 is now planned as part of the Euskotrain network, so a combined diagram with all urban-rail transit could be on the way.

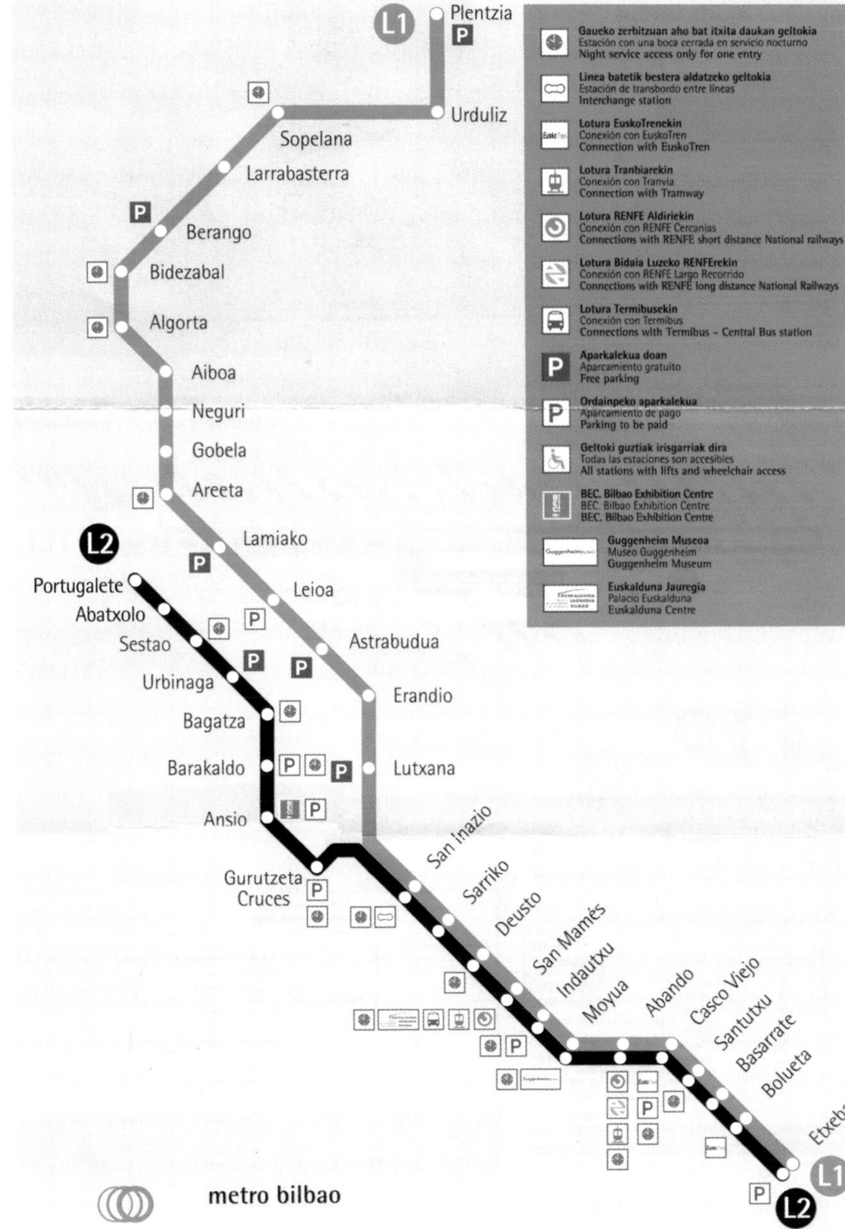

2.

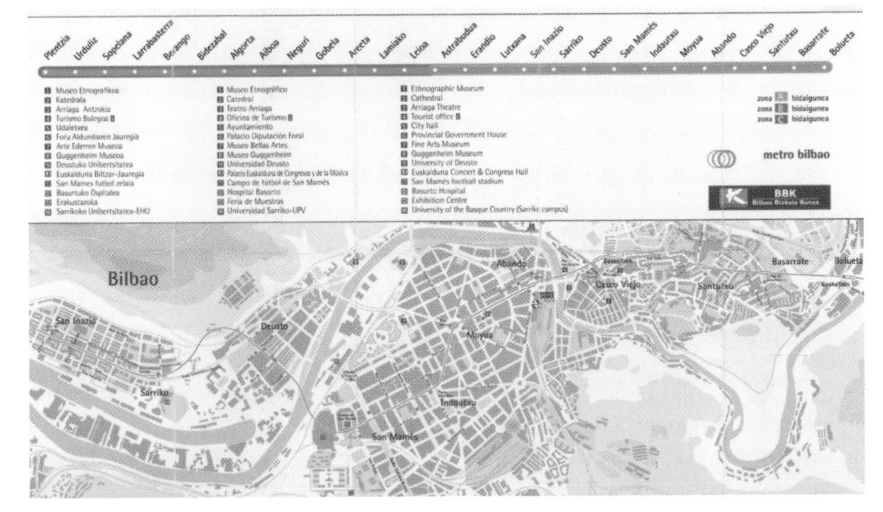

3.

Brussels

Urban population: 960,000. Map shows: Route length: 20.1 miles. Stations: 68. First section open: 1976 (1969). Underground: 100 percent.

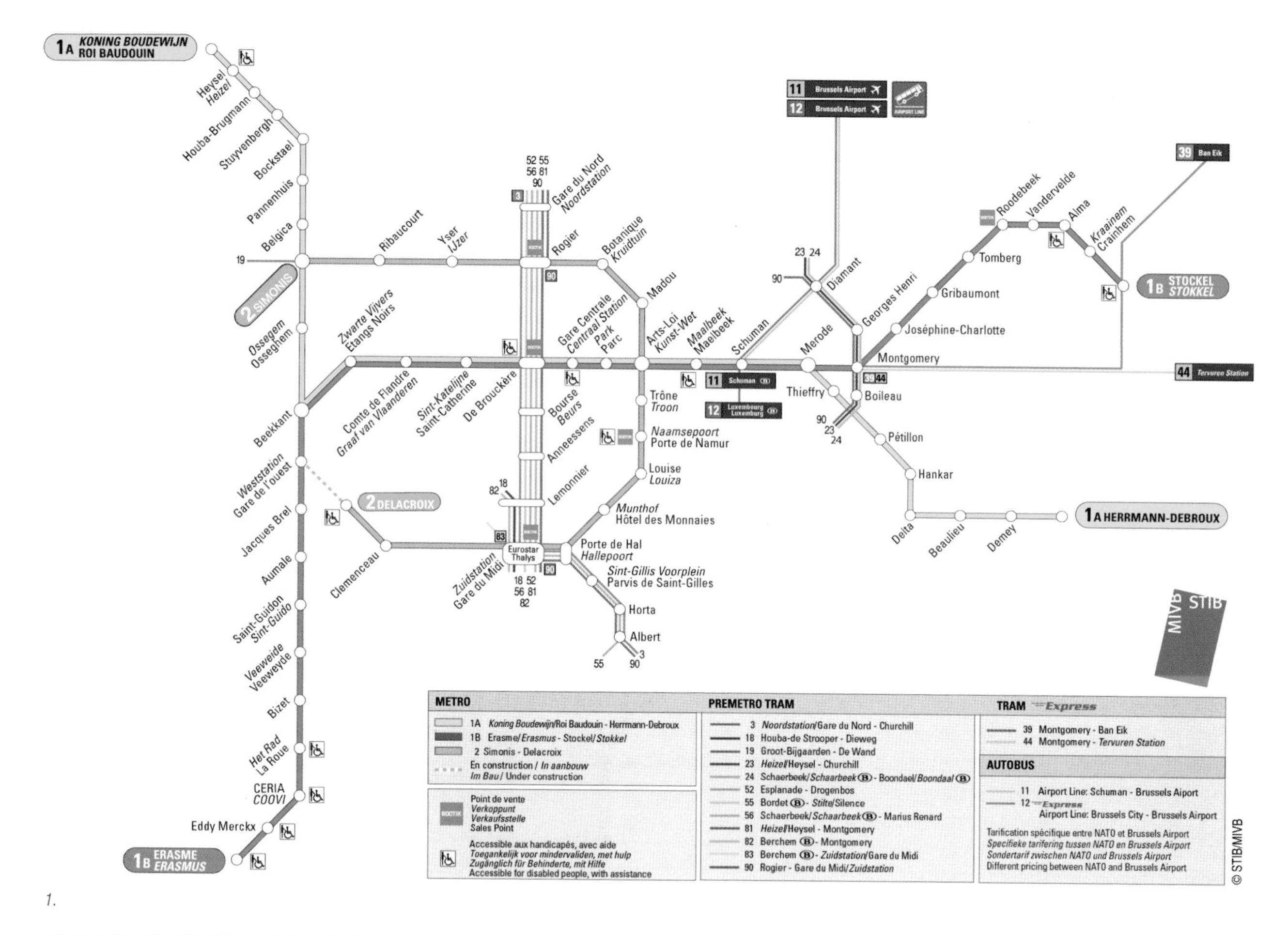

1.

Maps reproduced by kind permission of and copyright STIB/MIVE 2007.

2.

The Brussels diagram (1) has a slightly incomplete feel, as if it somehow forgot to show where the broad cluster of lines running north–south actually go. However, passengers soon discover that when traveling past Gare du Nord from Rogier, the line does not stop; it just surfaces. The diagram (and the figures given above for route length and stations) show only the *underground* sections of what is a large and effective rapid-transit system. As Belgium once had the densest network of streetcars anywhere on the planet, it did not get around to building any subways until relatively recently. In the mid-1960s the city finally decided to bury key downtown tram routes underground.

The first tunnel, in 1969, between De Brouckère and Schuman, had four intermediate stations. A year later a second tunnel crossed the first with an interchange at Arts-Loi. The infrastructure was built to handle full-size Metro trains but began by running the smaller, lighter streetcars alongside the specially lowered platforms. The concept was called "Prémétro" and has been subsequently tried elsewhere.

Two routes were converted from 1976 onward to full Metro operation by simply backfilling the platforms to bring their height up to the new rolling stock's doors. The initial east–west tunnel, regularly extended, now has two branches. Line 2, converted to full Metro from 1988 onward, will soon form a circle. The other tunnels are still called Prémétro and operated by through-running trams whose termini are in leafy suburbs. This evolution is what gives the unfinished air despite an otherwise neat look, with standard 45-degree diagonals, no station names crossing lines, and even bilingual names like Comte de Flandre/Graaf van Vlaanderen. The order has no language preference. To avoid any jealousies between the two groups, either Dutch or French is displayed first, which can be a little disconcerting for visitors. Slanting station names date back at least to the 1986 card folder (2), with its Morse code–style dots.

Bucharest

Urban population: 2.2 million. Route length: 38.7 miles. Stations: 40. First section open: 1979. Underground: 55 percent.

The Romanian capital is justifiably proud of its modern four-line metro system. An engineer, Dimitrie Leonida, first foresaw underground railways in the city in the 1930s, but in the lean post–Second World War years, when the country's status was that of a satellite state of the vast Communist Eastern Bloc, it took many years for Leonida's dream to become reality. Design, funding, and building eventually began in 1975.

Due to unfavorable geology for tunneling under Bucharest, much of the excavation had to be done by the cut-and-cover method, which required digging up the roads and then roofing over the trench to carry the reinstated streets. Thanks to the determination of the socialist Romanian government to improve the public transportation infrastructure, along with low car ownership in the "planned" economies, the system has been gradually expanded over the years.

Like many subways of its era, much of Bucharest's shows strong Soviet influences, which can still be seen in the style and architecture of the early stations, such as Republica. However, the city also eagerly embraced both construction and design influences from around the world.

The main diagram (2) is supremely clear and simple, with echoes of key devices used on other urban-rail schematics, including the use of the black circle with white center symbol, 45-degree diagonals, and strong, simple colors. Other graphic similarities include double-sided terminal ticks, equalization of distances between stations, and the simplification of waterways.

Due to the relatively small number of lines, this simple diagram has a light and airy feel. The kink approaching the end of the line M1 at Republica does not always appear on the diagram. The nomenclature of lines here, allied to the use of clear terminal designations, bears some resemblance to French and Spanish systems. An intriguing earlier diagram from 1987 (1) is both slavishly angular and includes many key streets and waterways at the same time.

1.

2.

Maps and photos reproduced by kind permission of and copyright Metrorex 2007.

Copenhagen

Urban population: 1.21 million. Route length: 13.7 miles. Stations: 22. First section open: 2002. Underground: 48 percent.

The Danish capital, Copenhagen, is one of the most recent cities to build a fully automatic mass-transit system. Before that, the region had been served by only a suburban commuter-train network marketed as S-Tog (1), which because it is so efficient has the ambience of a mass rapid-transit system and has an excellent diagram of its own in the classic urban-rail style.

Now the Danes are experiencing their first "real" Metro. At present only the first two lines are operating (2)—Line 2 reached the airport in 2007. The other diagram here (3), also based on 45-degree diagonals, even spacing of stations, and a little geographic distortion, shows how the full network should look in 2017, including a complete circle line, the M3, and a projected M4 from the S-Tog station at Valby to Gladsaxe and Brønshøj.

Tunneling under this ancient and beautiful city was a costly and arduous process, crossing under waterways and historic buildings as the aerial view of downtown shows (4).

The line that runs from Nørreport to Vestamager and Lergravsparken, the first section of the new system, has been one of the most impressive construction projects the country has seen. Known as the M1, the Vestamager branch serves a vast new business and residential development centered on Ørestad and is set to become the focus of Copenhagen's future expansion.

This is in anticipation of a major urban renewal effect and illustrates the role of improved transportation in economic redevelopment.

1.

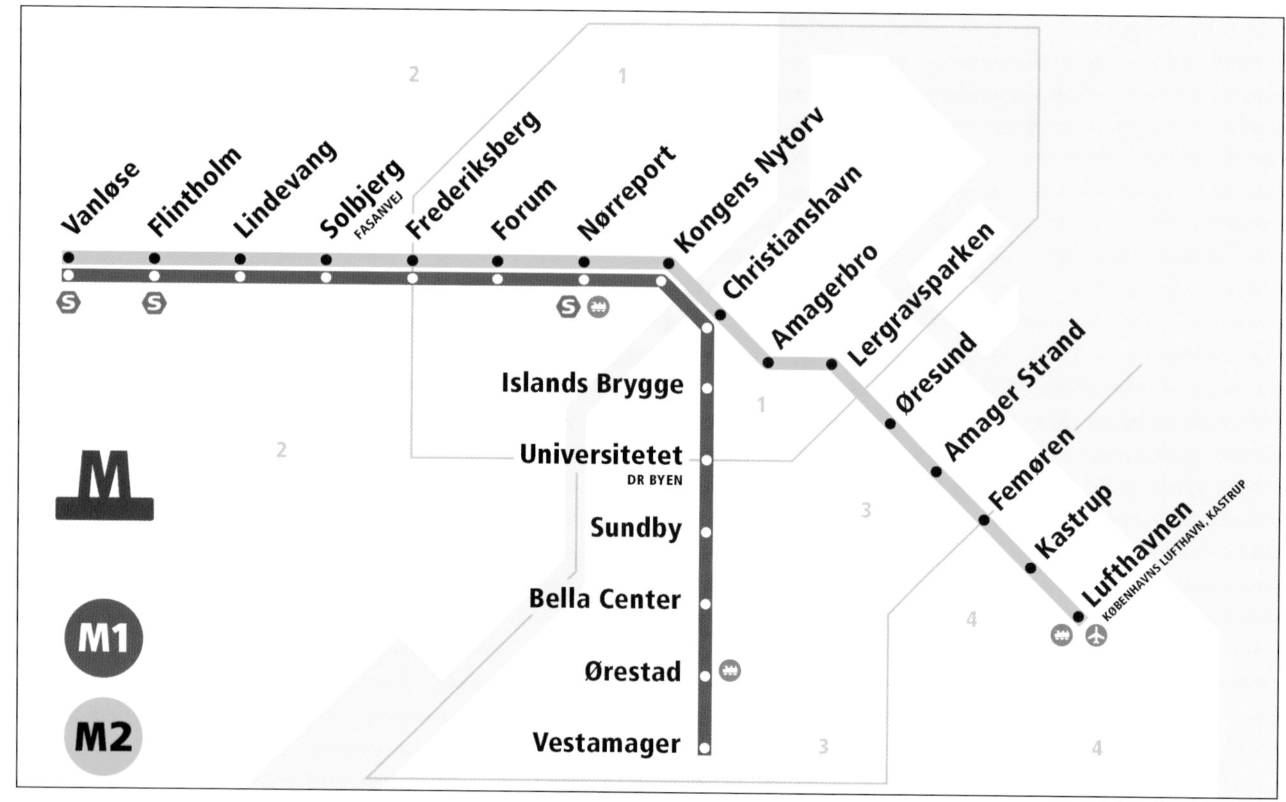

2.

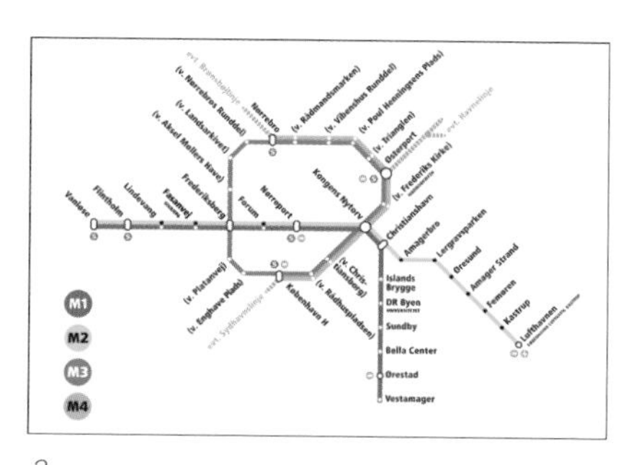

3.

4.

The new Delhi Metro is a full heavyweight system, with most of the yellow Line 2 underground. A huge planned network is being opened in phases, and the first of those is now complete.

The 2006 diagram (1) has a superbly clear and simple look, with bold primary colors, 45-degree diagonals, ticks for stations, and black circles with white centers for interchanges. The look screams tidiness and efficiency.

The current scheme dates from 1984, when years of studies proposed 125 miles of improved transportation networks, including the Metro, by 2021, to alleviate the Indian capital's chronic road congestion. Delhi then had more cars than Mumbai, Kolkata, and Chennai put together!

Over the years the plans have changed, giving rise to several different versions of the long-term view, including the highly detailed geographic version that was still on the company Web site in 2007 (2). By the speed of construction so far it looks like, at the very least, phases 1 and 2 will be open by 2010, consisting of 186 miles and 138 stations on three lines (3). A very impressive start to what could become a large network.

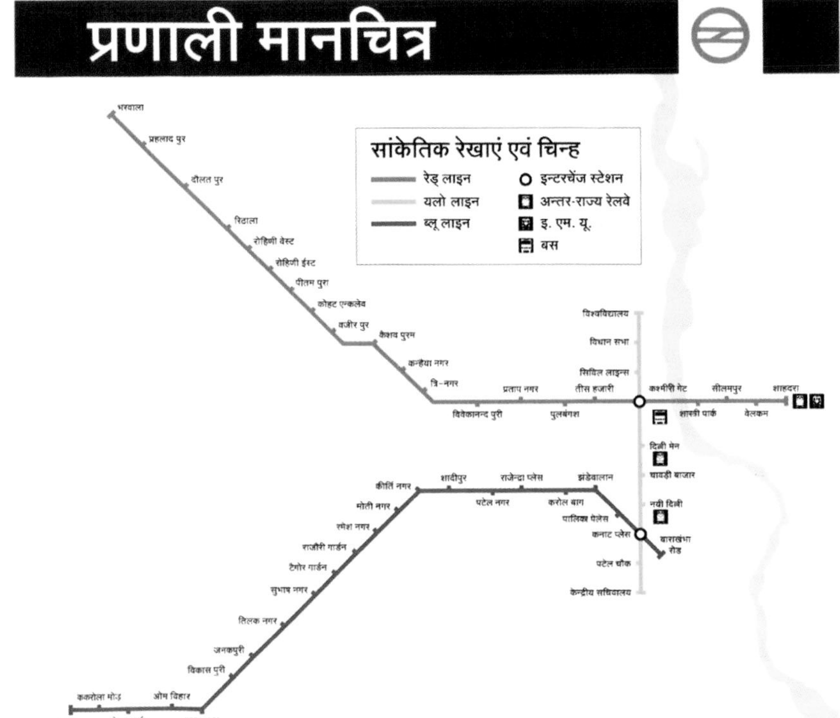

1.

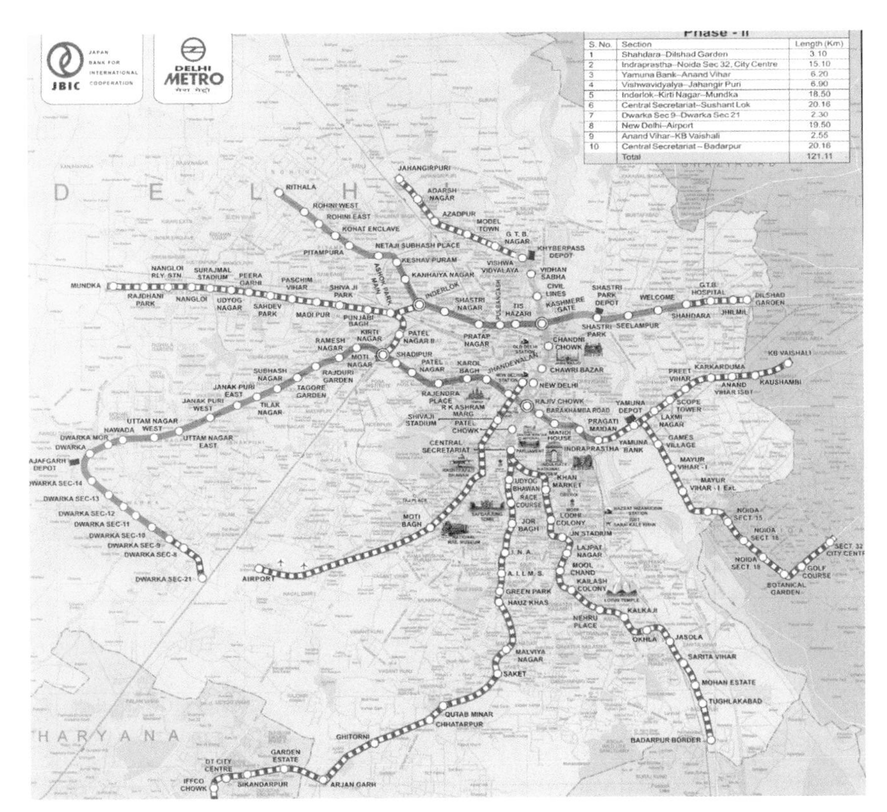

S. No.	Section	Length (Km)
1	Shahdara–Dilshad Garden	3.10
2	Indraprastha–Noida Sec 32, City Centre	15.10
3	Yamuna Bank–Anand Vihar	6.20
4	Vishwavidyalya–Jahangir Puri	6.90
5	Inderlok–Kirti Nagar–Mundka	18.50
6	Central Secretariat–Sushant Lok	20.16
7	Dwarka Sec 9–Dwarka Sec 21	2.30
8	New Delhi–Airport	19.50
9	Anand Vihar–KB Vaishali	2.55
10	Central Secretariat – Badarpur	20.16
	Total	121.11

2.

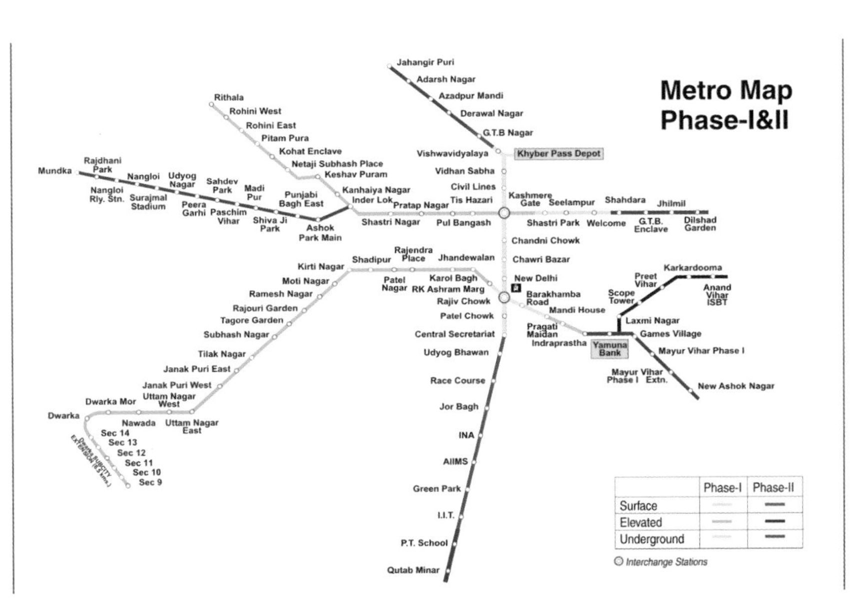

3.

Maps and photo reproduced by kind permission of and copyright Delhi Metro Corporation 2007.

Glasgow

Urban population: 2 million. Route length: 6.5 miles. Stations: 15. Entire line opened: 1896. Underground: 100 percent.

The main diagram (1) is of the entire region's rail network, with the little orange circle of the Subway almost lost in the center. It's the result of Strathclyde Passenger Transport Executive's persistence in promoting an integrated transportation policy. Though Glasgow uniquely in the United Kingdom calls its little circular line a subway, the diagram is very much in the classic style of the London diagram, with bold colors for lines, ticks for stations, and black circles with white centers for interchanges. The angle of the diagonal lines is 45 degrees, but in the 1980s and '90s the diagrams were square rather than oblong and used a 55-degree angle.

Stand-alone maps of the Subway are less common and, given the relative compactness of the line, considered almost unnecessary. Thus for some years publicity consisted of a leaflet with strong emphasis on the timetable. On this was found a Subway diagram similar to that currently displayed in the cars (3), with the rounded rectangle sitting over a broad blue band of the River Clyde. In contrast, early in-car versions (2 and 4) included many more topographic features, like key roads, docks, parks, and mainline stations.

Despite the city's having Britain's only true "Underground" outside London, Glasgow's publicity material lacked the thoroughly integrated graphic design approach of its English sister. Since the total rebuild in 1977–79, however, a much clearer corporate identity has emerged: functional brown and orange tiling, white station-wall flanking panels, and sans-serif fonts have set the tone. The initial livery was bright orange, and the line soon became affectionately known as the "Clockwork Orange."

Scotland's largest urban area has had, along with the most extensive suburban rail system in Britain outside of London, the benefit of a subway since the end of the nineteenth century, effectively the third after London and Budapest. The line is a simple two-track circle circumnavigating downtown.

Until electrification in 1935, it utilized cable haulage. The cars themselves are quite small, operating on a track gauge of 4 feet, and the older island platforms could become perilously cramped at peak hours. Strangely, the circuit managed to avoid both the majestic Central Station and also some of the downtown area. In fact, the one mainline station it did serve, St. Enoch, is now a shopping mall. However, the line does continue to efficiently connect, as the promoters hoped, the thriving city center with bustling suburbs such as Hillhead.

There have been a few plans to build branches over the years, but all have come to nothing because of a lack of funding, as well as the limiting gauge of the tunnels. There are now some plans for new light-rail schemes in Glasgow.

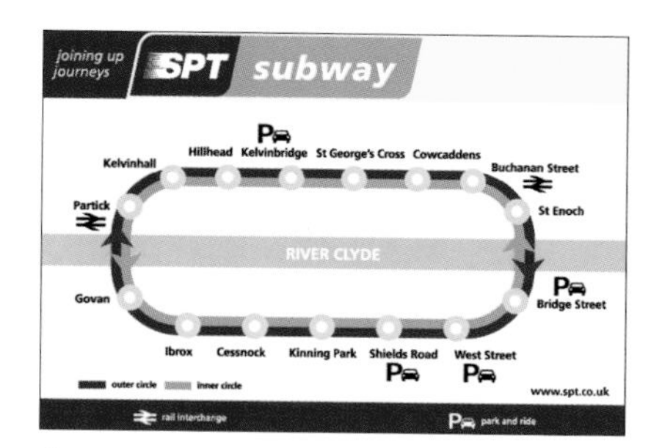

3.

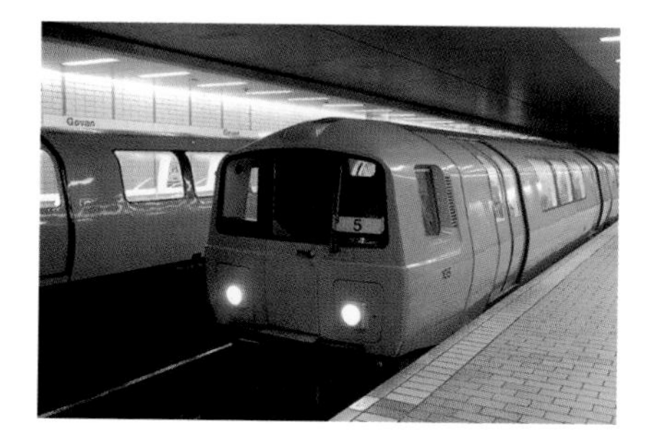

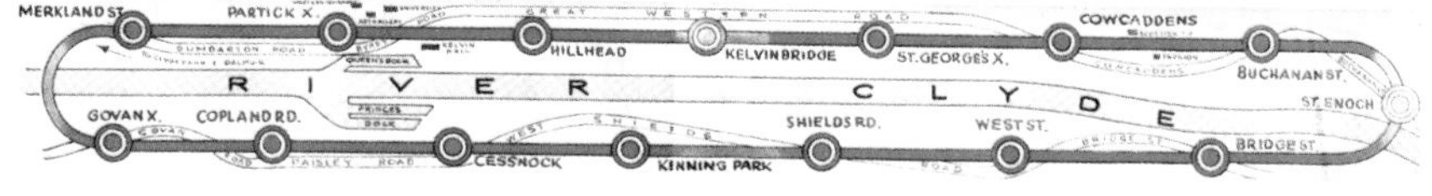

4.

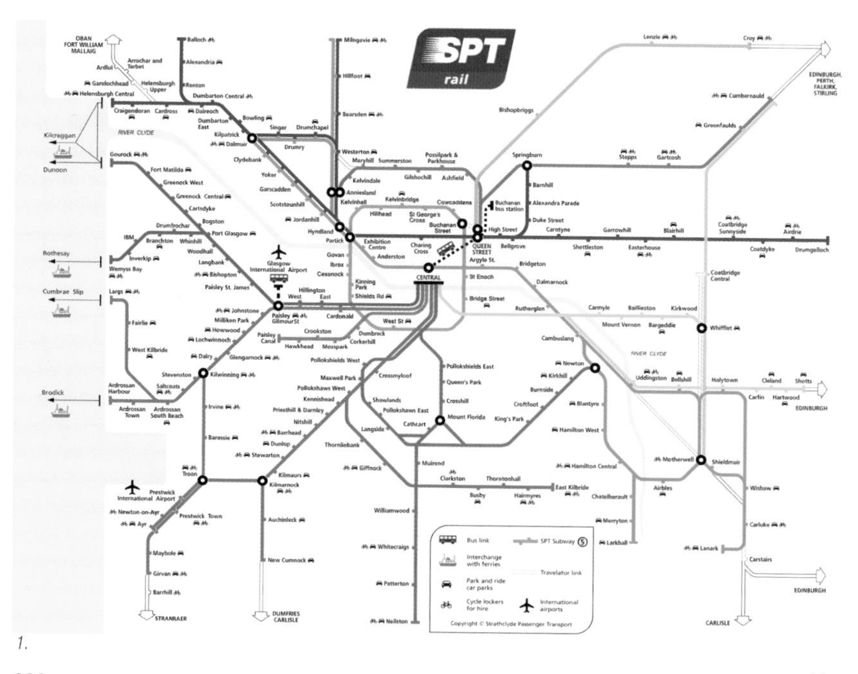

1.

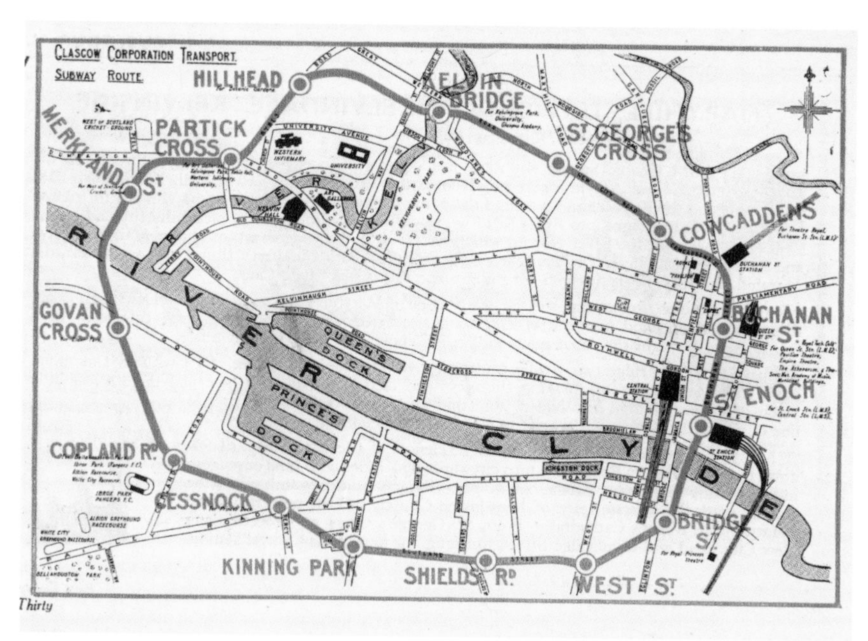

2.

Urban population: 2.6 million. Route length: 34.8 miles. Stations: 43. First opened: 1960. Underground: 90 percent.

There were plans for a subway in Kiev as early as 1916, and these surfaced again in 1936, but building only started in 1949 after terrible war damage to the historic city. The first 3 miles, with traditional Soviet-style palatial stations, between Vokzal'na (Central Station) and Dnipro, did not open until 1960, when a basic map appeared (1).

The next section rose from belowground to cross the river and continue on the surface. This was extended at either end throughout the next thirty years. In 1976 the Kurenivs'ko–Chervonoarmiys'ka Line (the second meaning "Red Army") opened between Kalinina (which is now an interchange with Line 1 and renamed Maidan Nezalezhnosti) and Pl. Krasnaya (now Kontraktova Pl.). The line was further extended during the 1980s (2). In 1989 the Syrets'ko–Pechers'ka Line entered service in the city center (3) and has been extended toward the southeast.

The current Kiev diagram (4) is typical of a classic three-line system, common across the entire former Eastern Bloc. The airbrushed texture of the station markers is another common theme (5). The version from 2000 (6) is an amalgamation of both former Eastern Bloc and more commonplace metro map design: massive simplification of the lines, roughly equal spacing between stations, and standard 45-degree angles. Here the artistic elastic has stretched the system to an almost symmetrical pincer shape.

Compare it to a geographic version from 2007 showing some planned extensions (7).

Yet more extensions, including fourth and fifth lines, were planned for 2020, to give a total network of over 60 miles, but at least one proposed subway line, entirely in eastern Kiev, has been built as a fast-tram light-transit line to save money.

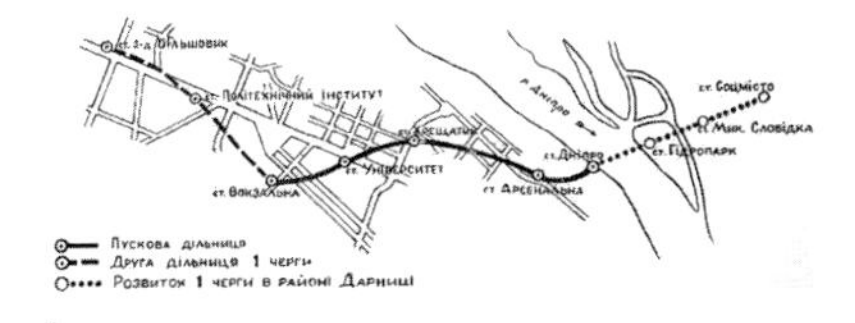

1.

2.

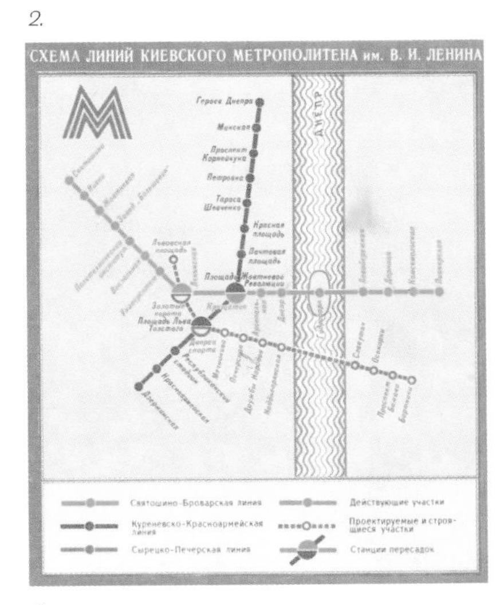

3.

4.

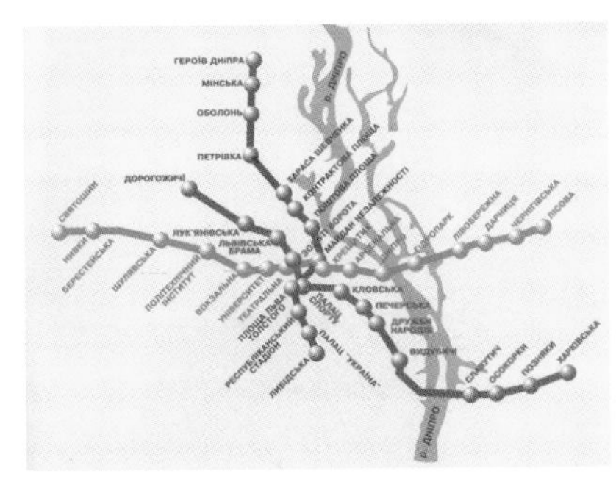

5.

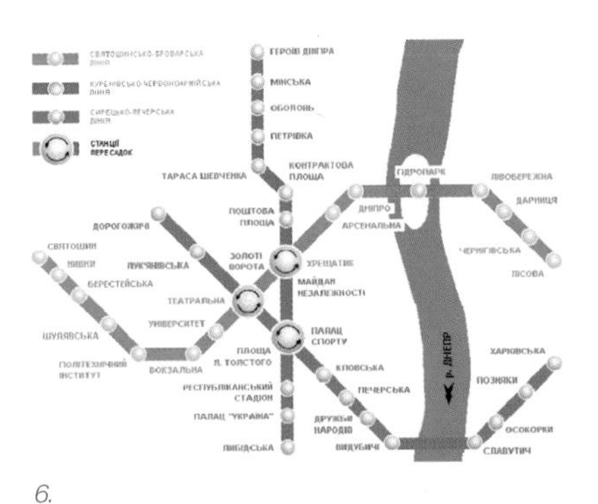

6.

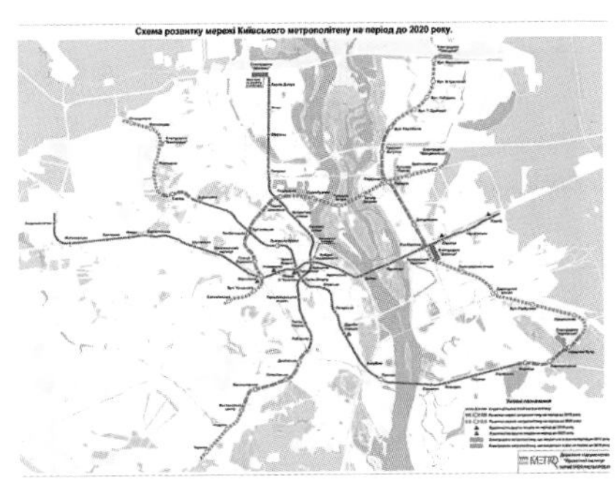

7.

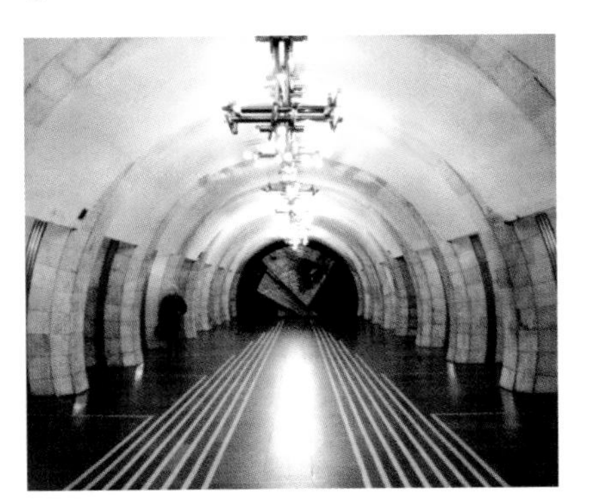

Maps and photos reproduced by kind permission of and copyright Kyivsky Metropoliten 2007.

Kuala Lumpur

Urban population: 6.5 million. Lines C, D, E and F: Route length: 44.7 miles. Stations: 54. First section open: 1996. Underground: 6 percent.

Kuala Lumpur, one of Asia's new supercities, is home to the magnificent Petronas Twin Towers, the world's tallest buildings until 2004. There are five independent rail companies here: two light-rail systems that operate as three lines, and a monorail plus high-speed and airport express services. Each produces its own maps, some of which overemphasize their own service, so a collective has emerged that aims to promote the whole system as a network. Despite this, there is still not yet much through-ticketing.

LRT 1, the Ampang, and the Sri Petaling lines C and D are elevated. The diagram of the entire planned network from 2000 (1), when it was known as STAR, made a good effort at clarity with black circle and white-eye center symbols for stations and 45-degree angles.

The PUTRA light-rail system (Line E) is also mainly elevated but with 2.5 miles underground. It was one of the world's first full-scale driverless mass-transit lines. A cubist diagram made by Stesen Sentral (Central Station) land developers in 2001 (2) shows PUTRA in purple. An integrated diagram of 2003 (3) from KTM Komuter, which runs lines A and B, has its services in blue and the PRT Monorail, Line F, shown in red.

The most recent integrated diagram (4), used to varying extent by all the networks, is perhaps the least proficient geometrically, with a few odd angles creeping in especially on lines E and F, but it does at least attempt to show all routes with equal prominence.

Extensions are continuing gradually, and neighboring Putrajaya, which was to have been the country's future capital city, may yet build an 11-mile, twenty-three-station monorail of its own.

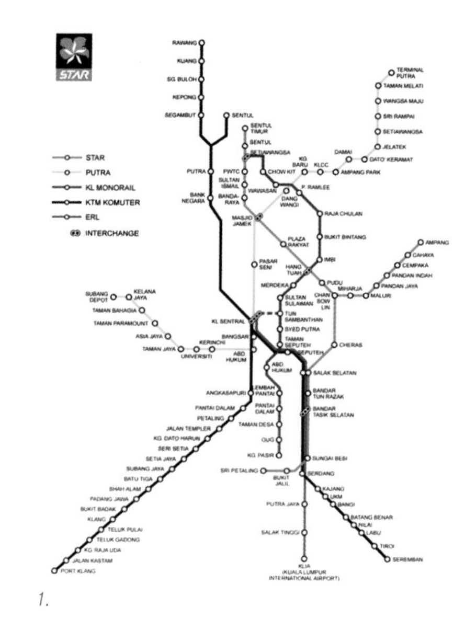

1.

2.

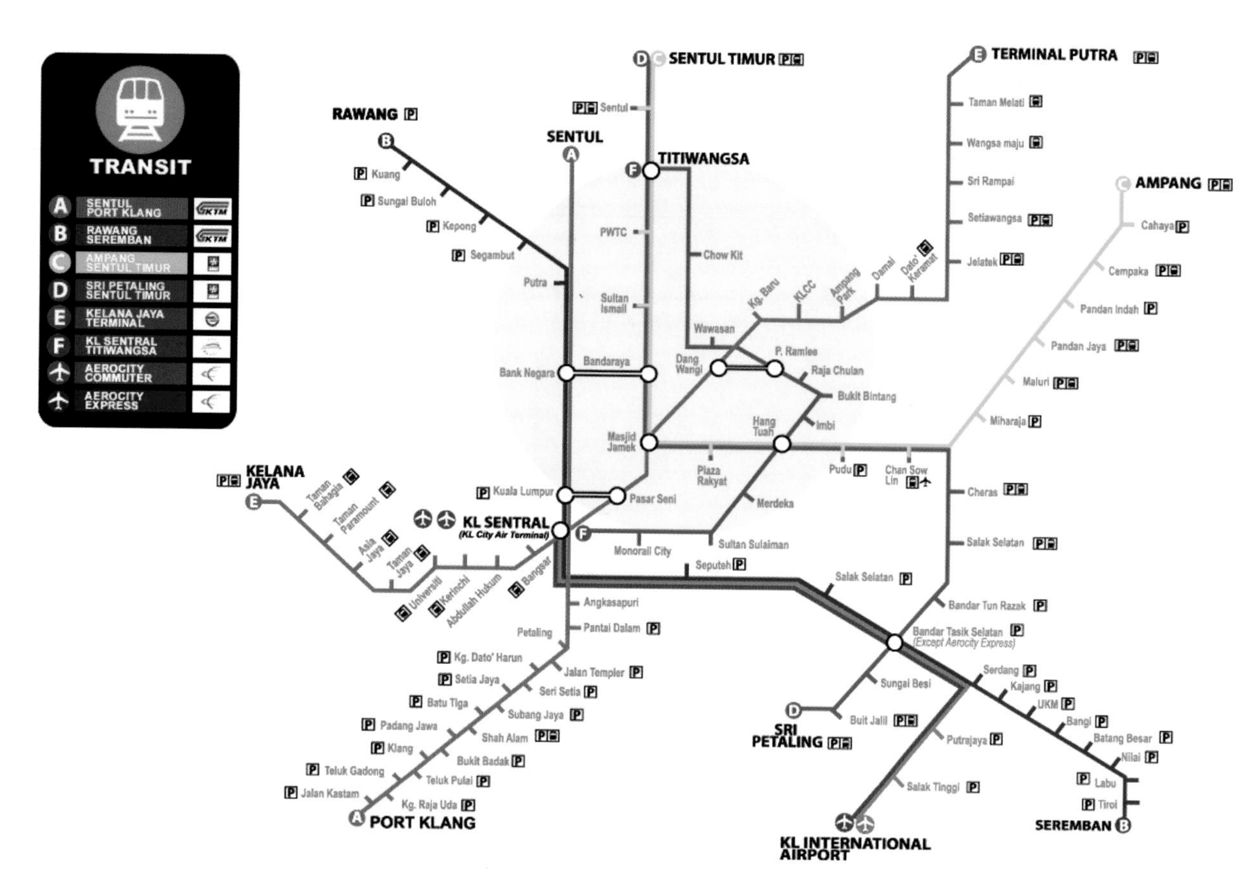

4.

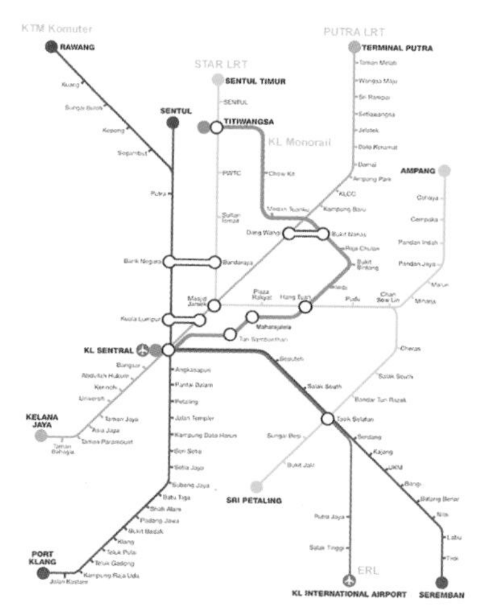

3.

This awesome, if at times not always pleasurable, urban landscape is often perceived as the ultimate car city. That has not always been the case. Early urban development was inexorably linked with the streetcars, whose 1,100 miles of "Big Red" track in the Pacific Electric system (see p. 9) laid the framework for the city's massive scale. The basic premise behind the movie *Who Framed Roger Rabbit?*, in which highway-building oil interests drove the trolleys out of existence, may be part of the story of their decline.

Since the 1950s car-exhaust pollution had increased to near suffocation levels. Sanity now prevails. Environmentally savvy Californians are rebuilding the mass-transit network with the subway Metro Red Line and light-rail Metro Blue, Green, and Gold lines.

The current diagram, with its neat station ticks, white space, and mix of 60/35-degree angles is both clear and confident. It will soon show more of the MTA's impressive expansion plans to East LA (Gold Line) and Washington/National (Exposition Line). The "Transitway" shown here is a bus rapid-transit route called the Orange Line.

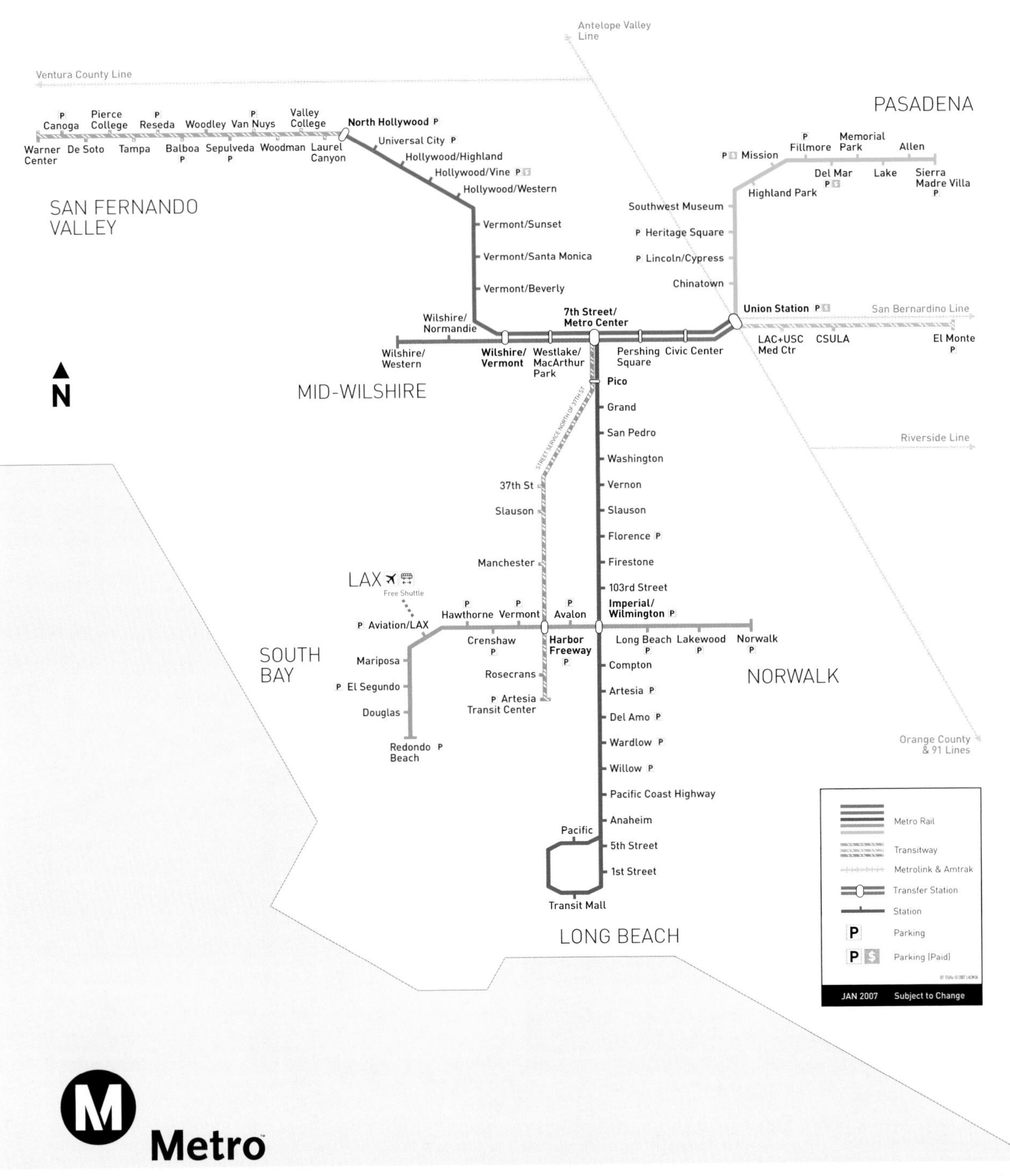

Maps and top photo reproduced by kind permission of and copyright MTA 2007. Train photo: author.

Lyon

Urban population: 1.3 million. Route length: 18.5 miles. Stations: 39. First section open: 1978. Underground: 80 percent.

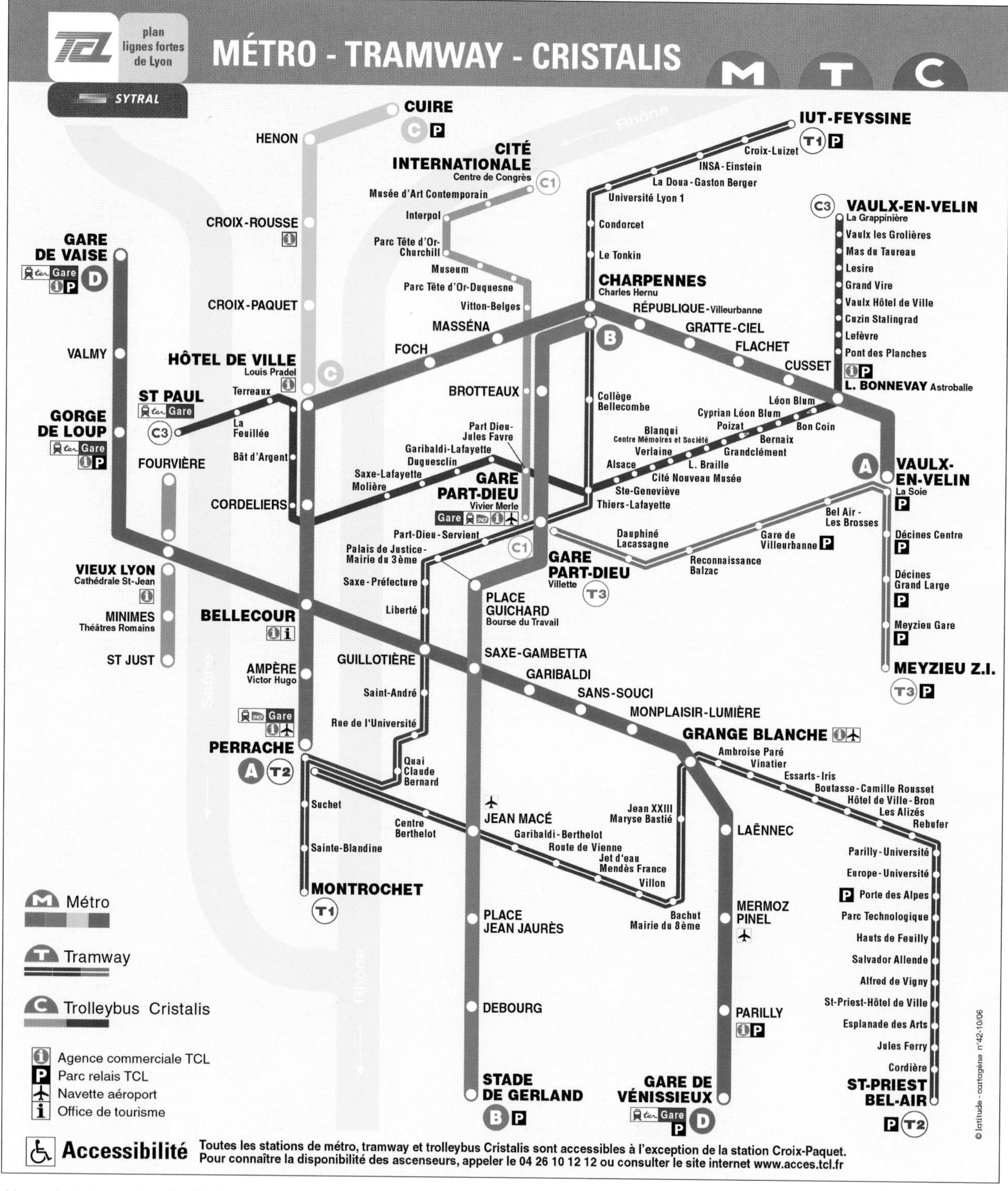

Map and photo reproduced by kind permission of and copyright TCL 2007.

An effective diagram for France's second urban-transit city, the layout gives a good balance of weight between the positioning of each line and the background. This is almost the only major urban-rail diagram using diagonals at 20 degrees, and it works well. Earlier versions used a slightly steeper 30/45–degree mix of diagonals or, back in the mid-1990s, the traditional 45-degree angle.

Lyon's distinctive and stylish trams (T1, T2 and T3) and rebranded "Cristalis" trolleybuses (C1 & C2) have been the most recent additions to the urban-transit scene, but Métro Line B could be extended in the future.

The diagram shows all modes fully integrated with the Métro system, almost as if they were continuations of the subway lines themselves. Line C is an old funicular railway, absorbed by the Métro in 1974, and is a rare occurrence of the "rack" system of wheels and cogs, which assists in pulling the train up a steep incline in a tunnel toward Cuire. Besides this, two funiculars, one dating from 1862, run from Vieux Lyon.

All other lines operate on the rubber-tired system. In 1997, Lyon opened a fully automated heavy-metro route, Line D, which has infrared detectors on the platform edges to check for obstructions, and whisks along driverless at a thrilling speed for excited children young and old.

Architects Jourda and Perraud designed several award-winning stations here.

Urban population: 2.2 million. Route length: 54 miles. Stations: 109. First section open: 1957. Underground: 90 percent.

Nagoya can trace its urban-rail roots back to the first electric railway of 1898, which became a subway in the early 1930s but not a proper metro until the late 1950s. The 1995 diagram on the back of a ticket (3) is almost devoid of angles, yet the 2002 version (2) introduced a curious mixture of diagonals.

With the completed the Meijo circle line, the current diagram (1) now has a beautiful eye-catching oval in the middle of the network. Although angles are not standard, station spacing is even and names do not cross subway lines. The boxes on mainline rail interchanges; just visible on the 2005 diagram (2), have now disappeared in favor of the East Asian tendency to number stations. These appear inside spherical or oval station markers.

Following the opening in 2005 of the maglev-powered Linimo Line, there is yet more construction due for an extension of the red Sakura-dori Line, with Nagoya embracing forward-thinking new technology for subway construction.

2.

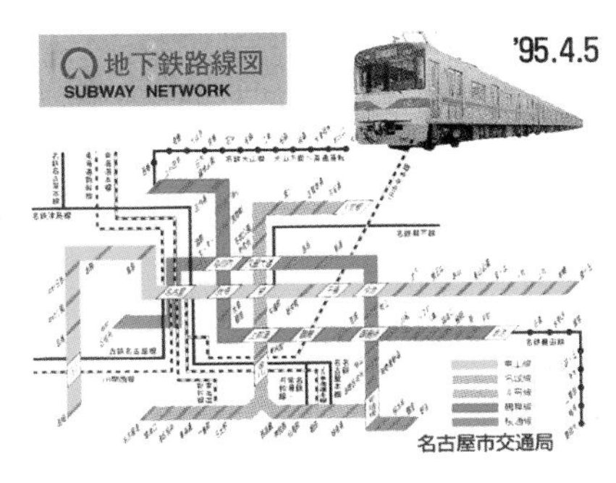

3.

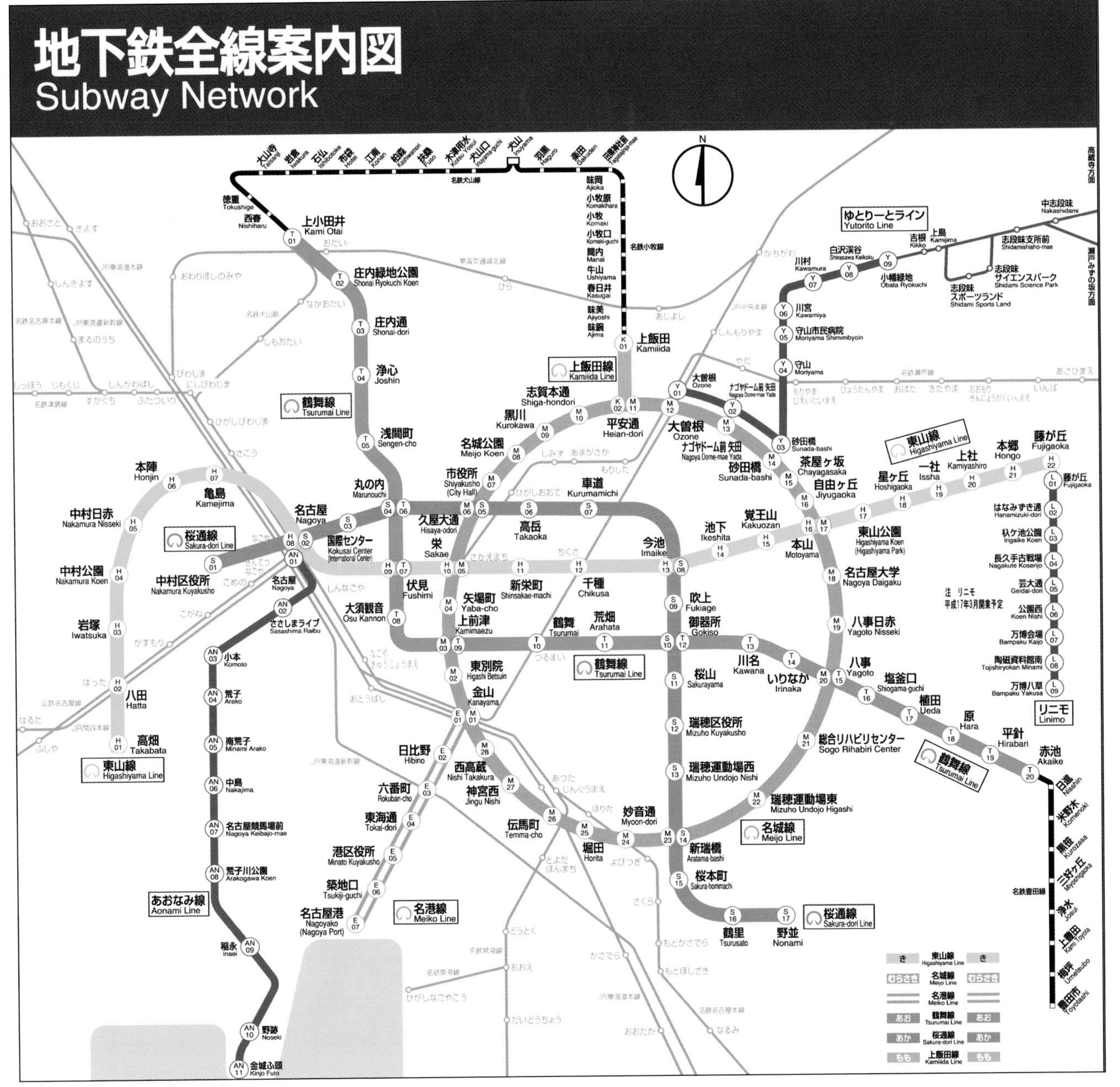

1.

Maps reproduced by kind permission of and copyright City of Nagoya Transportation Bureau 2007. Photo: Mark Kavanagh.

Newcastle

Urban population: 1.1 million. Route length: 48.1 miles. Stations: 60. First section open: 1980. Underground: 12 percent.

When power over fragmented public-transport organization in Britain's major conurbations was handed to the new Passenger Transport Executives after 1968, Tyneside took its requirement to develop an integrated transport plan very seriously. In just over a decade, the dream of converting 25 miles of underutilized and disused British Rail (BR) routes into the United Kingdom's first modern urban-transit system finally became a reality when in 1980 the Queen opened the Tyne and Wear Metro to huge crowds.

With 3.8 miles of new tunnel through the heart of Newcastle, and a brand-new bridge over the Tyne, this was a monumental milestone in the revival of mass transit in the United Kingdom—the first purpose-built new network outside London for over eighty years!

The Metro diagrams have a distinct character but with an appreciation of their British cousins. The system had its own font, designed by the darling of British signage, Margaret Calvert (who with Jock Kinneir designed the Transport font for U.K. roads and the Rail Alphabet for BR).

The first Metro diagrams were brightly colored and had ticks for stations and black circles with white centers at interchanges (1). The central spine was set at 60 rather than the traditional 45 degrees, with simple curves, standardized angles, and equal spacing between stations all in evidence. Unlike other British systems, major bus interchange hubs were highlighted—a firm intention of the Metro plan. Unique was a diamond for the interchange between all Metro lines at Monument. It appeared as if the system had four lines, when in fact colors were used to differentiate between service patterns using the same tracks.

The Metro saw some small-scale extensions, such as that from Bankfoot to the regional airport that opened in 1991. However, the new millennium brought a large extension to the system, again partially over converted heavy-rail lines, to Sunderland in the south of the conurbation, giving rise to a redesign of the diagram (2).

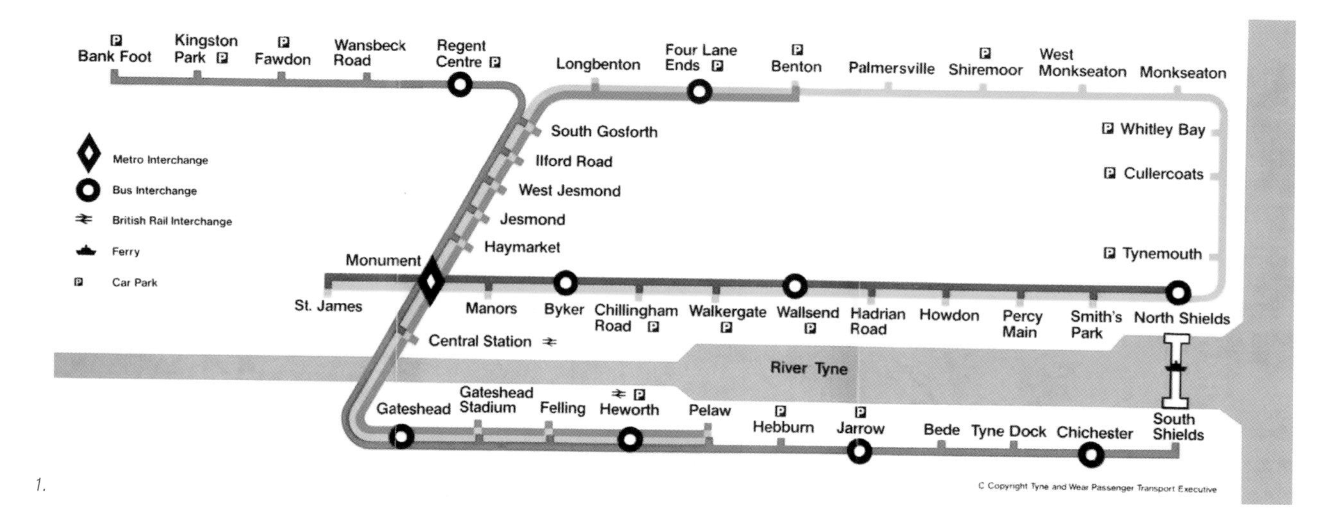

1.

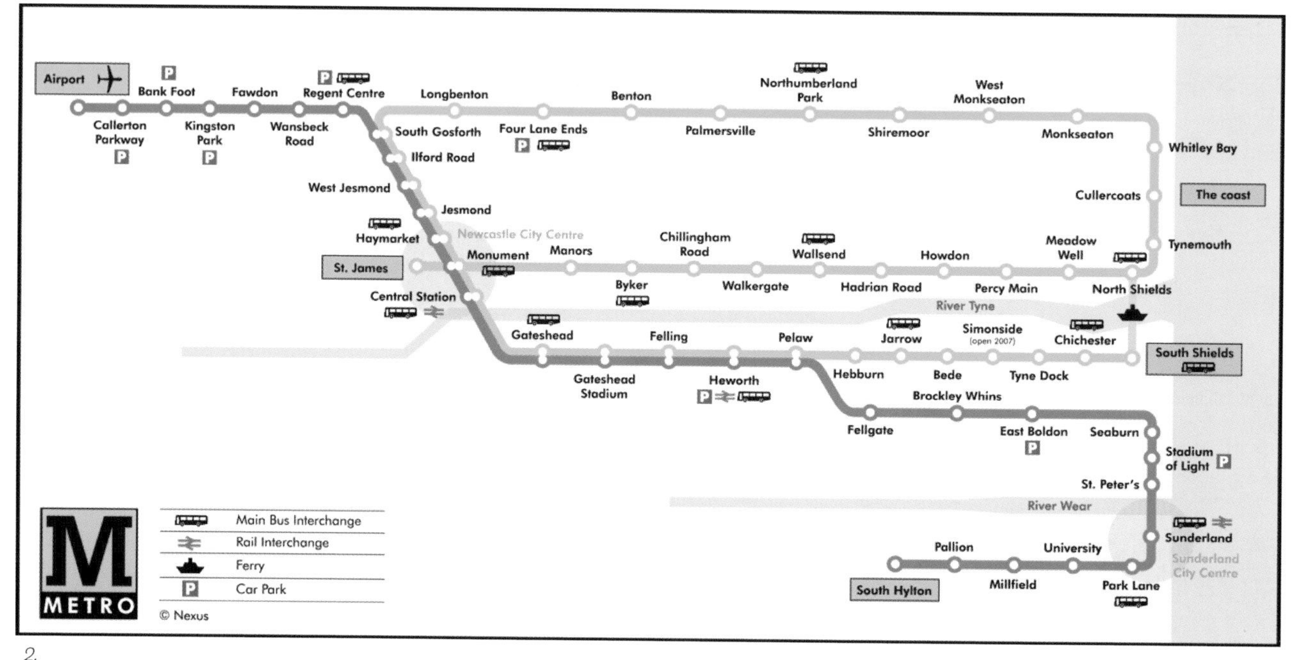

2.

Maps reproduced by kind permission of and copyright NEXUS 2007.

Oslo

Urban population: 900,700. Route length: 42.3 miles. Stations: 104. First section open: 1966 (1928). Underground: 22 percent.

Maps of the Oslo T-bane, trams, and railways have always been clear and distinct, the current diagram (1) being all in bright, bold colors, with 45-degree diagonals, equalization of spaces between stations, and distortion of the topography to fit. This and the full rail-system diagram (2) are equally well spaced. The 1973 Forstadsbanene diagram (3) illustrates how Oslo is another example of originally unconnected streetcar and suburban rail lines evolving into a full-scale metro.

As early as 1898, Scandinavia's first urban-rail line terminated at Majorstuen, on the western edge of the city center. As in many other cities, the station's location was unsatisfyingly distant from the true heart of the city. So in 1928 a tunnel from Majorstuen to Nationaltheatret created Scandinavia's first subway. On the eastern side, Jernbanetorget, which lies adjacent to the city's central station, gradually became the focus for a group of both heavy- and light-rail routes.

As work continued during the 1970s and '80s on a tunnel between Nationaltheatret via Stortinget to connect the two halves of the system, these lines were converted into rapid transit. Thus linked, there was now only one problem: as the power supply systems to the trains differed on either side of the city, passengers had to change trains at Stortinget right up until the mid-1990s!

Now, with new dual-powered trains, Lines 1–5 pass right through the central tunnel without passengers' being inconvenienced. The full circle line recently opened, and the Kolsås branch was closed in 2006 for upgrading to full T-bane standard.

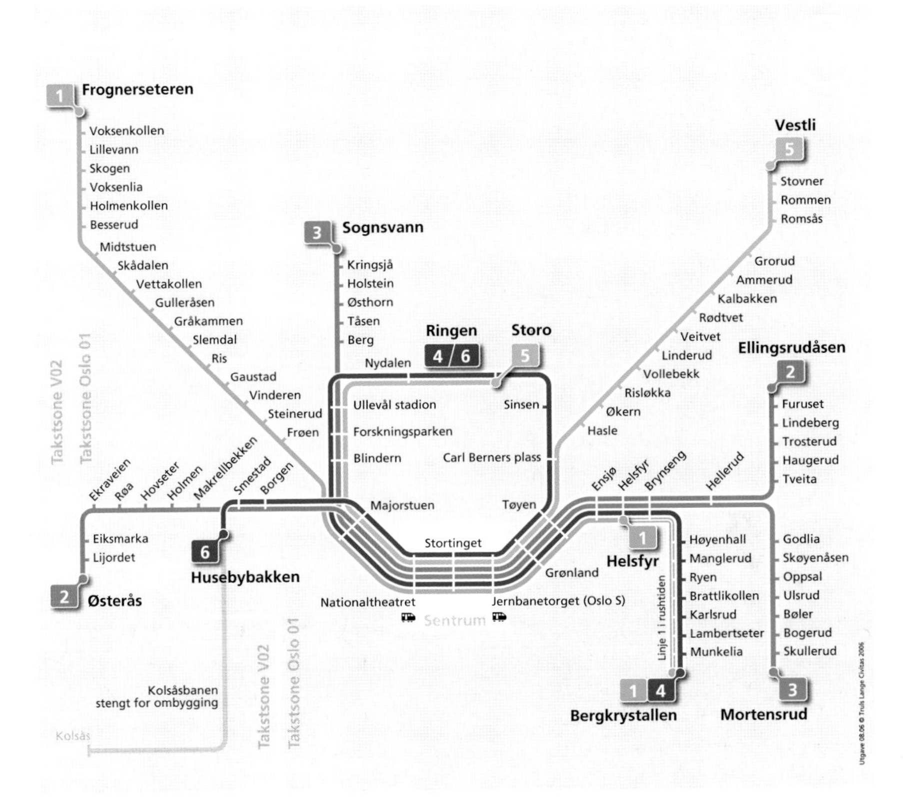

1.

2.

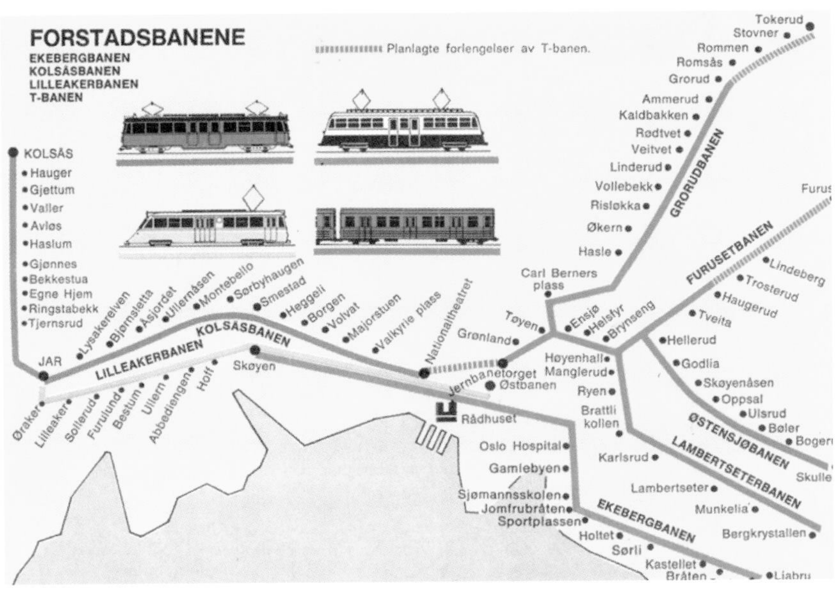

3.

Maps reproduced by kind permission of and copyright Trafikanten 2007.

Philadelphia

Urban population: 5.8 million. Route length: 43.5 miles. Stations: 62. First section open: 1907. Underground: 33 percent.

Given its pivotal role as one of America's first cities, Philadelphia has a beguiling mixture of mass rapid transit, ranging from trolleys running underground downtown and surfacing in the suburbs to the beautiful original tiling of the Broad Street and PATCO lines. However, most of the Market–Frankford Line lost its early ceramics in a recent "renovation," and in places the system is rather showing its age.

There are two main operators, SEPTA (Southeastern Pennsylvania Transit Authority) and PATCO (Port Authority Transit Corporation), which runs the connection to New Jersey. As you might guess by looking at the neat, stylish, but slightly busy SEPTA diagram (1), getting from place to place in Philadelphia can be quite a complicated affair! Not only is the full diversity of operations shown, from streetcars to light and heavy rail, but services depicted on the diagram also run from street to subway to elevated! Add to the mix the complex politics and funding issues that abound in this city, and you could be forgiven for being a little confused.

We can be thankful that this most pleasing of schematics offers some relief. It rests on solid urban-transit diagram traditions of 45-degree diagonals, bright colors, equally spaced stations, and the black circle with white (and idiosyncratically here, yellow) centers denoting free interchange.

There is an incredibly well-thought-out amount of measure and balance between the lines and stations. Look for example at the spacing between the Regional Rail lines R6, R7, and R8. This is geographic distortion, removal of topographical features, and equalization of gaps between stations at their best, and not a single station name here breaks over a line.

The PATCO Speedline–only map of 2004 (2) has a more topographic outlook for one of the world's few twenty-four-hour mass-transit routes.

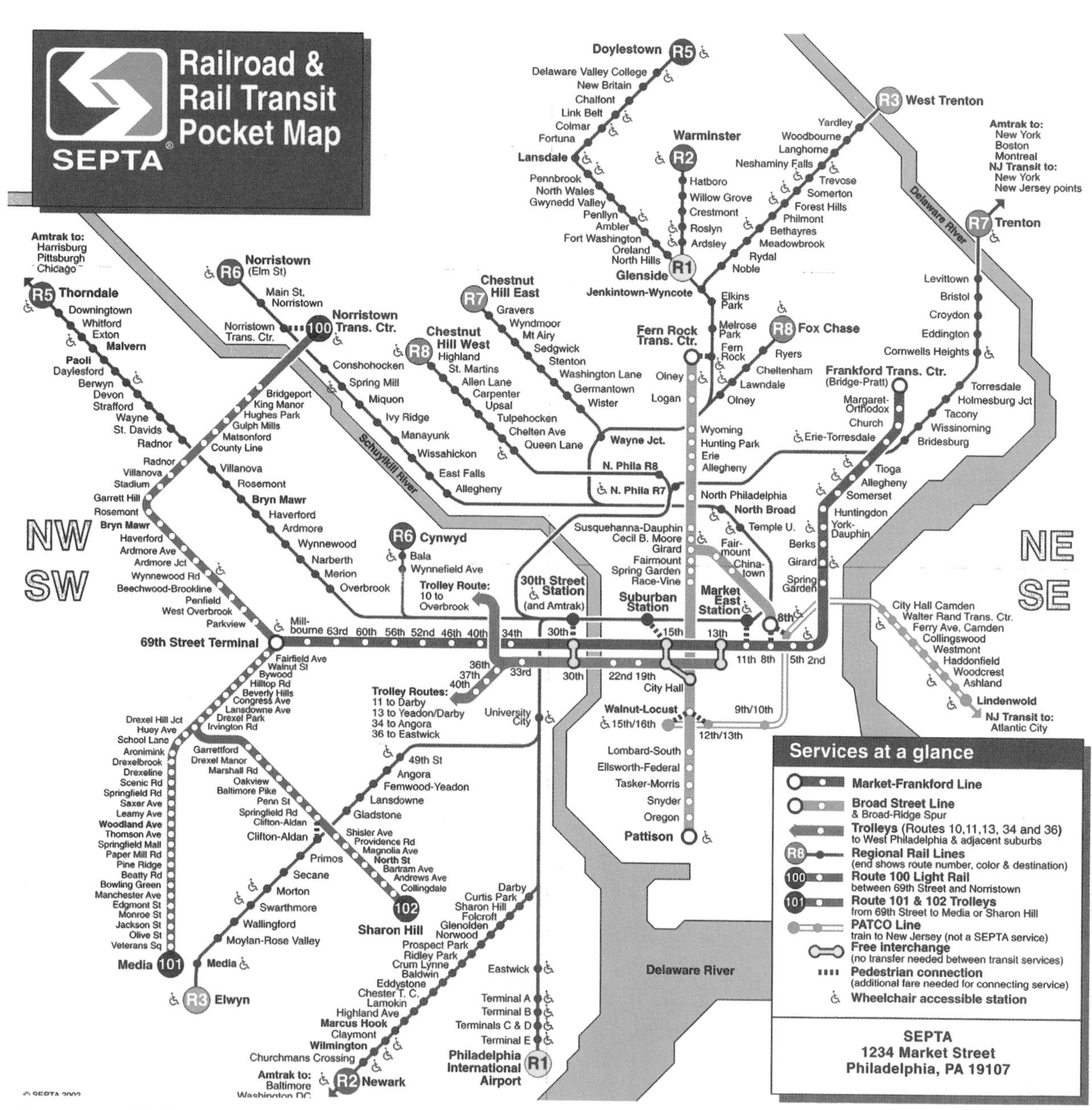

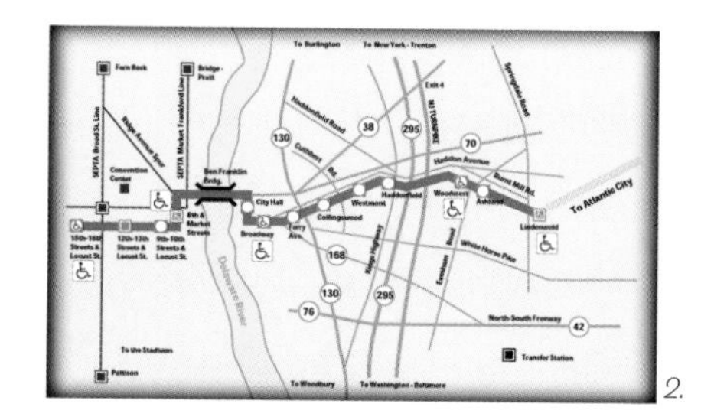

Maps reproduced by kind permission of and copyright PATCO 2004 and SEPTA 2007. Photo: Richard Brome.

1.

2.

Prague

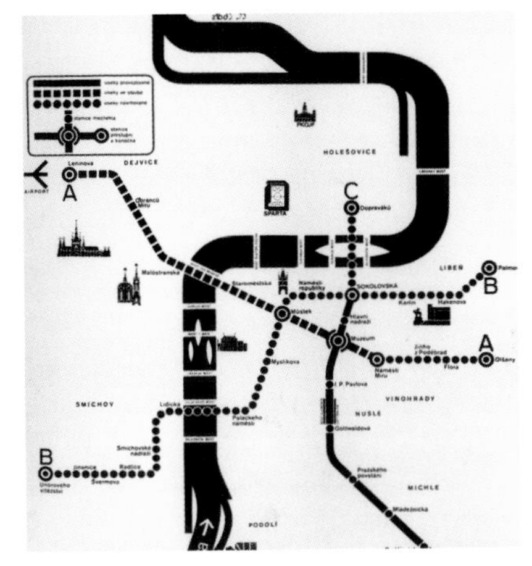

1.

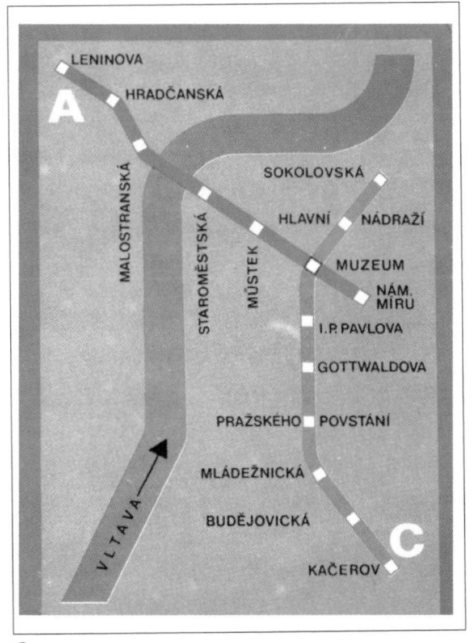

2.

METRO PRAHA UNDERGROUND U-BAHN

3.

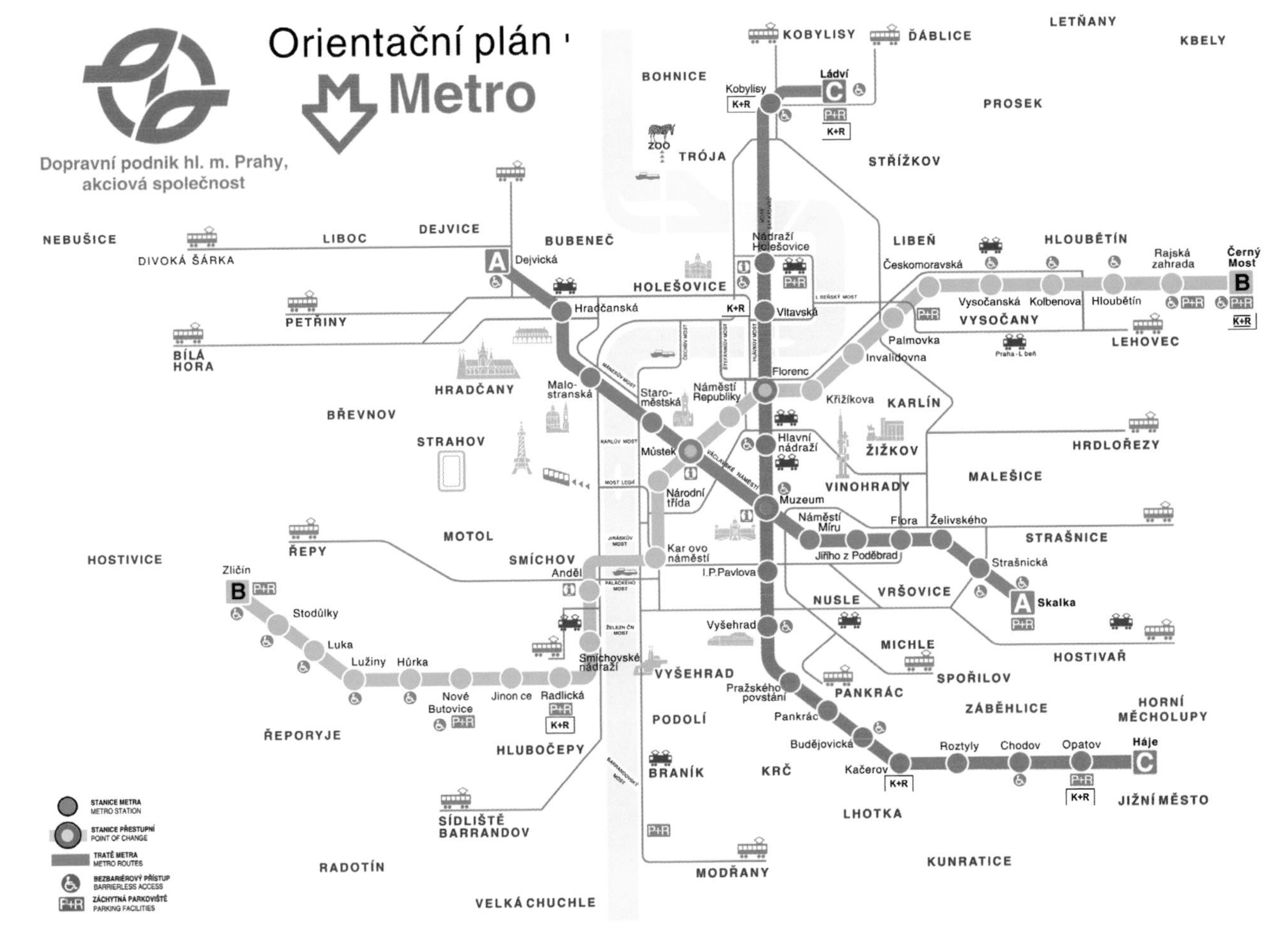

4.

Prague's relatively modern Metro disguises the fact that there were calls for a subway here as early as 1898 and again in 1926. Between the wars a plan for tram tunnels gained popularity, and in 1967 work even started on the first one at Hlavní nádraží (which was incorporated into the first section of the Metro, to open on what is now Line C), but plans were changed and the first portion of full metro opened in 1974, with heavy Eastern Bloc support.

The emphasis on design was exemplary, leading to some of the most distinctive subway interiors in the world (see photo). The first of Prague's many highly stylized maps produced for the opening also showed the proposed extensions, all of which have now been realized (1). An example from after the opening of Line A in 1978 (2) kept things more basic.

By the late nineties Prague was gaining tourist-destination status, as this black-background version with landmarks shows (3). The latest diagram (4) also depicts the city's excellent tram network, a feature that has been on and off the Metro map since its inception. Although there are still just three lines—among the heaviest-used in Europe—plans exist for extensions and a new north–south line.

Rio de Janeiro

Urban population: 10.8 million. Route length: 21.7 miles. Stations: 32. First section open: 1979. Underground: 55 percent.

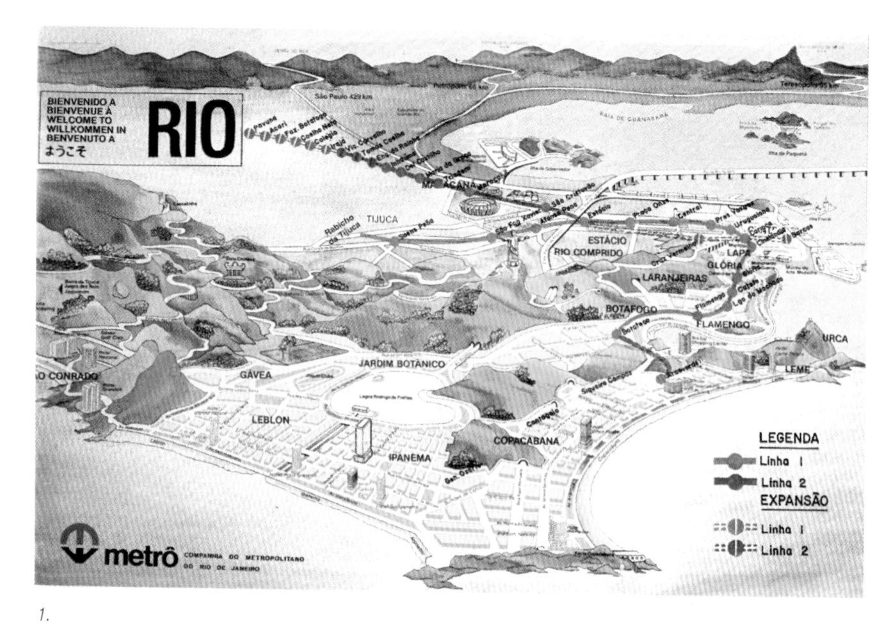

1.

For such a gigantic city, Rio has a relatively small Metro system. Abstract diagrams are sometimes seen as unnecessary when a city is trying to establish the location of a new system—Rio's maps, using the *M* system logo as station markers, focus on geography so people can see where the relatively new system actually runs.

Thus a beautiful representation of scenic Rio from 1995 is still on display in many parts of the system (1). The current map also takes an aerial view (2). Other recent maps retain key geographic features. The central detail (3) is from a pocket map released when Linha 2 fully opened. It began life in 1983 as a pre-metro and was progressively extended north toward Pavuna, with conversion to full metro operation in 1998. The station at Cardeal Arcoverde on Linha 1, from where a new extension to Siqueira Campos opened in 2003, is right by the famous Copacabana beach, and further extensions to the Ipanema district are under construction, including the Metro's latest opening at Cantagalo. Several other big extensions, including an airport branch, are on the drawing board.

2.

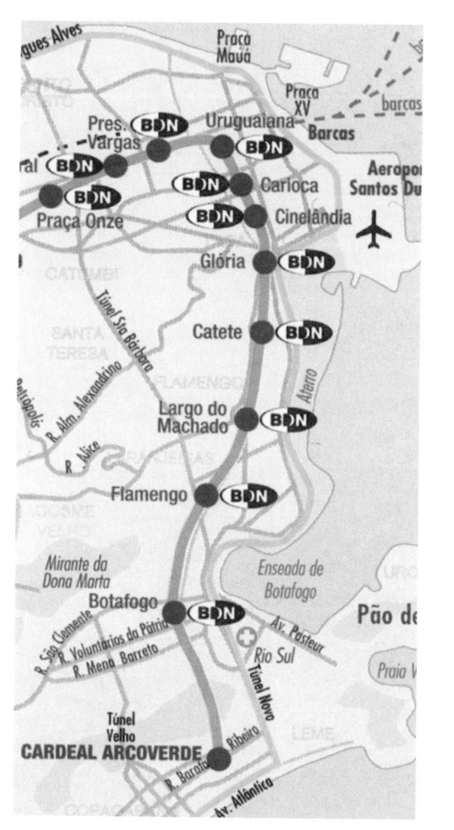

3.

Maps and photo reproduced by kind permission of and copyright Metro Rio 2007.

By virtue of its history, Rome proved to be one of the world's most complex places archaeologically to build a subway, a task made even more delicate by a network of natural caves that riddle the city. Utilizing traditional urban-rail diagram diagonals at 45 degrees, creators Steer Davis Gleave have crystallized the result of these intricate excavations into an effective and simple representation of Rome's entire urban-rail network (1). The diagram is in itself a minor work of art.

This has not always been the case. Some earlier maps (2), difficult to track down, were far cruder efforts.

For the future, Rome is embarking on major extensions, building the "new backbone," line C, which will run from the Vatican in the northwest of the city out to the suburbs of the southeast. This will be one of Italy's biggest construction projects, with complex tunnel boring under some of the most treasured architectural monuments on the face of the globe. There is also a plan for a new line, D.

Diagrams showing just the two existing Metro lines are featured on some tourist guides and maps, but official versions show all urban rail, including Roma–Lido, Roma–Pantano and Roma–Viterbo, which operate a metro-like service, the Leonardo Express (fast train to the airport), and the FR lines (Ferrovia Regionale), which offer suburban commuter services. In addition, ATAC also runs a number of modern light-rail streetcars. The two existing Metro lines are shown on the diagram as thickened blue and orange.

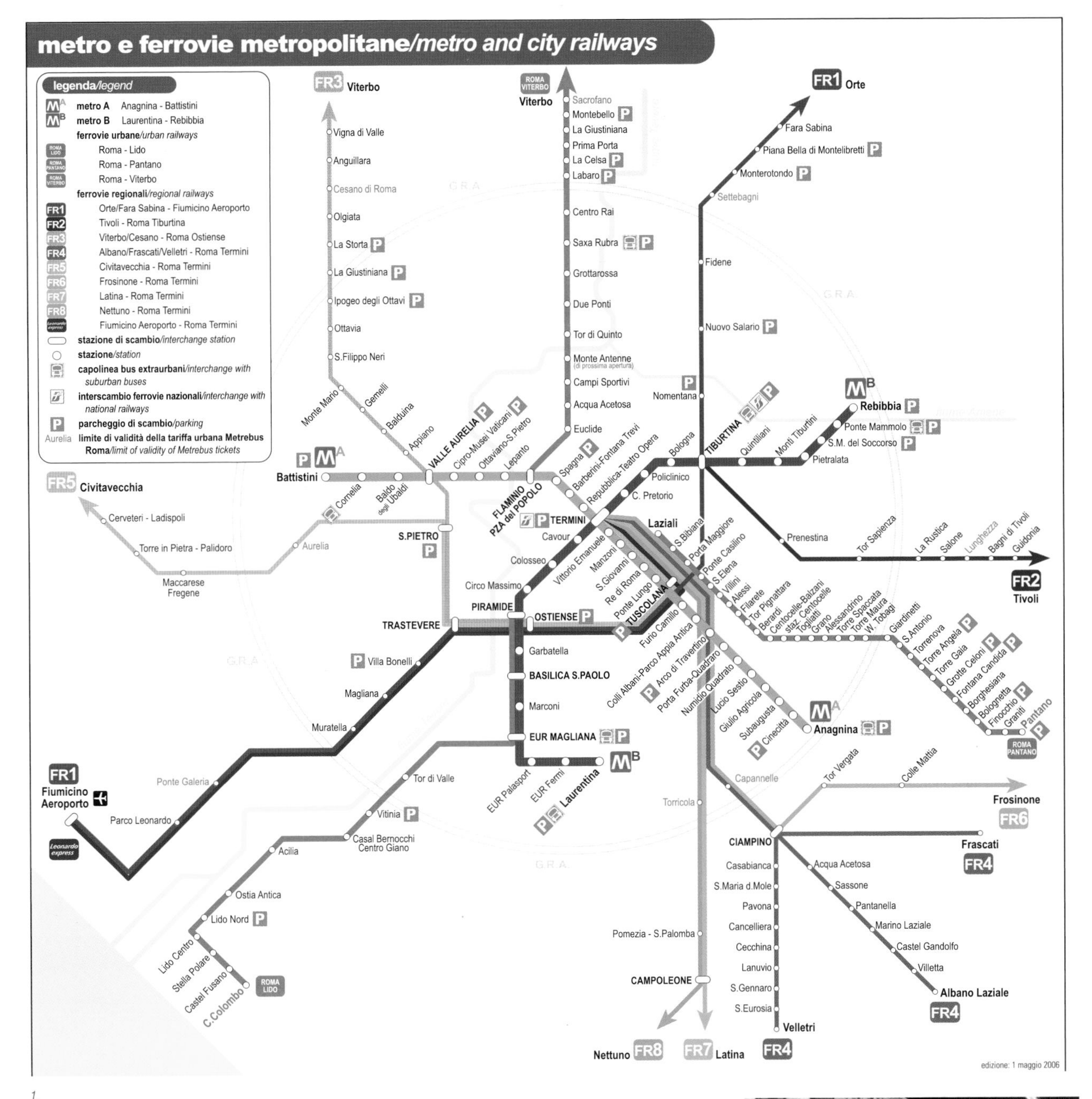

1.

2.

Maps reproduced by kind permission of and copyright ATAC 2007. Current map originally designed by Steer Davis Gleave.

Rotterdam

Urban population: 1.2 million. Route length: 34.1 miles. Stations: 36. First section open: 1968. Underground: 75 percent.

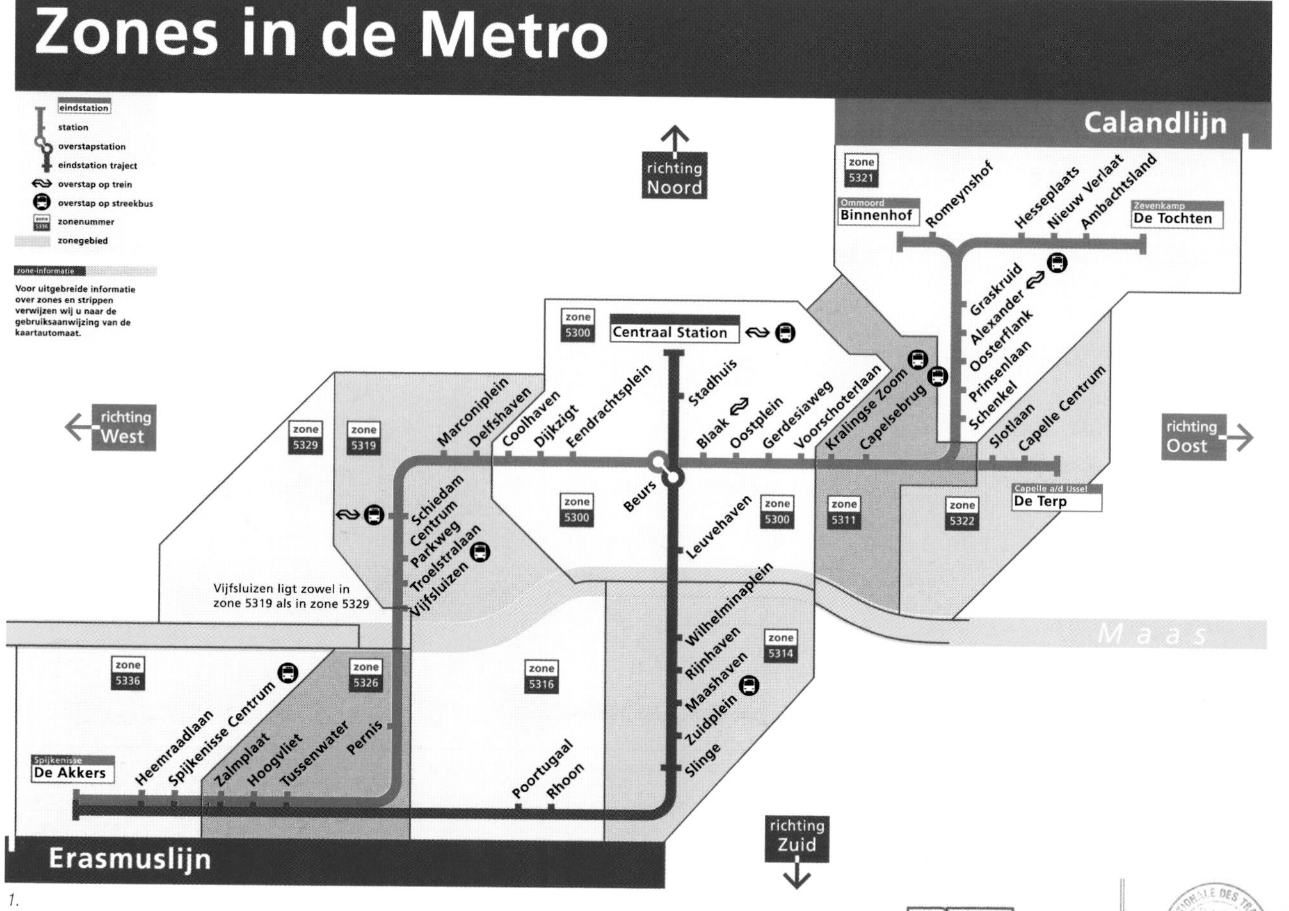

1.

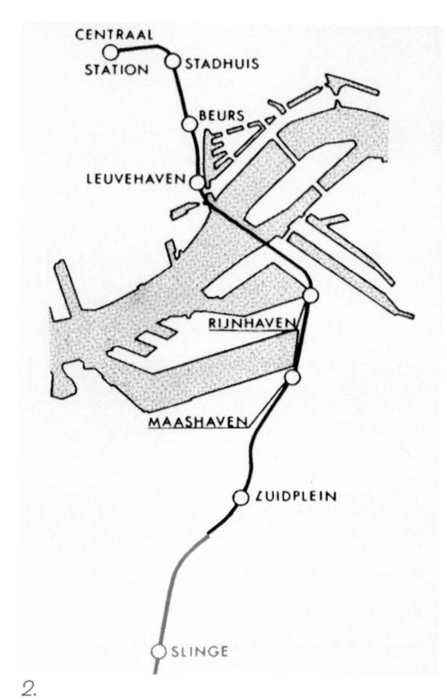

2.

3.

Europe's largest port, the Netherlands' second city boasts a heavily London-inspired diagram (1) from Bureau Mijksenaar, the same people who designed the Amsterdam one (see p. 78). Here we have ticks for stations, double-sided ticks for terminals, and white-line connectors at Beurs interchange. An interesting variant is the slanting of station names, giving a hint of the slender geographic nature of this inland port city.

The 1970 map (2) shows the planned extension to Slinge in red. The 1982 version (3) exhibits an early interest in diagrammatic form and the proposed extension to Marconiplein.

When building the first tunnels, engineers overcame a number of difficulties in the swampy ground near the river Nieuwe Maas, using the "caisson method." They sank tunnel parts in the river, covered them, and drained them. They dug canals for the city-center underground sections, filled them with water, then sank the tunnel parts in those canals. After pumping the water out, they covered them over and remade the land above.

A new service called RanstadRail has opened between here and The Hague, for which there will soon be a direct link into the Rotterdam Metro with a new station at Blijdorp.

São Paulo

Urban population: 20.2 million. Route length: 37.6 miles. Stations: 54. First section open: 1974. Underground: 45 percent.

With more than 20 million people in the metropolitan area, São Paulo is truly a monster-sized city. It has South America's largest concentration of people and is Brazil's most important economic and financial center, being responsible for almost a fifth of the entire GDP! A flier in general circulation folds out into a large-scale city map with Metro routes depicted as thickly colored meandering lines, and includes major roads (2). On the reverse is a smaller, neater diagram covering a wider region, based on 45-degree diagonals (1).

One of the earliest maps, dating from 1978, uses open circles for stations and an unusal crossover at Sé (3), but this device could not have worked for more than two lines.

Construction of the subway—Brazil's first system—started in 1968, and it opened from Tiradentes to Jabaquara in 1974. The network, now approaching 40 miles, could total 100 when all extensions are ready. Lines are colored and numbered, 1 Blue, 2 Green, and 3 Red, each of which is fully integrated with other modes of transit. Construction started in 2004 on Line 4 Yellow, between Luz and Morumbi.

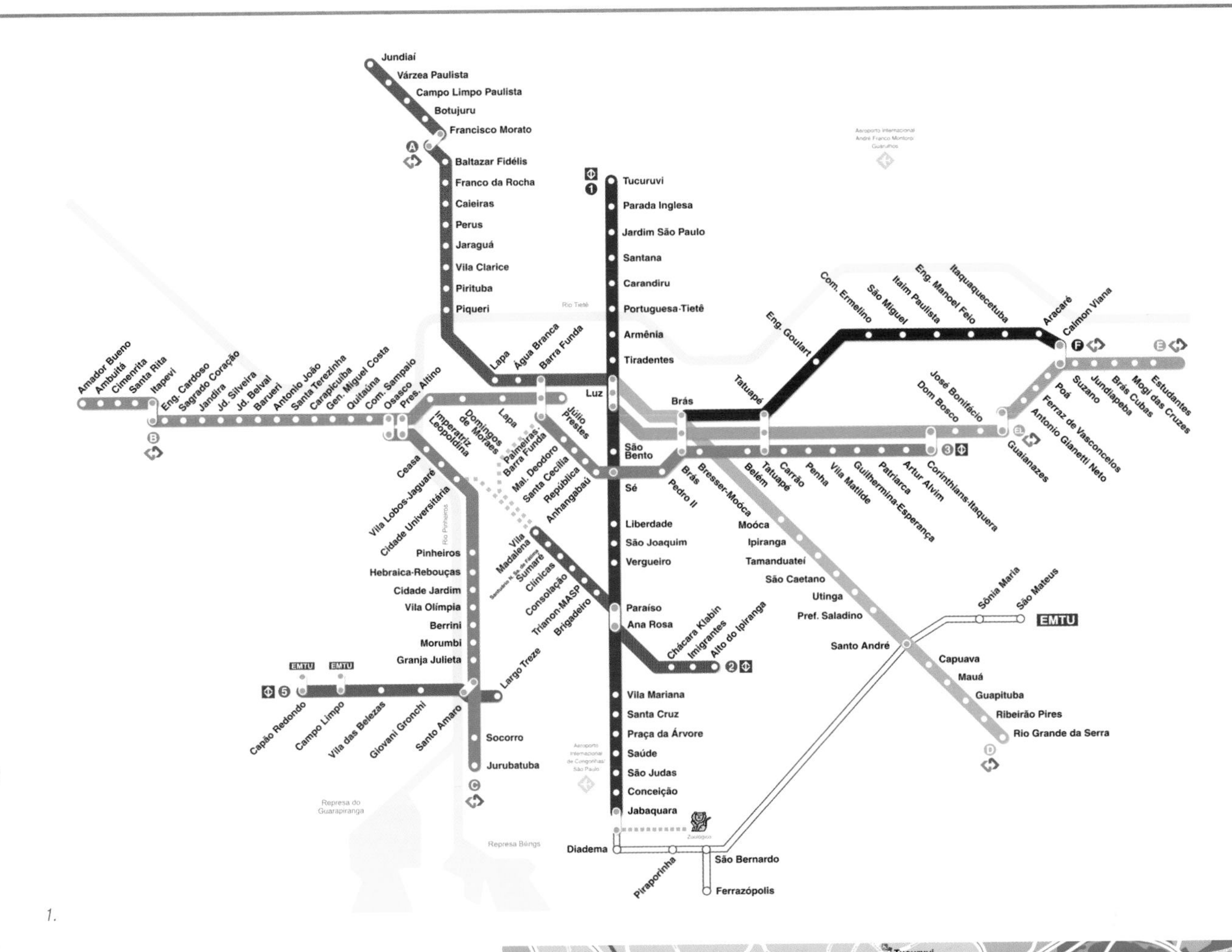

1.

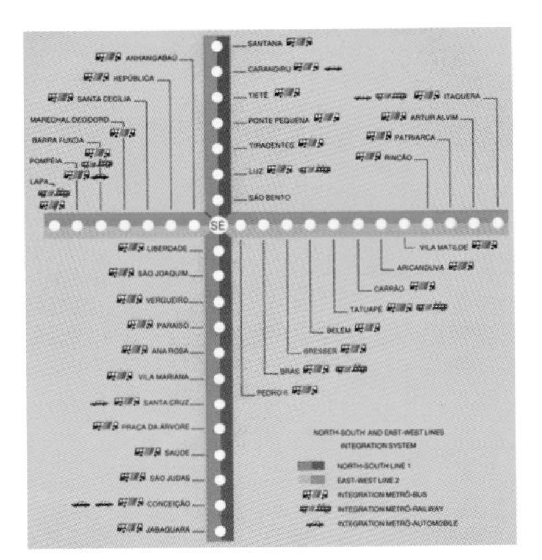

3.

Paralleling this is work on the second stage of Line 5 Lilac, designed to add initially ten stations to the network, but in time interchanging with Line 1 at Santa Cruz and Chácara Klabin on Line 2.

The São Paulo Metro is one of the world's most heavily loaded, being used by 2.6 million passengers daily. The citizens are particularly proud of its efficiency, without which a city of this size would doubtless grind to a halt.

The Metro operator places a strong emphasis on the display of contemporary art. Nearly 100 works by well-known artists are dotted around the central stations.

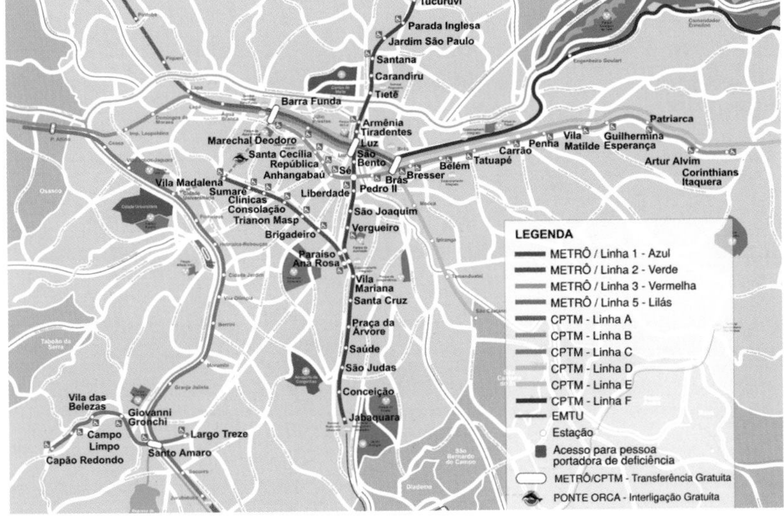

2.

Maps by kind permission of and copyright Metro São Paulo 2007.

Singapore

Urban population: 4.3 million. Route length: 68 miles. Stations: 64. First section open: 1987. Underground: 26 percent.

The Singapore diagram (1) is a triumph of clarity and chic for a city that takes mass rapid transit very seriously! Diagrams do not need to be flamboyant, but at first glance this has an air of meticulous sterility—until you hold it at arm's length and see that the perfectly balanced undulating logo hiding behind the routes is both a stylized *S* and also the shape of the island of Singapore. Most ingenious.

In graphic terms the diagram exhibits many classic influences, for example the black circle with white center symbol at interchanges and the use of white-line connectors at Dhoby Ghaut and elsewhere. There are also the station-marker ticks, with double-ended ticks at terminals, the 45-degree diagonals, and the gentle curves into the vertical and horizontal.

Uniquely, the Singapore system uses destination numbers instead of line numbers. A reminder of colonial times are station names such as Admiralty and Commonwealth. The diagram from 1988 (2), resembling an exotic snake, and the version from just four years later (3) show how quickly the system is growing.

A whole new circle line (yellow on diagram 1) is currently under construction, and more extensions of the automated LRT feeders are planned. Much like the island metropolis it serves, Singapore Metro is one of the most modern, clean, and efficient in the world. It was one of the first to employ full platform-edge sliding doors to aid ventilation and stop people from falling onto the tracks, and it is among the world's busiest systems.

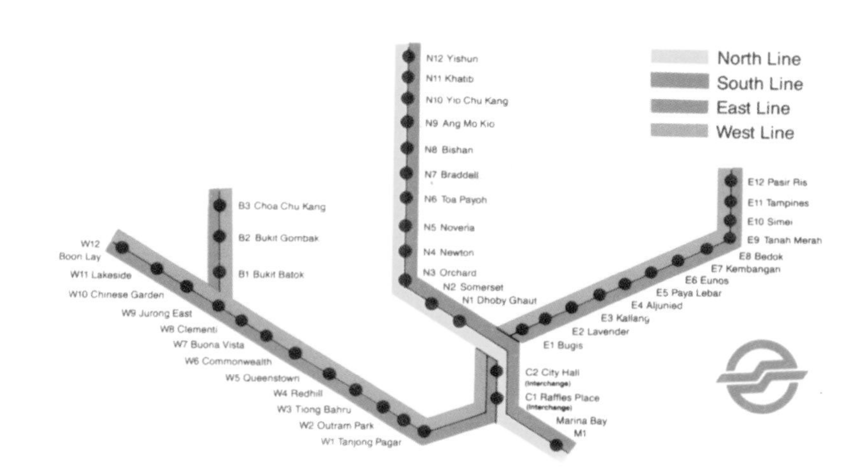

3.

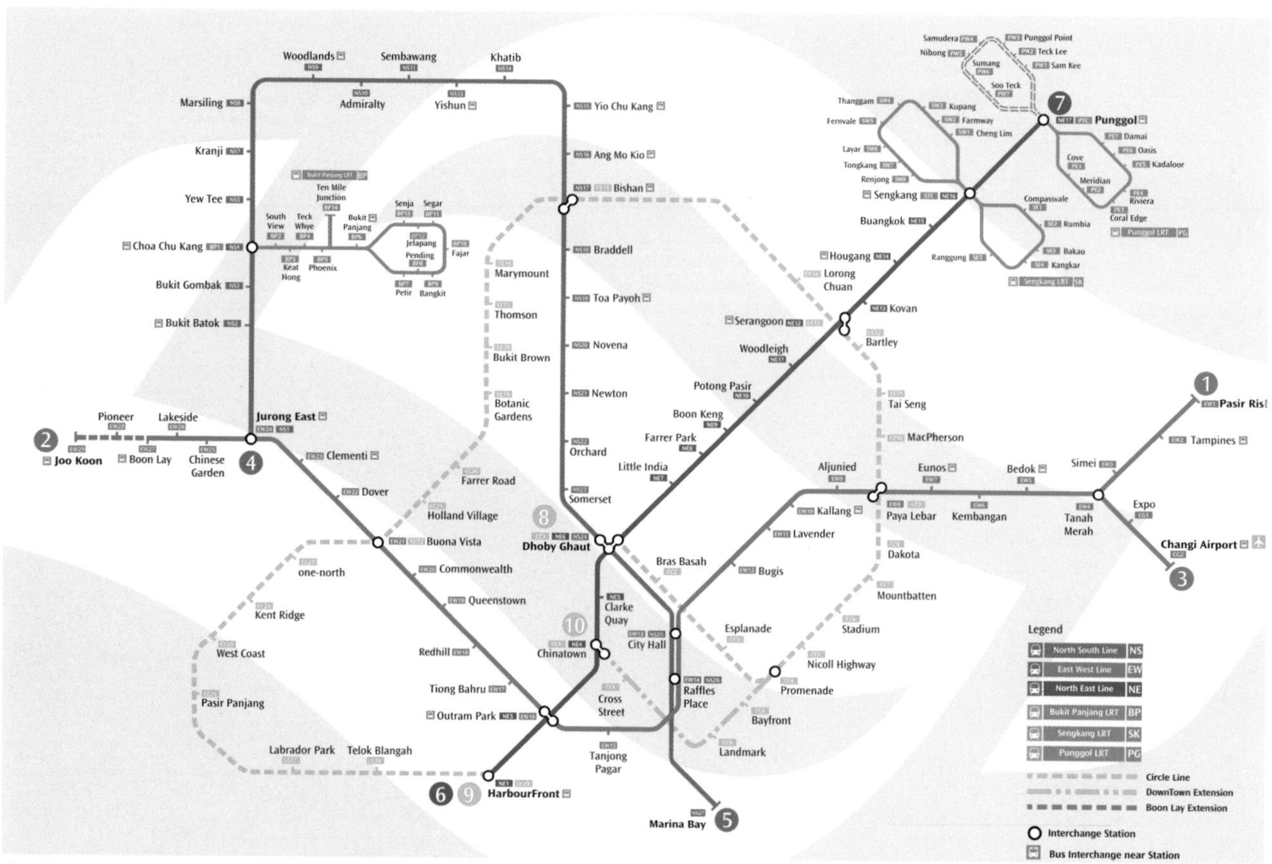

1. *Maps and photo reproduced by kind permission of and copyright SMRT 2007.*

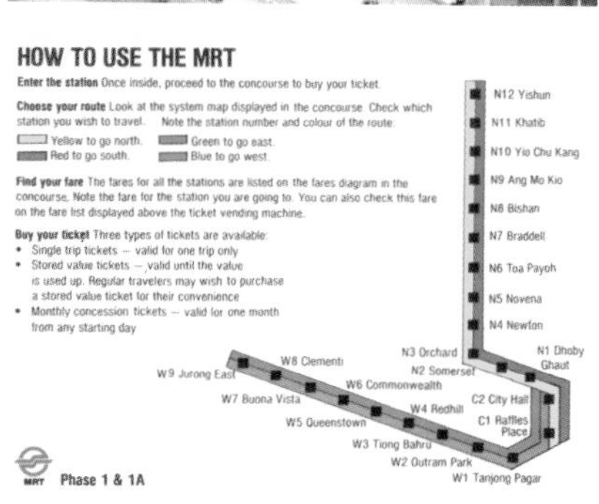

2.

Stockholm

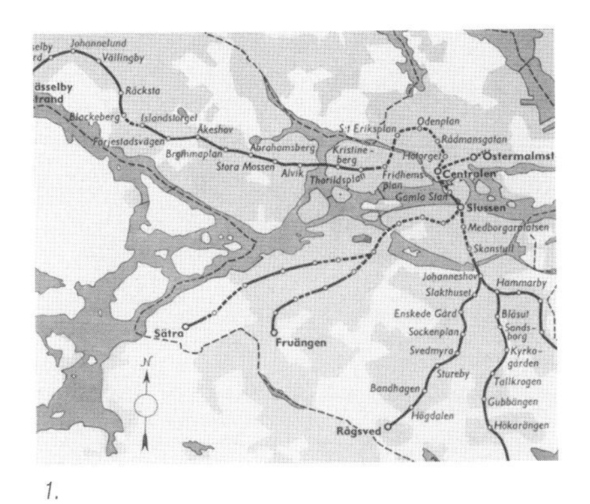

1.

2.

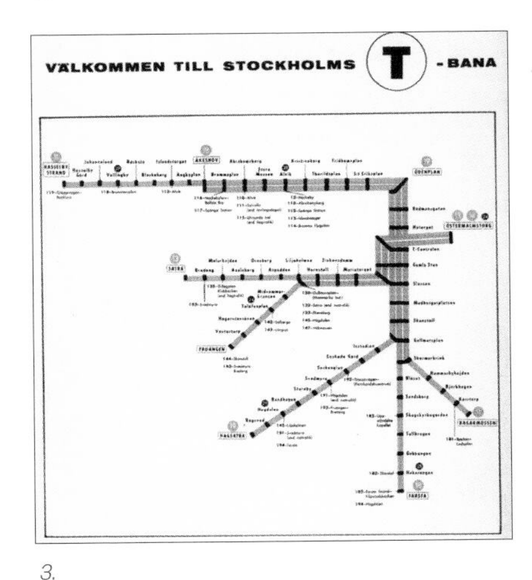

3.

4.

Stockholm is by no means a large city but is probably one of the best served by public transport in the world, with a dense network of efficient services penetrating much of the built-up area. The current schematic of the Tunnelbana (5) should be contrasted with the more geographic version from 1957 (1), which illustrates how greatly diagrams distort geography. It also demonstrates how much had been built by the late 1950s and how extensive the coverage of the system is now.

The lack of topography on the contemporary diagram, however, hides the complexity of the network, which burrows beneath the waters no fewer than nine times. This is a country where geology is a formidable barrier for engineers to overcome. The rock is literally blasted away, rather than bored (in fact a technical advantage over boring), and so has turned some cavernous station openings into natural canvases for artists to cover in original works.

Stockholm's earliest diagrams (2 and 3, from the mid-1960s) had a rather stark and utilitarian feel, with little of the lightness and grace injected in later years; for example, in 1983 (4).

There are a number of plans to extend the system, including a branch of the Green Line from Hagsätra to Älvsjö, but no work is currently under way. A visit to Stockholm's Transport Museum provides insight into the achievements of a well-planned system.

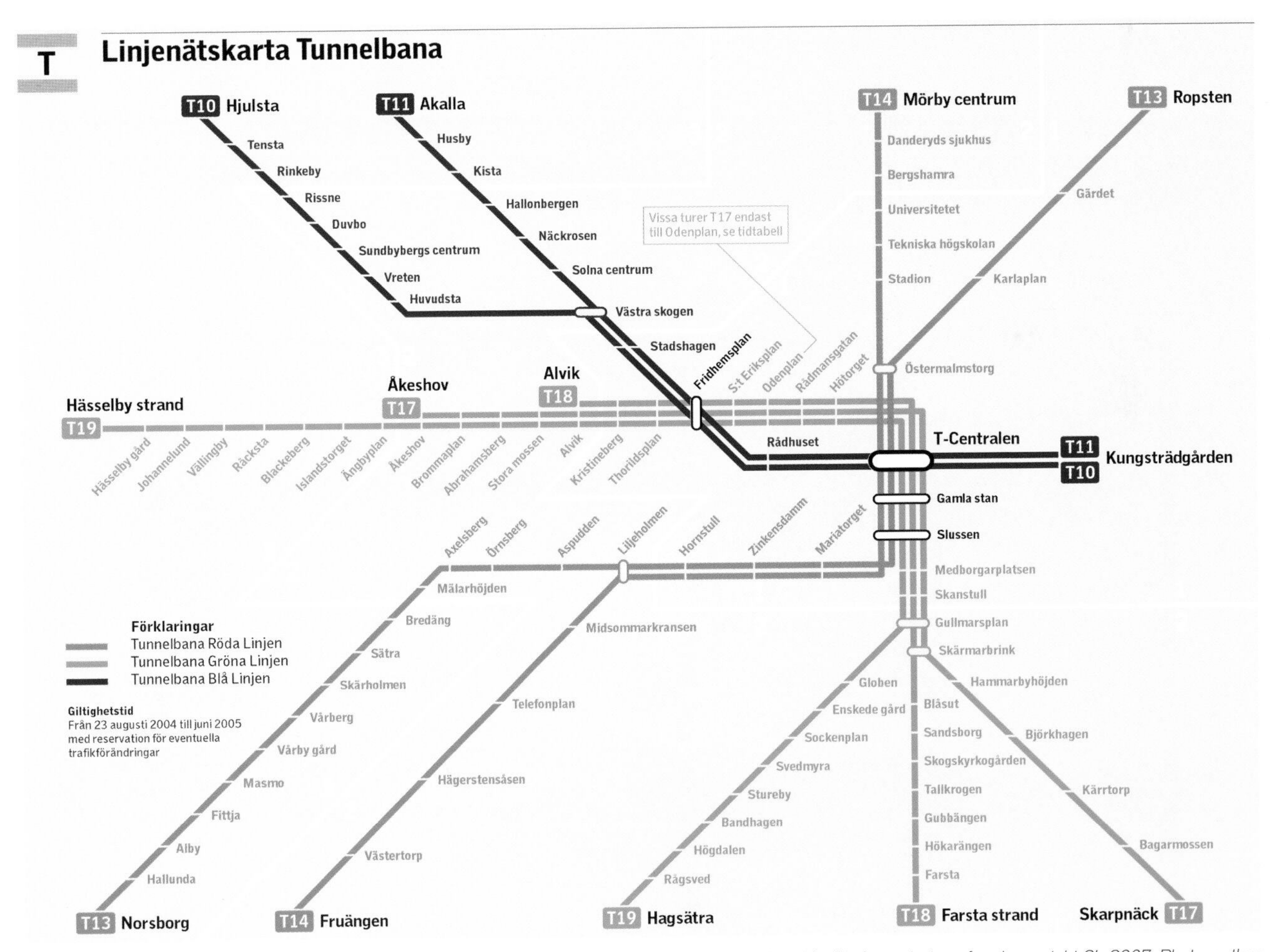

5.

Maps reproduced by kind permission of and copyright SL 2007. Photo: author.

Taipei

Urban population: 6.7 million. Route length: 50 miles. Stations: 61. First opened: 1996. Underground: 70 percent.

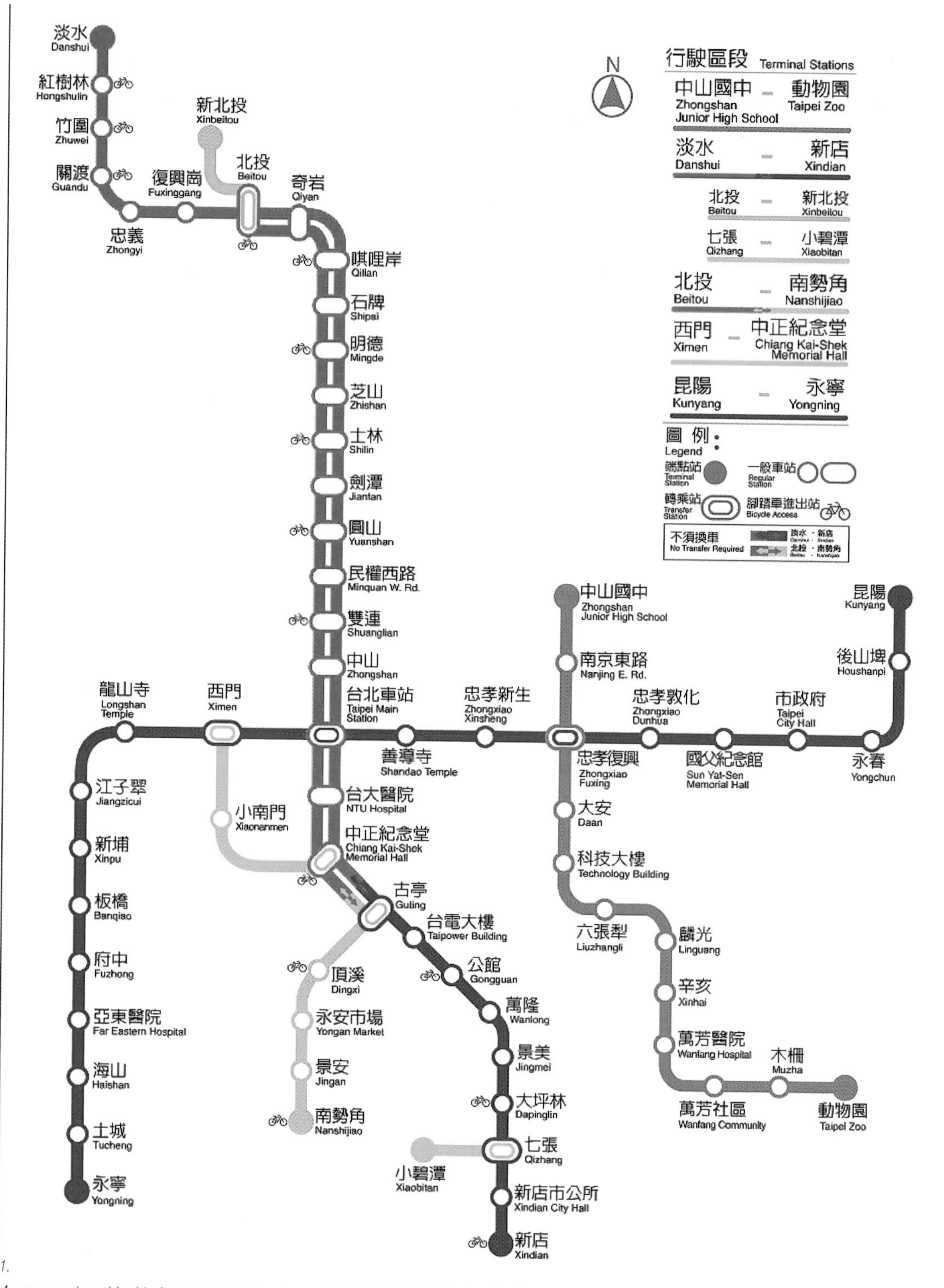

1.

Six new lines in a decade is incredible in urban-transit construction terms. Although three of the routes are little more than shuttles right now, two more major lines and many extensions are on the way. Four of the lines are mostly underground and one mostly elevated, operating an automated service of the French VAL type on rubber tires.

Green, Orange and Red lines on the current diagram (1) are now pretty much operated as one, with different routings. Both the current and planned network diagrams use the same model of extreme simplification and geographic distortion, with very few diagonals.

Notice how the stations are all evenly spaced, the lines are straightened out, and no station name breaks over a line despite being written in two languages. White-centered ovals, doubled up for transfer stations, are replaced by open circles on the planned system diagram (2). A system to watch.

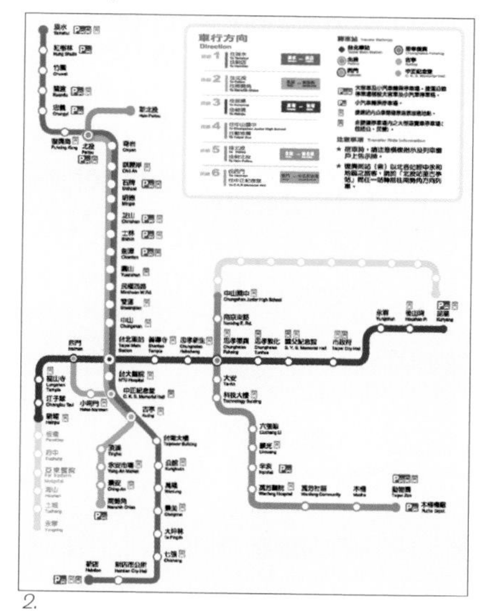

2.

Maps reproduced by kind permission of and copyright TRTS 2007. Photo: Chaffee Yiu.

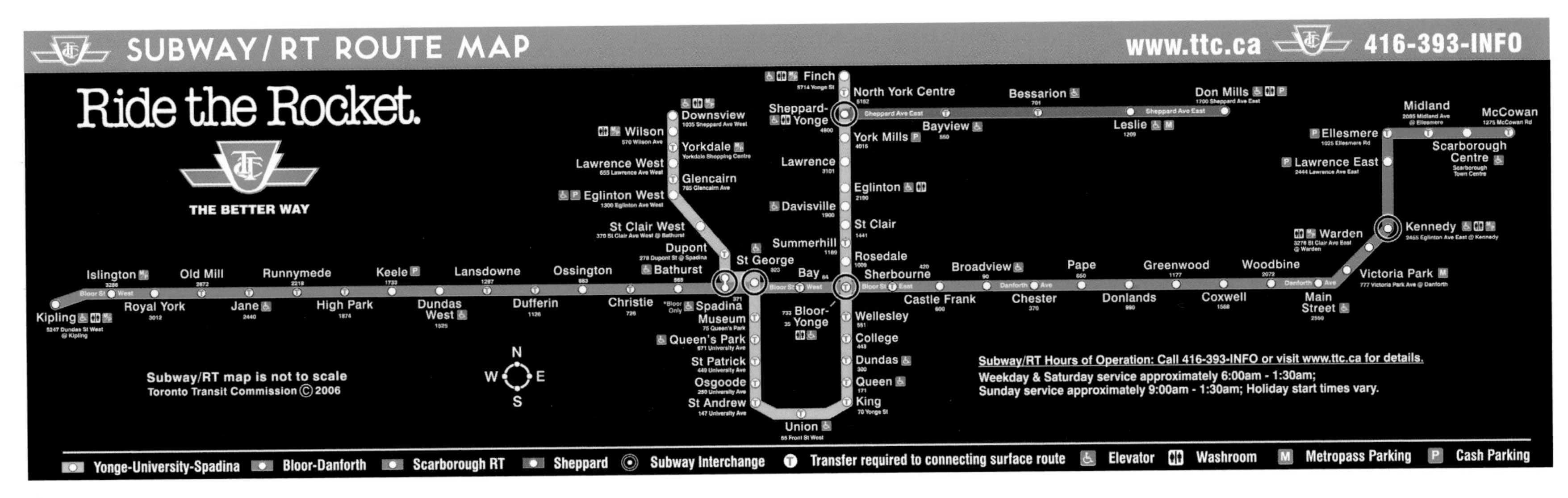

1.

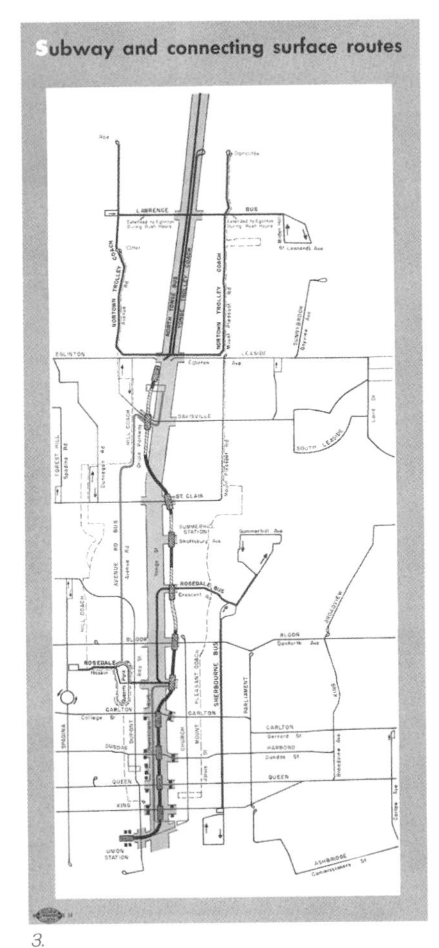

3.

Toronto was the first Canadian city to get mass transit. The Yonge Subway, running from Union to Eglinton Avenue, opened in 1954. Although the system initially used British-built trains, it is very much in the North American style. The network is now marketed as the "Rocket." Its diagram, with a black background (1) similar to Montreal's (see p. 62), neatly compresses all the stations at regular intervals, with just the diagonals adding quirkiness—each at a different incline! Stations are marked with a fine-rimmed black circle with white-eye center, and at interchanges they are larger with a transparent center.

A foldout card pocket map from 1981 (2), with black dots for station markers, shows the eastern end of the Bloor–Danforth Subway in a rather graceful curving flourish. An earlier plan shows how far the modern diagrams have come (3).

The $875-million Sheppard Subway opened in 2002. It was built under Sheppard Avenue East by cut-and-cover, the same construction technique that caused much disruption during the building of the city's first subway in the early 1950s.

Along with improvement and expansion of the light-rail network, Toronto also has long-term plans for the subway, including completion of the Sheppard Subway right through from Downsview to Scarborough, and the Yonge–University line will probably grow four or six stations northward on its shorter western arm.

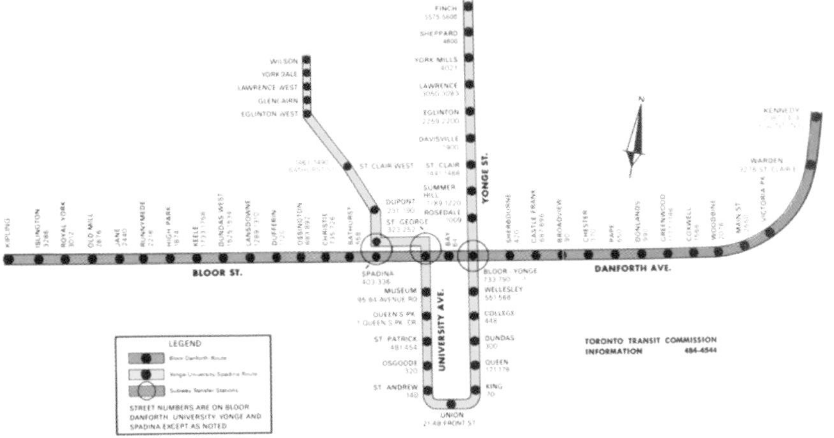

2.

Maps reproduced by kind permission of and copyright TTC 2007. Photo: author.

Valencia

Urban population: 1.8 million. Route length: 83.7 miles. Stations: 123. First opened: 1988 (1888). Underground: 20 percent.

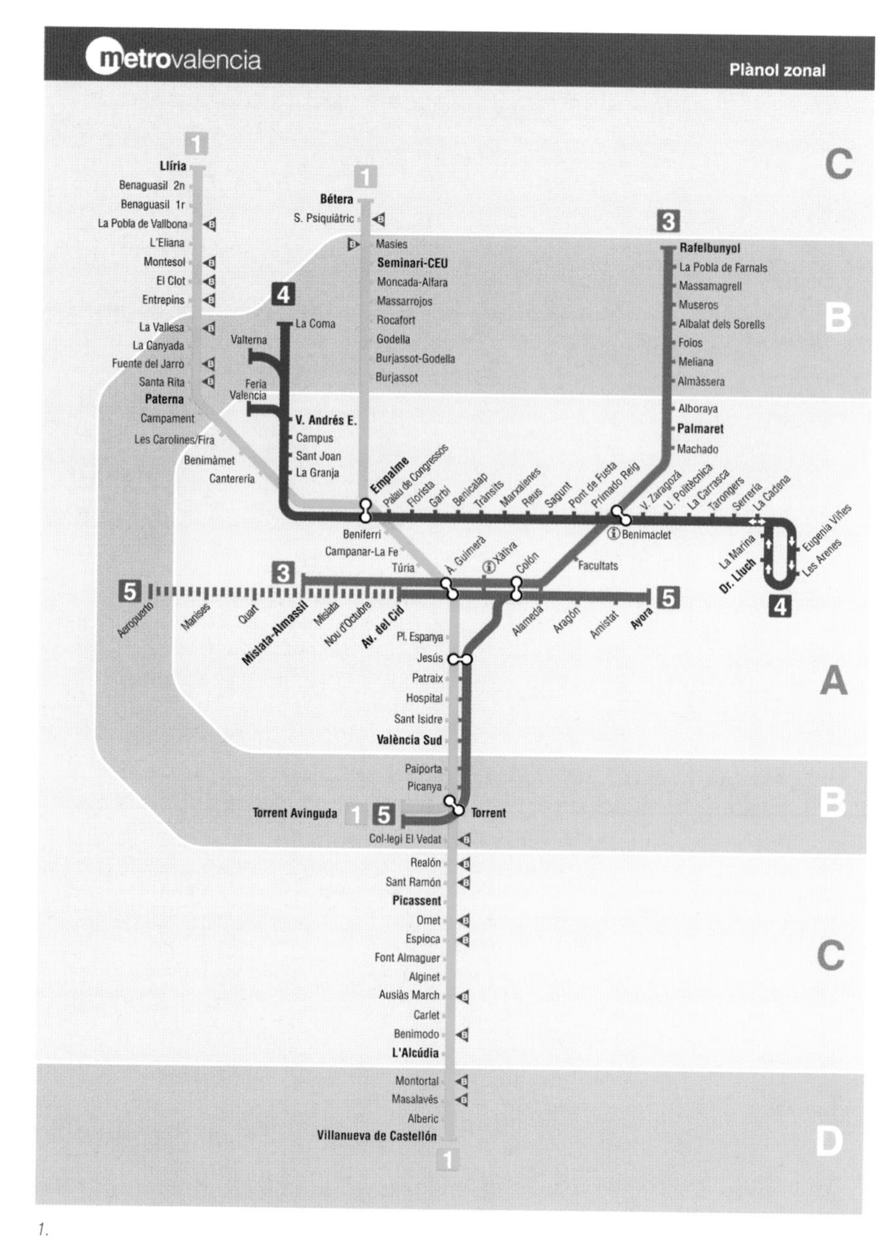

1.

Valencia now has a modern mass-transit system, although it has been built on the shoulders of a much older network: a group of typically Spanish narrow-gauge railways assembled by the local authority and partially converted to carry bigger metro trains or high-floor trams. In the suburbs the system can still seem strangely rural and quaint.

The main diagram (1) reflects the aim of the Metro system to convey ease of use and simplicity of connections between the various modes. Even a cursory glance at the geographic system map (2) shows the degree of space saved by using a diagram, distorting true geography and aiding clarity. Again, simple, strong colors such as reds and blues are used along with perpendiculars linked by diagonals at 45 degrees, ticks for evenly spaced stations, and standard interchange symbols. The roots of these colorings and the black circle with white-eye center device were evident on the more topographic 1999 map (3).

Although there has been rail-based transport in the city since 1888, it took 100 years for the amalgamation of all railways under the Ferrocarrils de la Generalitat Valenciana (FGV) to take place. With the building of new tunnel sections in the center, the system was christened Metro de Valencia.

All of the central-area stations are in typically exuberant, modern Spanish architectural style. Extensions are now planned that will turn part of what is Line 3 into Line 5, with a direct link to the airport, and place the central spine of the new T2 streetcar in a tunnel under the downtown area. In Valencia streetcars have even numbers and Metros are odd, but all 123 stations on the varying modes are marketed as being part of the "Metro" system.

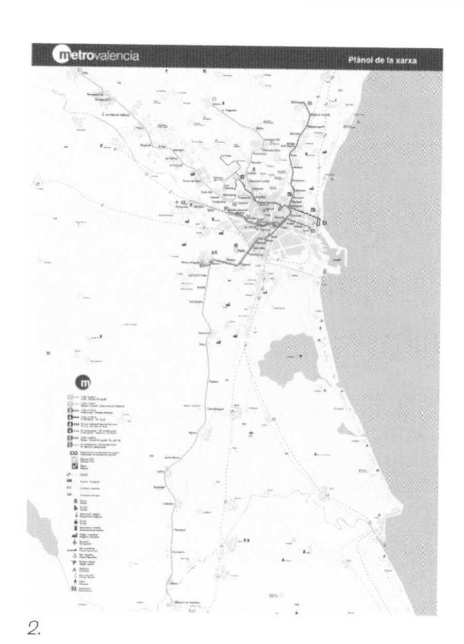

2.

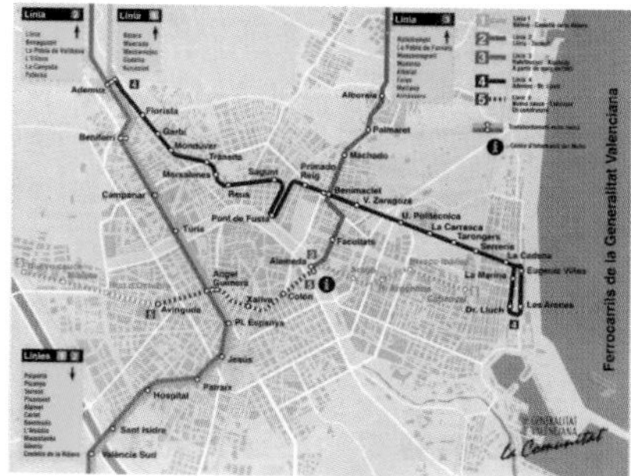

3.

1.

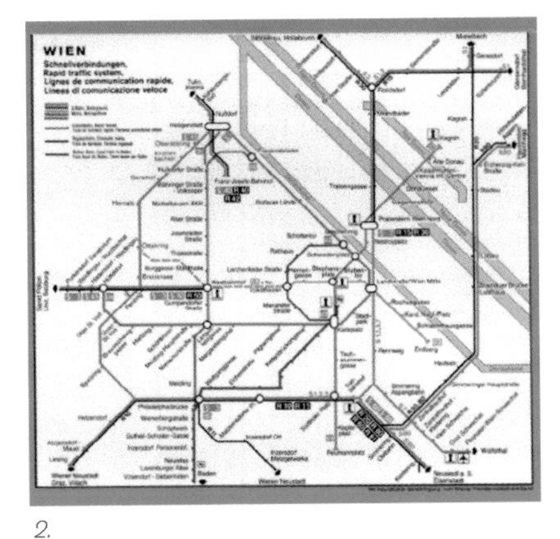

2.

The simplicity of the Vienna U-Bahn diagram probably makes it one of Europe's most straightforward. The current version (3) is a direct graphic descendant of the city's design tradition of straightforward, clean-cut maps and plans like the 1972 edition (1). Although this is quirky, it shares many common metro map themes. Every diagram is unique, of course, and the contemporary one has a variation on the black circle with white-eye center interchange symbol, here appearing as an elongated rounded oblong. An earlier version was this pocket plan (2).

The system's evolution dates from 1898/9, when the kaiser opened three Stadtbahn lines, literally translated as "town railway," to serve what was then the capital of the Austro-Hungarian Empire. Historians trace the idea of building an underground railway in this beautiful city back to 1843, when a plan was first mooted to build subterranean railways rather than risk destroying sections of the ancient city center, with its numerous historic buildings. The old Stadtbahn lines now form the core of the modern network.

An eastward extension of Line U2 toward Stadion opens in 2008, and in the long term a completely new U5 is also on the drawing board.

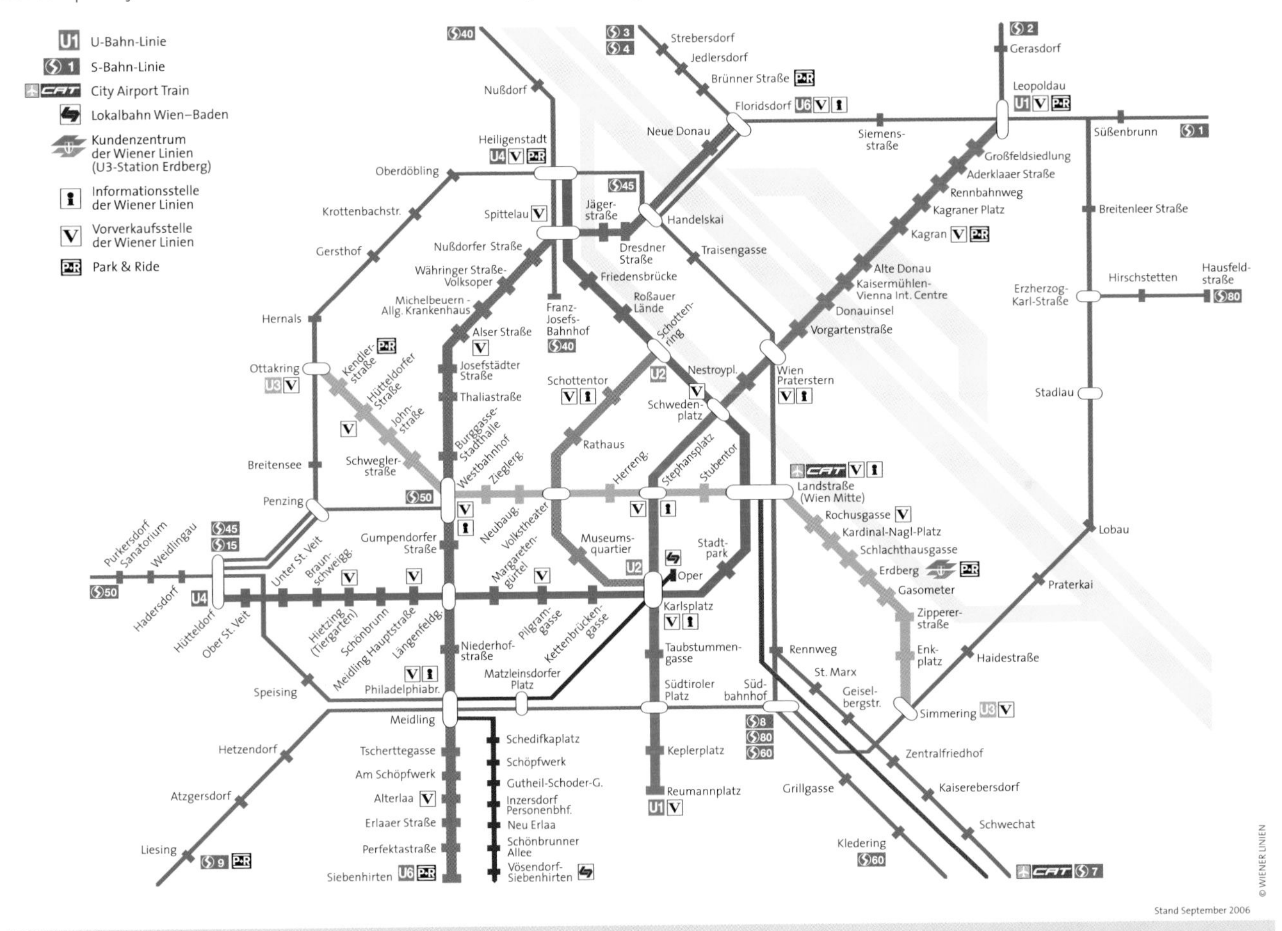

SCHNELLVERBINDUNGEN IN WIEN

3.

Maps reproduced by kind permission of and copyright Wiener Linien 2007. Photos: lower left, Robert Schwandl; upper middle, Capital Transport.

BTS

Zone 4

Maps are just one part of the graphic package presented to travelers. System logos should be easy to recognize in a crowded streetscape. They must convey safety, speed, and authority. Here is a selection of logos used by the systems in this zone, almost all of which are relative newcomers to the urban-transit field and, consequently, have produced only a small number of map variations.

Atlanta	108
Baltimore	108
Bangkok	109
Cairo	109
Guangzhou	110
Kharkiv	110
Lille	111
Marseille	111
Milan	112
Naples	112
Newark	113
Nizhniy Novgorod	113
Recife	114
Santiago	114
Shanghai	115
Warsaw	115

Atlanta

4.9 million. 1979. 49 miles. 38 stations. 20 percent underground.

A simple, neatly executed diagram, complete with major highways, conveys with ease Atlanta's straightforward two-line system. It employs geographic distortion and evenly spaced stations, diagonals at 45 degrees, and an interesting deviation at Dunwoody and Sandy Springs. Each station marker is formed by an open circle, filled with a *P* where parking facilities are available alongside the station. At the central interchange of Five Points is a larger black-rimmed open circle with a white-eye center. The coming years may see construction of two extensions.

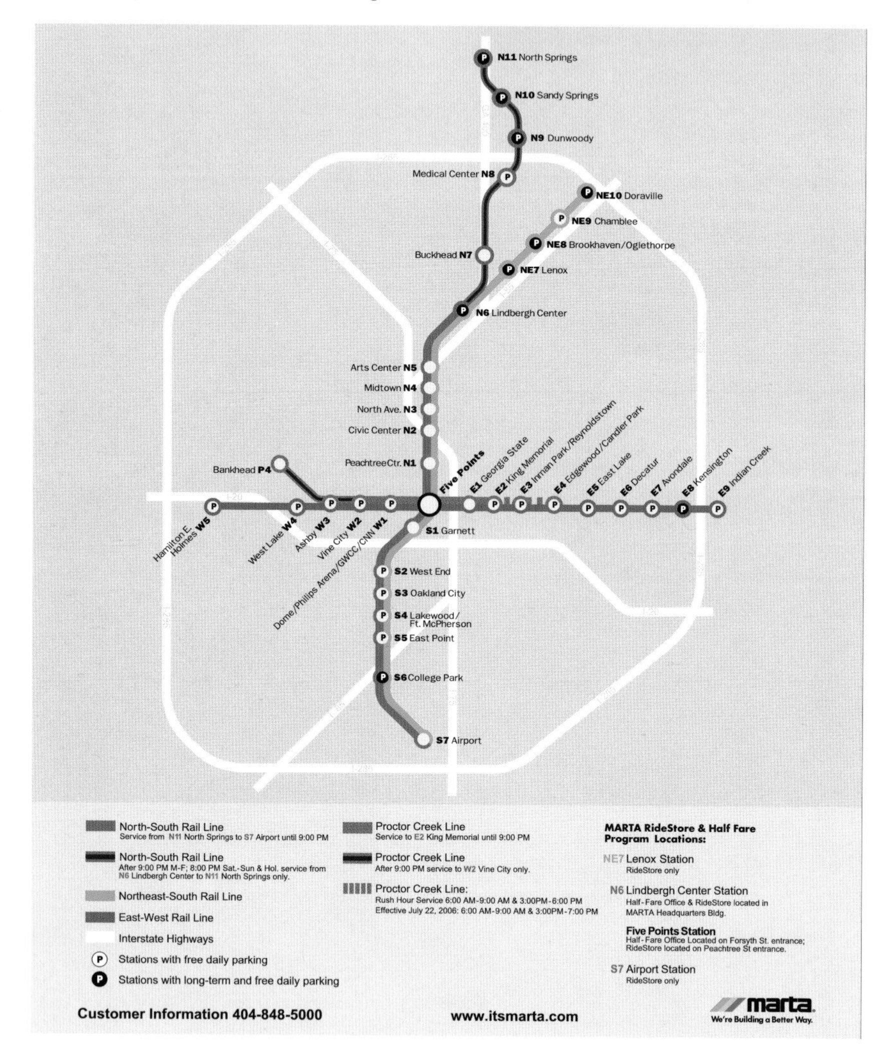

Baltimore

2.6 million. 1983. 15.2 miles. 14 stations. 45 percent underground.

The Baltimore Subway is currently a single line (1). Plans for a 30-mile system (1), first mooted in the early 1970s, have been revised by the Baltimore Region Rail System Plan, but since its adoption in 2002, with the venerable aim of bringing 850,000 residents within 1 mile of rail-based service, costs may prohibit it from becoming a reality. The most likely extension will be from Johns Hopkins Medical Campus toward the northeast of the city. It is interesting to note that the diagram of proposals strongly resembles the style successfully used by the neighboring Washington Metro (see p. 74).

1.

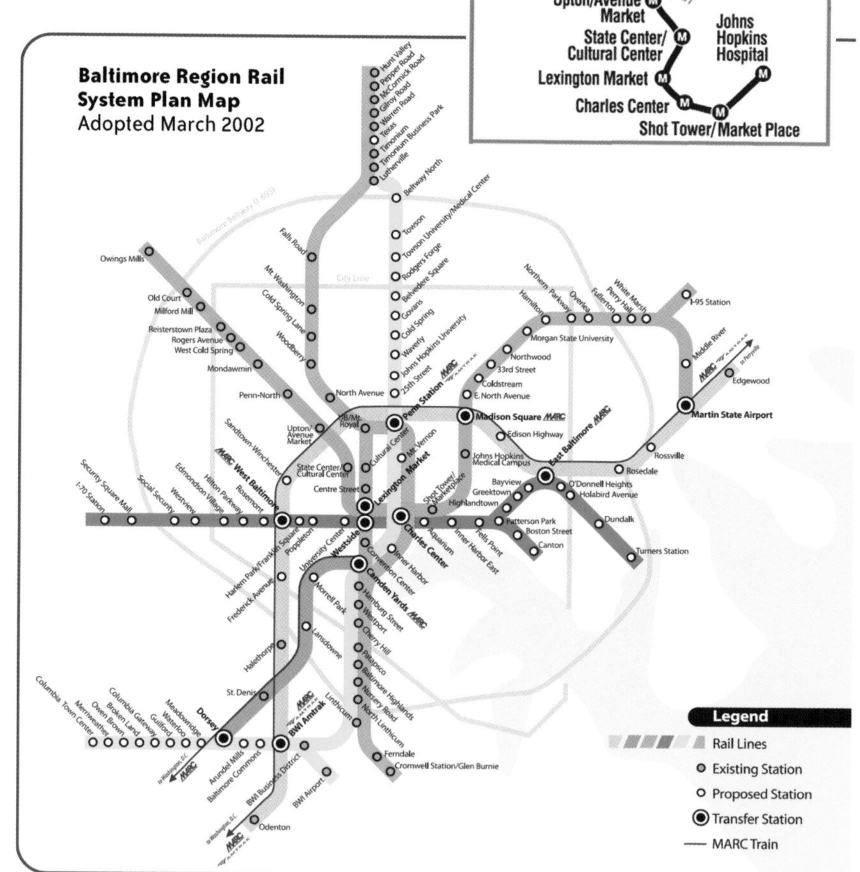

2.

Unless stated, all maps by kind permission of and copyright the system operator © 2007.

Bangkok

6.5 million. 1999. 27.3 miles.
40 stations. 40 percent underground.

Bangkok's recently opened rapid transit consists of a subway, Bangkok Metro, run by BMCL (1), and an automated, elevated Skytrain run by BTS, which has two lines (2). A third, partially elevated system (BERTS) began construction in 1990 but was abandoned due to costs. Its route is now emerging as a light-rail option for an airport express, among other proposals to give Thailand's capital over 150 miles of mass rapid transit.

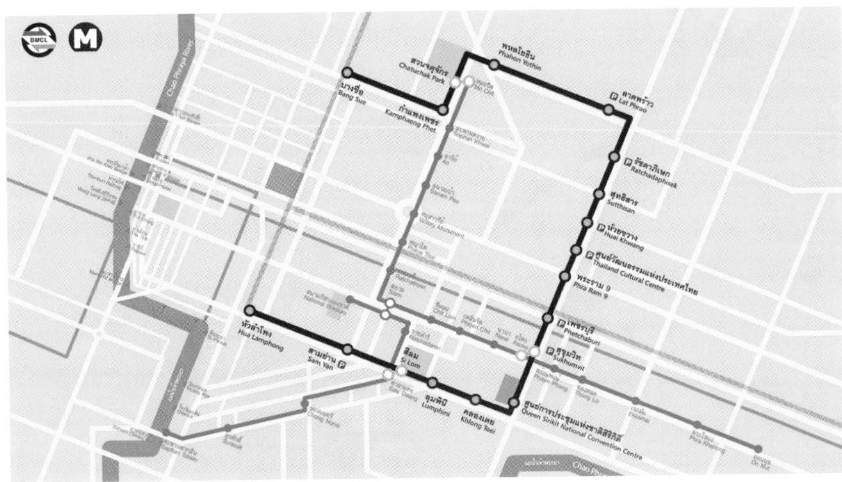

1.

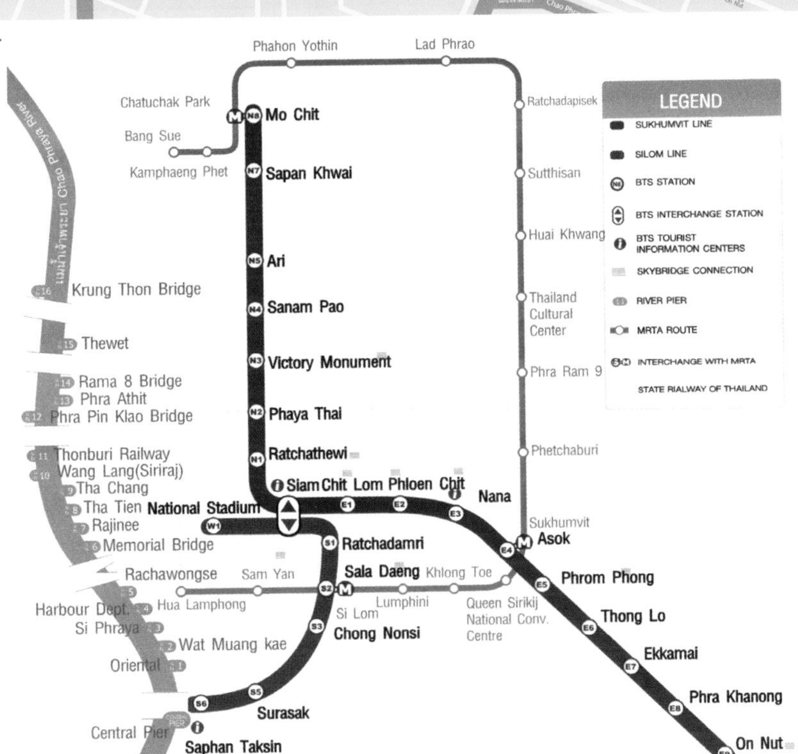

2.

Cairo

11.1 million. 1987. 38.5 miles. 50 stations. 30 percent underground.

Line 1 of Africa's first metro system grew out of two commuter lines that were connected through the city center by a new tunnel. All the downtown stations are named after Egypt's leaders: Nasser, Mobarak, Sadat, etc.

The second line now reaches Giza's huge residential area and is close to the ancient pyramids. A third line to the airport is part of a six-line plan. The signage and logo are strong, but maps are rare, most, like this one below, resembling engineering drawings.

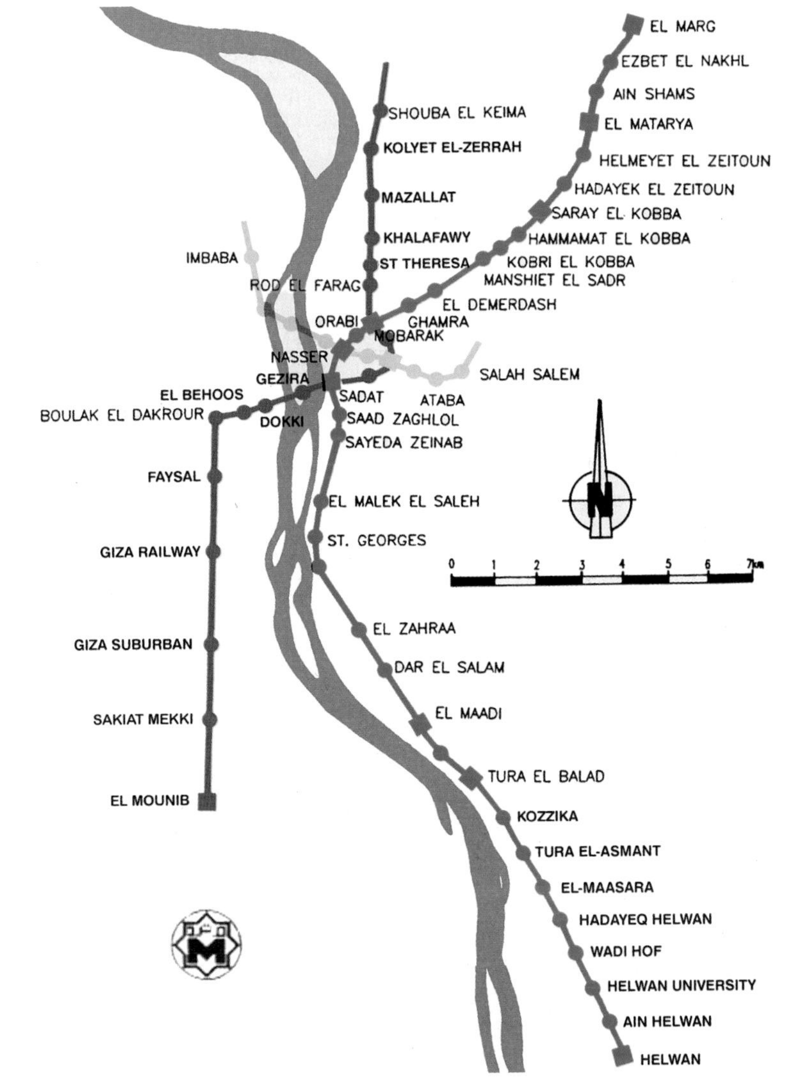

Unless stated, all maps by kind permission of and copyright the system operator © 2007.

Guangzhou

12 million. 1998. 34 miles. 57 stations.
90 percent underground.

The first three lines of the Metro in Canton, as the city was formerly known, are just the tip of the iceberg for this massive city in the booming Hong Kong/Pearl River delta region. The first parts of Line 4 have just opened, and construction is under way on *fourteen* other sections of what is set to become a network around 160 miles long by 2010. Some of this incredible growth will come from a connection with the neighboring city of Foshan, via the Guangfo subway due to be opened soon, but the majority of the seven lines will be in Guangzhou itself. The main diagram (1) shows the current network, but just look at the planned system (2).

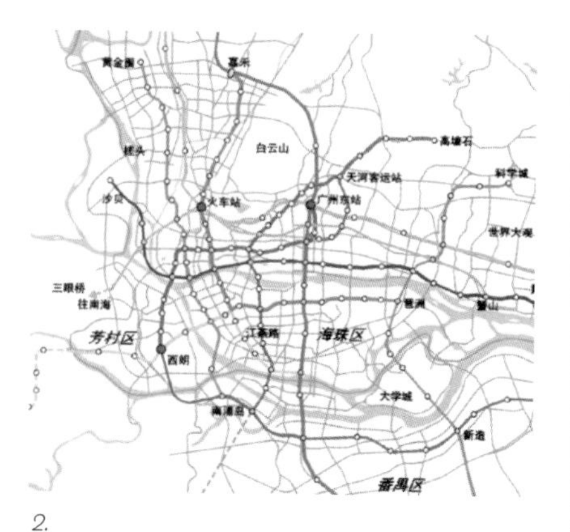

2.

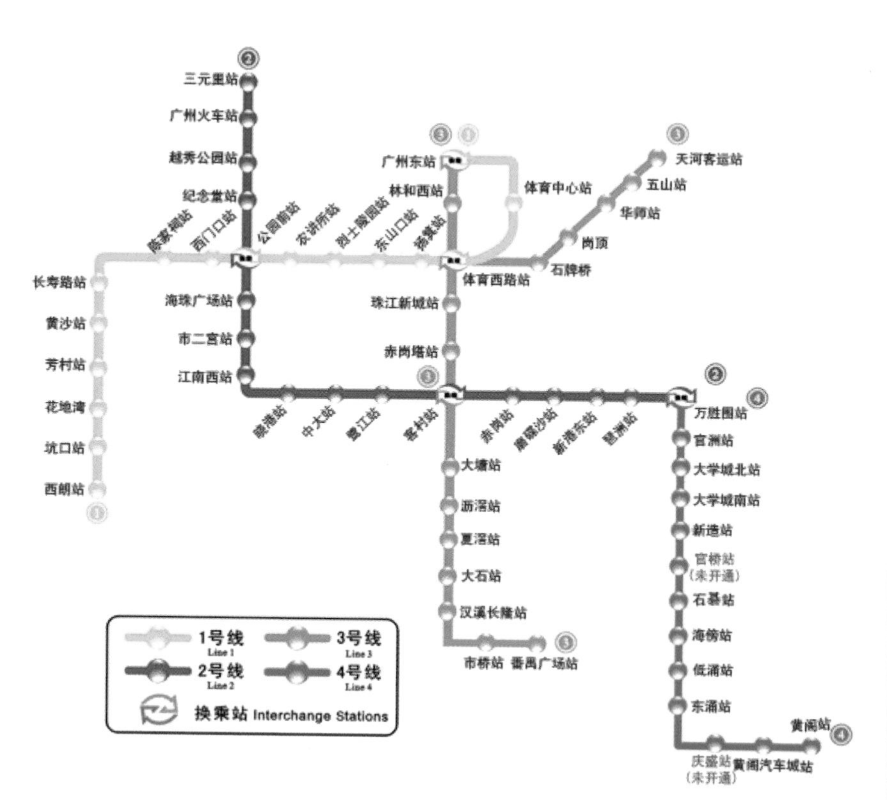

1.

Kharkiv

1.9 million. 1975. 22 miles. 25 stations.
100 percent underground.

2.

The main diagram (1) involves some line straightening and the distinctive, somewhat Asian feature of coloring wards or districts, but lots of key streets and district names are also shown. This is in stark contrast to the austere monochrome diagram from 1986 (2). Despite its relatively small size, the system moves about a million people a day, which means that over half the population takes the Metro! By 2025 this important Ukrainian city could have 31 miles of subway serving thirty stations. In keeping with most of the other former Eastern Bloc–style subways, it's a simple three-line network, and the stations are ornate and elegant palaces of subterranean style, as a postcard from the early 1980s shows (3).

3.

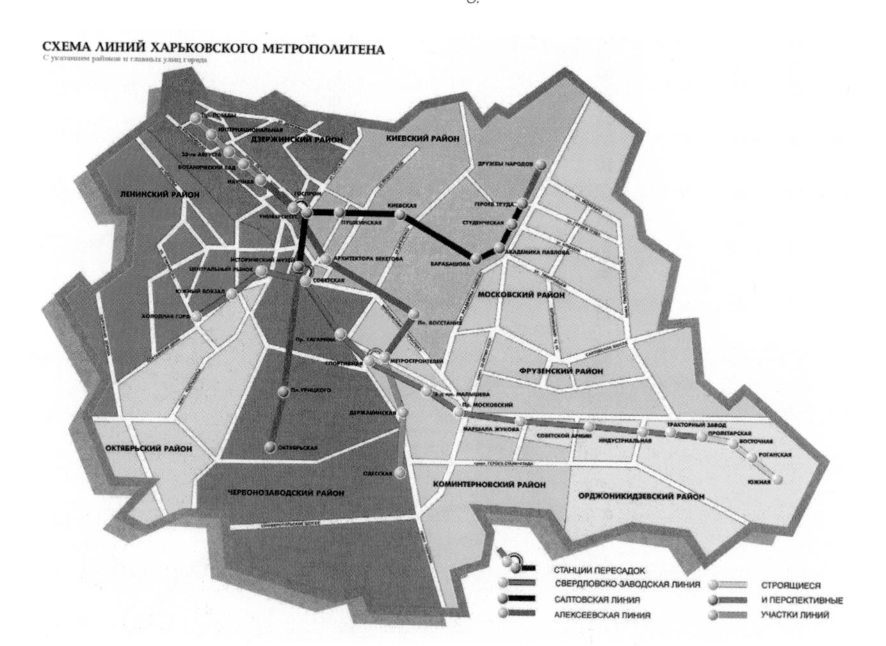

1.

Unless stated, all maps by kind permission of and copyright the system operator © 2007.

Lille

1.1million. 1983. 28.3 miles. 59 stations. 30 percent underground.

Marseille

1.5 million. 1978. 11.8 miles. 24 stations. 77 percent underground.

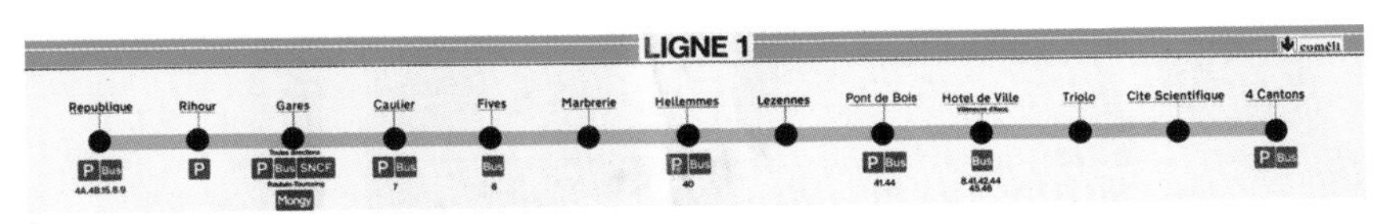

1.

Lille is synonymous with the VAL (Véhicule Automatique Léger) system, the first fully automated line in Europe. The linear diagram (2) is distorted from the geographic deviations the line takes but is partly representative of the long, sinuous nature of the urban area the system serves, though C.H. Dron is actually the most northerly station!

The in-car line diagram (1) comes from the opening year. An extension from C.H. Dron into Belgium may make this the only cross-border mass transit! A large modern tram system (with four underground stations) links up with the Metro at two points.

2 St Philibert
Bourg
Maison des Enfants
Mitterie
Pont Supérieur
Lomme - Lambersart
Canteleu
Bois Blancs
Port de Lille
Cormontaigne
Montebello
1 C.H.R. B - Calmette
C.H.R. Oscar Lambret
Porte des Postes
Porte d'Arras
Porte de Douai
Porte de Valenciennes
Lille Grand Palais
Mairie de Lille
Wazemmes
Gambetta
République - Beaux Arts
Rihour
Gare Lille Flandres
Caulier
Fives
Marbrerie
Hellemmes
Lezennes
Pont de Bois
Villeneuve d'Ascq - Hôtel de Ville
Triolo
Cité Scientifique
4 Cantons 1
Gare Lille Europe
Saint Maurice Pellevoisin
Mons Sarts
Mairie de Mons
Fort de Mons
Les Prés
Jean Jaurès
Wasquehal - Pavé de Lille
Wasquehal - Hôtel de Ville
Croix - Centre
Croix - Mairie
Epeule - Montesquieu
Roubaix - Charles de Gaulle
Eurotéléport
Roubaix - Grand Place
Gare - Jean Lebas
Alsace
Mercure
Carliers
Tourcoing - Sébastopol
Tourcoing - Centre
Colbert
Phalempins
Pont de Neuville
Bourgogne
2 C.H. Dron

Transpole

Correspondances
T Tramway Lille - Tourcoing
R Tramway Lille - Roubaix
Liaison Inter'Val
S SNCF
Terminus
Terminus partiel
(Repérez sur le quai la destination de votre rame).
P Parkings métro gratuits et surveillés
T Taxi

2.

France's third-largest metropolitan area is expanding its mass rapid transit. The Métro itself is set to grow by four stations when the blue Line 1 reaches La Fourragère in 2009, and Line 2 should be getting five more stops after that, but the big transit development is the new tramway, which opened in 2007 and has three lines interlacing with the Métro.

The diagram of the Métro has a profusion of interesting angles, with barely one that matches another, but a clarity that outweighs its lack of uniformity. The black circle station markers have an ultra-fine rim and massive white-eye centers, and rounded oblong sisters at interchanges. Though the rubber-tired Métro is clean and effective, it oddly closes at 9 pm.

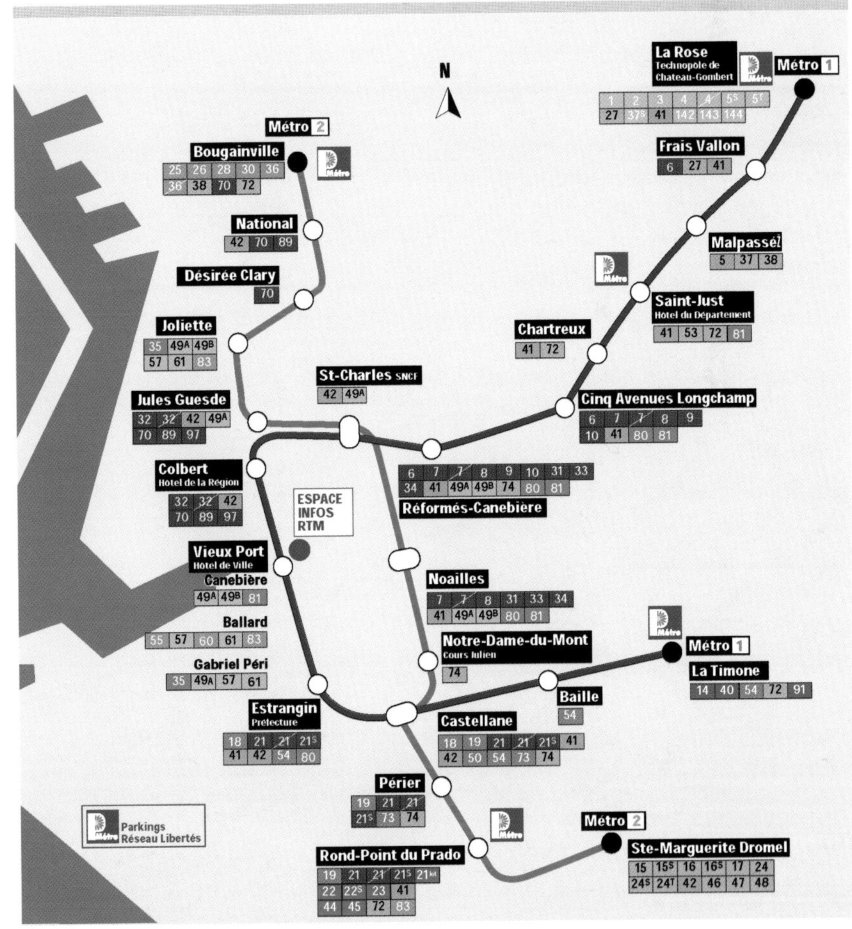

Unless stated, all maps by kind permission of and copyright the system operator © 2007.

Milan

4.2 million. 1964. 46 miles. 87 stations. 80 percent underground.

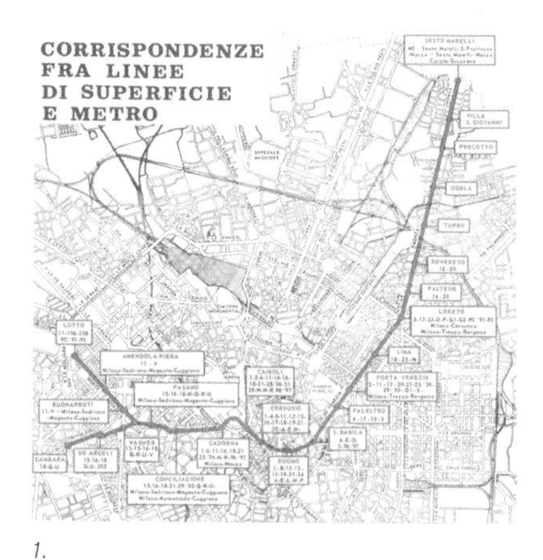

1.

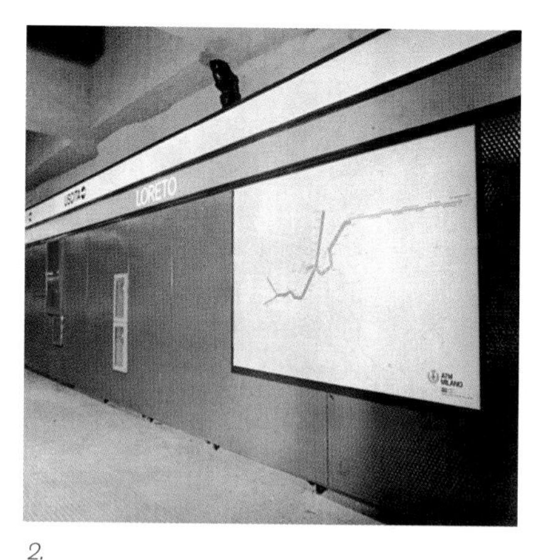

2.

Italy's second (but largest) city, and home to its most extensive Metro system, has an appropriately impressive diagram (3). It bears all the hallmarks of good Italian design, on a traditional urban-transit map: bold, strong colors, 45-degree diagonals, geographic distortion, and equally spaced stations. Only where station names crash over the lines, for example at Lanza and Montenapoleone, does the map feel in any way cluttered.

The current network followed three abortive plans to bring underground railways to Milan—one as early as 1848 and a huge projected seven-line system in 1938.

The 1970 map (1) showed surrounding streets. By the mid-1970s wall maps were diagrams (2), which demonstrates the evolution to the current one. At least three more full lines are projected, with many other improvements due.

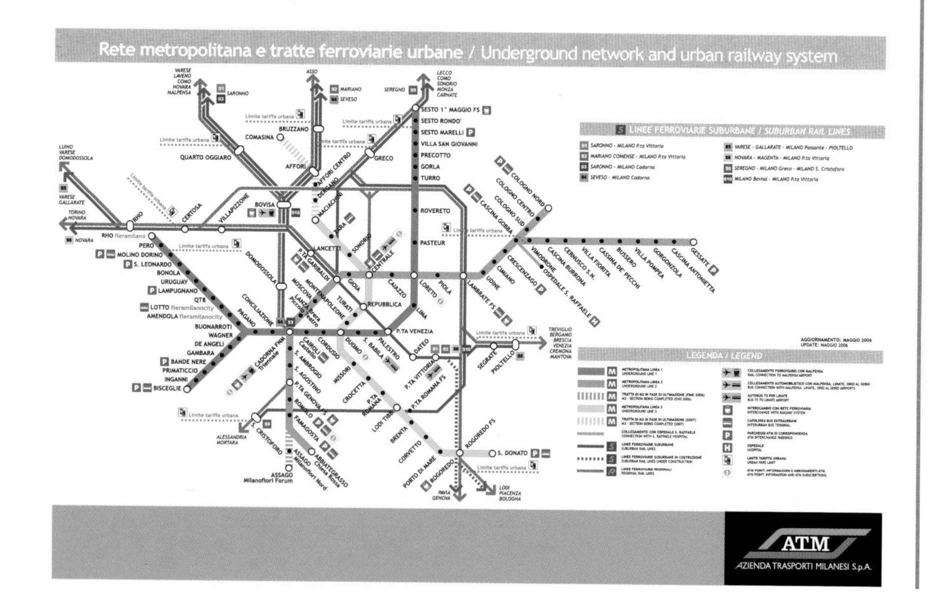

Naples

3 million. 1993 (1925). 17.4 miles. 25 stations. 60 percent underground.

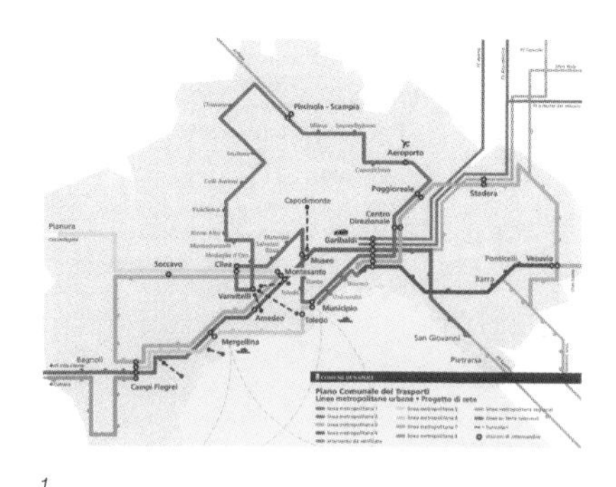

1.

The first diagram (1) shows that the Naples administration has some very ambitious plans to expand and convert the two existing Metro lines and the region's complex arrangement of heavy- and commuter rail routes into a new, fully integrated network of ten mass rapid transit lines.

The current network diagram of what has so far been achieved (2) is a striking and bold effort with good Italian design concepts, and its dark background certainly aids legibility.

The 45-degree diagonals, even station spacing, and dark hexagonals with thick blue rims and white centers work well as a variant on the traditional interchange symbol. However, the small nicks in the line indicating stations might not survive when all the new lines open.

The network is built around, through, and under the rocky volcanic terrain of this beautiful port. The central portion of Linea 1 opened in 1993, while FS (Italian National Railways) operates a metro-like service through a 1920s tunnel, labeled Linea 2. Small portions of some other lines are already in operation on former rail alignments, but the majority will not be open until 2011.

2.

Unless stated, all maps by kind permission of and copyright the system operator © 2007.

Newark

4 million. PATH: 1908. 13.7 miles. 13 stations. 70 percent underground.
Newark Subway: 1934. 5.3 miles. 12 stations. 23 percent underground.

Newark is linked to Manhattan by a subway known as Port Authority Trans-Hudson, or PATH. Its system diagram (1) is simple and clear, and the front cover of a 1974 guide (3) shows an alternate version. Within Newark, the old City Subway has been renovated and absorbed into part of a modern light-rail service that uses a geographic map (2).

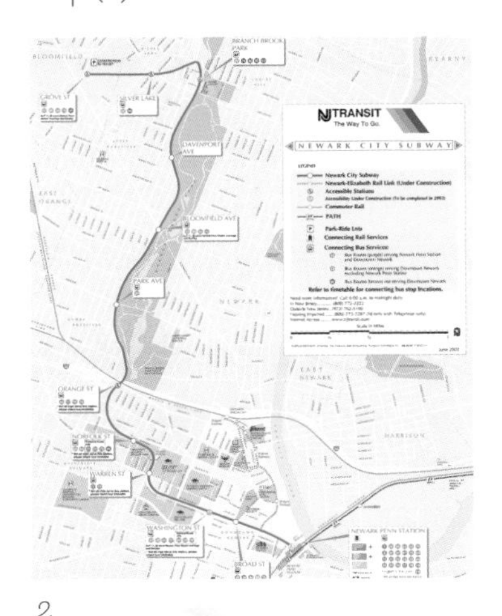

2.

PATH Service Guide

Daylight Saving Time

Effective April 28, 1974

NEW YORK

33rd St.
23rd St.
14th St.
9th St.
Christopher St.
World Trade Center
Hoboken
Hudson River
Pavonia Ave.
Exchange Pl.
Grove St.
Journal Square
Penn Central
Erie Lackawanna
NEW JERSEY
Harrison
Newark
NJ NY

Amtrak / Penn Central / Central of New Jersey / Reading

3.

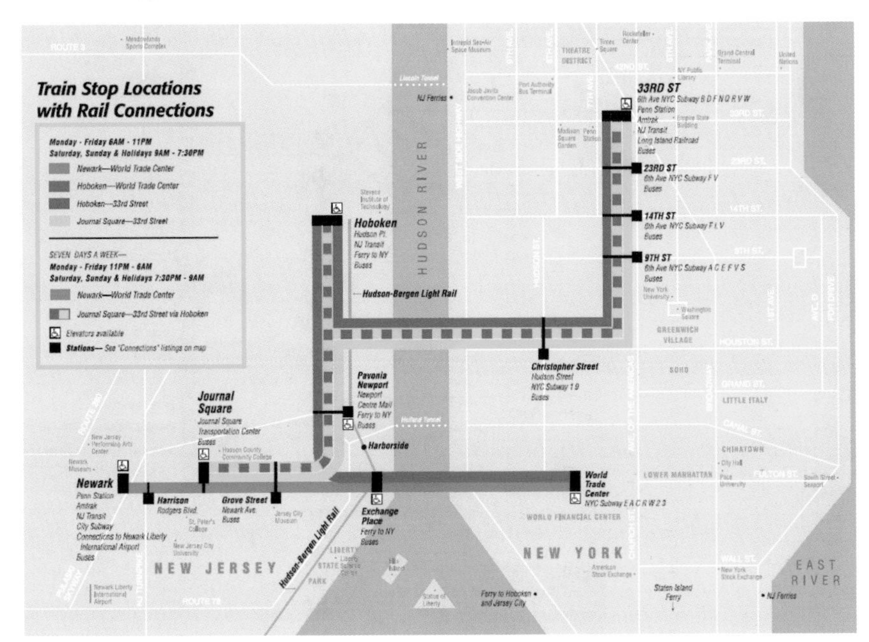

1.

Nizhniy Novgorod

1.2 million. 1985. 9.6 miles. 13 stations. 100 percent underground.

1.

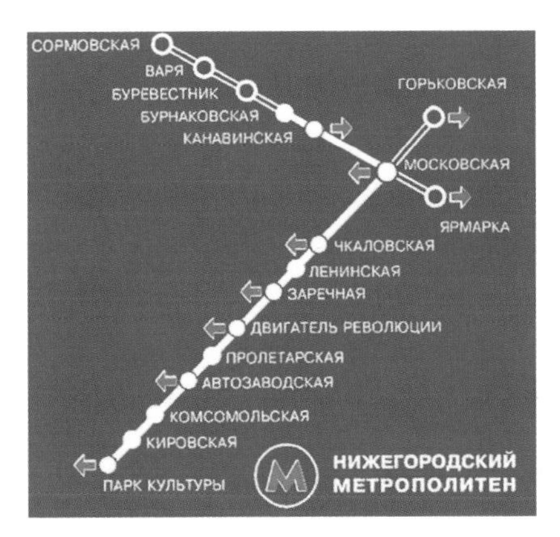

2.

The former Soviet city of Gorky, where political dissidents were once exiled, only recently opened to tourists. Its typical Russian station architecture is beautiful, with huge chandeliers hanging from platform ceilings, and marble statues and artworks gracing the walls.

Progress with extending the system and with the maps themselves has been less impressive. A stylized X shape was first seen on a poster from 1996 (1) celebrating 100 years since the All-Russia Art Expo and on a small card diagram from 2001 (2), both of which proudly predict new extensions.

The recent backlit wall map (3), which was erected when the Sormoskaya line finally reached one stop farther, to Burevestnik, in 2002, shows that expansion has been slow, and the line remains truncated and distant from the current downtown area. The proposed river crossing may yet happen before 2010 if funds become available.

3.

Unless stated, all maps by kind permission of and copyright the system operator © 2007.

Recife

3.6 million. 1985. 18.2 miles. 20 stations. 0 percent underground.

Like Spain and China, Brazil is engaged in some major mass rapid transit construction. In Recife, Metrorec currently runs the Central Line, which was built on an old railway alignment, although the line may be transferred to state control after a 3-mile extension on the Rodoviária branch to Timbi is completed in the next few years.

The first 7.4-mile section of the suburban line from Recife to Cabo is being upgraded to metro standard from Joana Bezerra to Cajueiro Seco, with the aim of providing an 8.9-mile metro service from Central to Seco and ten new stations.

The remaining 7 miles of single track on the 11.5-mile line will then be doubled, and the five stations south of Cajueiro Seco will be rebuilt. This will give Recife a 25-mile metro network, plus an 11.5-mile suburban line.

Metro access is being improved by a connection to Rodoviária bus terminal, a moving walkway from Guararapes Airport to the station, and a monorail between the shopping mall and a Metro station named simply Shopping.

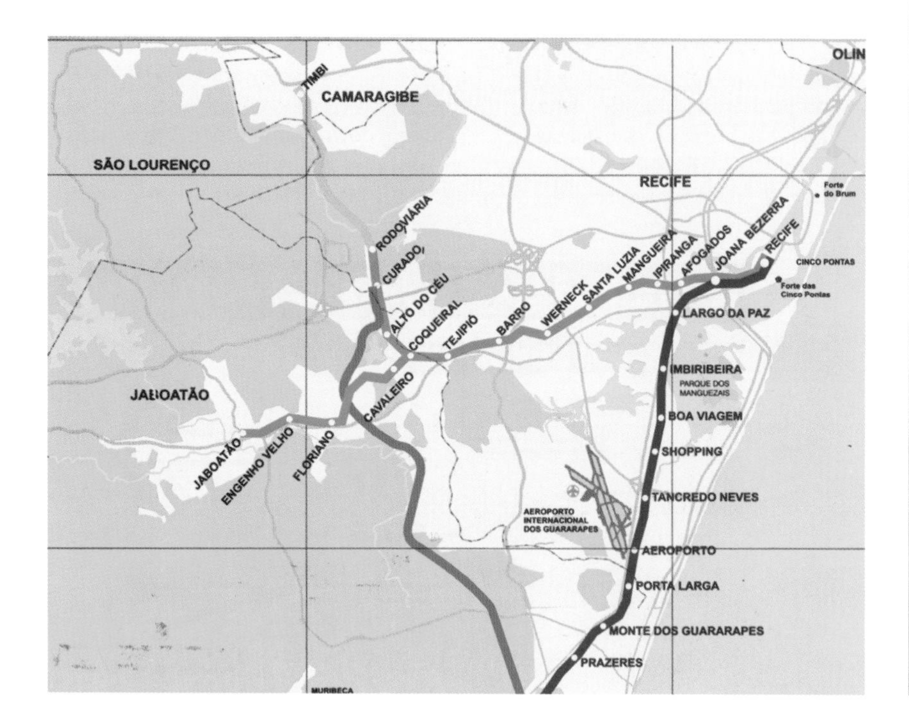

Santiago

5.4 million. 1975. 45.3 miles. 78 stations. 70 percent underground.

A clear diagram with slight geographical distortion shows the system that runs on rubber tires under Chile's capital. Diagonals are on a 45-degree angle, and also used here are the black circle with white center at interchanges, with stations marked as an open circle.

The whole 20.5-mile extension of the new Line 4 is now operating, but the diagram excludes the intended Line 3, planned to run from the north of the city to the southwest. Eventually, there could be a 65-mile network by 2009.

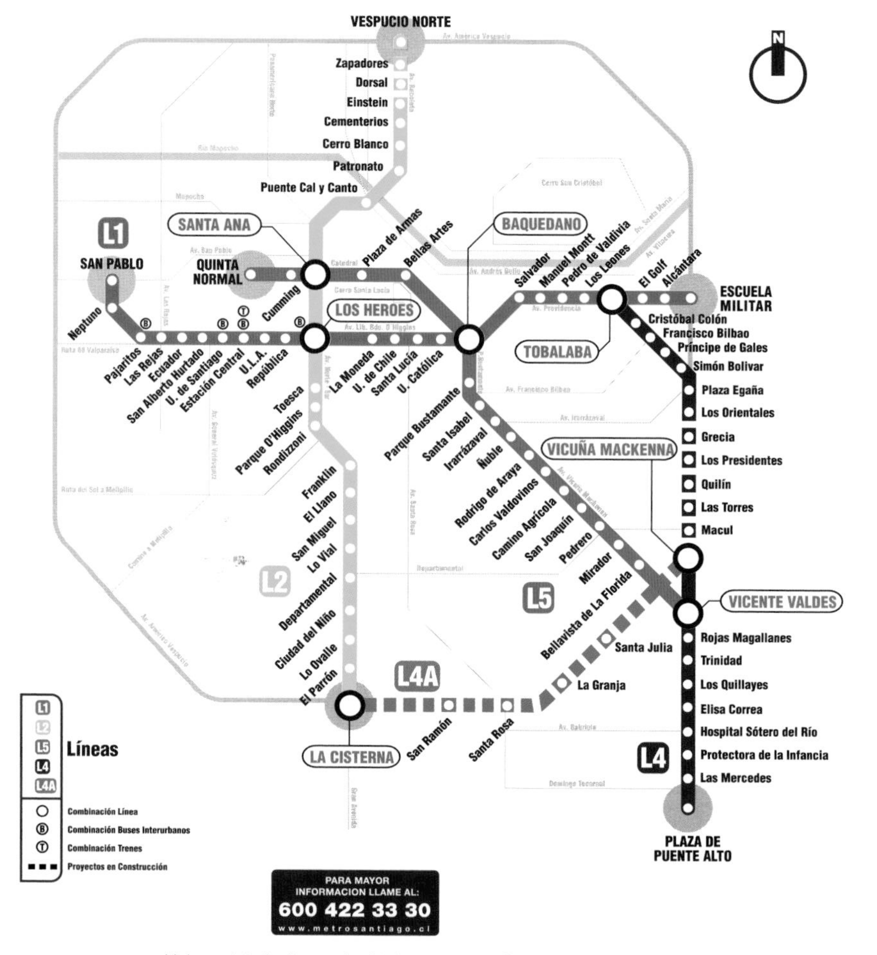

Unless stated, all maps by kind permission of and copyright the system operator © 2007.

Shanghai

18.6 million. 1995. 51.5 miles.
73 stations (w/ Pearl Line). 80 percent underground.

With arguably the world's most ambitious rapid-transit growth plan, the city wants to build *eighteen* new lines! By 2025 there could be at least eleven lines, covering almost 190 miles. Some older maps showed the network with all its curvilinear meanderings alongside ward areas or districts colored in traditional Asian style (1). Others, like the excellent current example, have dispensed with all topography to produce an effective and clear diagram using open circles for station markers, doubled at interchanges (2). Like some other Asian maps, it resembles a new character in the alphabet.

1.

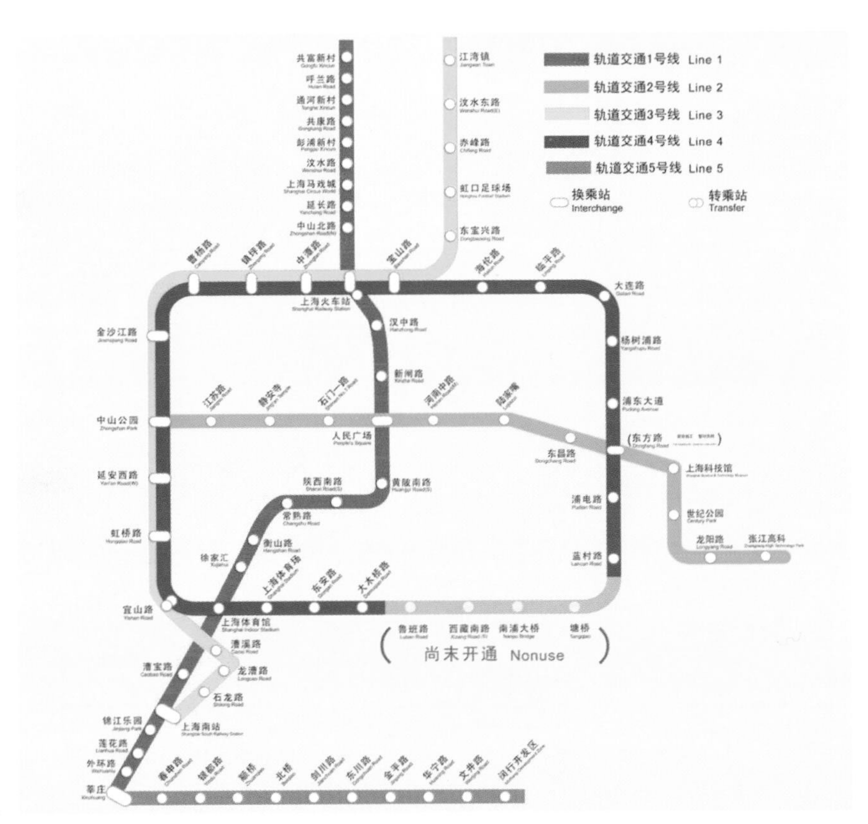

2.

Warsaw

2.8 million. 1995. 11.3 miles. 16 stations. 100 percent underground.

The relatively new system in Warsaw currently has just one line. It has been shown on maps in a fairly topographical manner to help the city's inhabitants place it in context with existing road and rail routes (1). However, two new lines are in the planning stage, so there may yet be a diagram to come out of the Polish capital in a style worthy of the distinctive red and yellow *M* logo and bold, clean station signage, like the platform wall diagram from before the line was extended to Marymont (2).

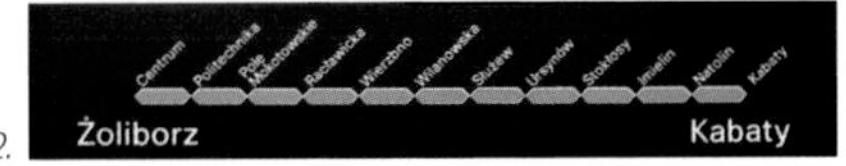

2.

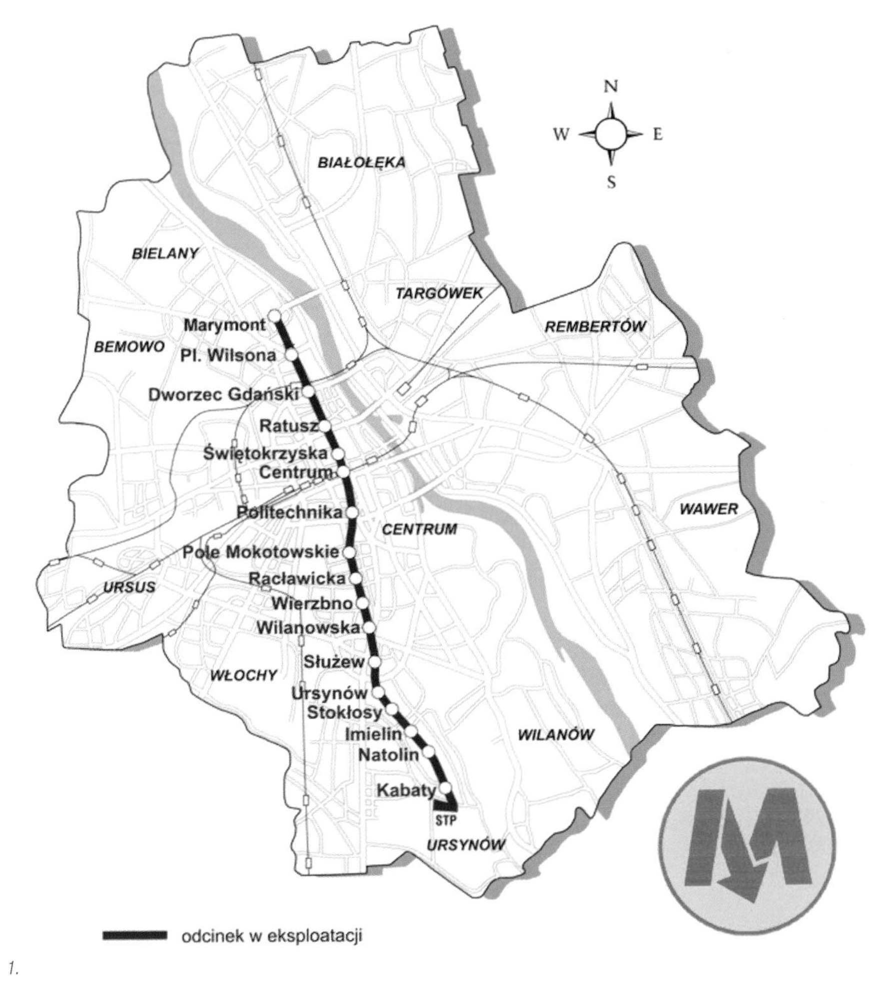

1.

Unless stated, all maps by kind permission of and copyright the system operator © 2007.

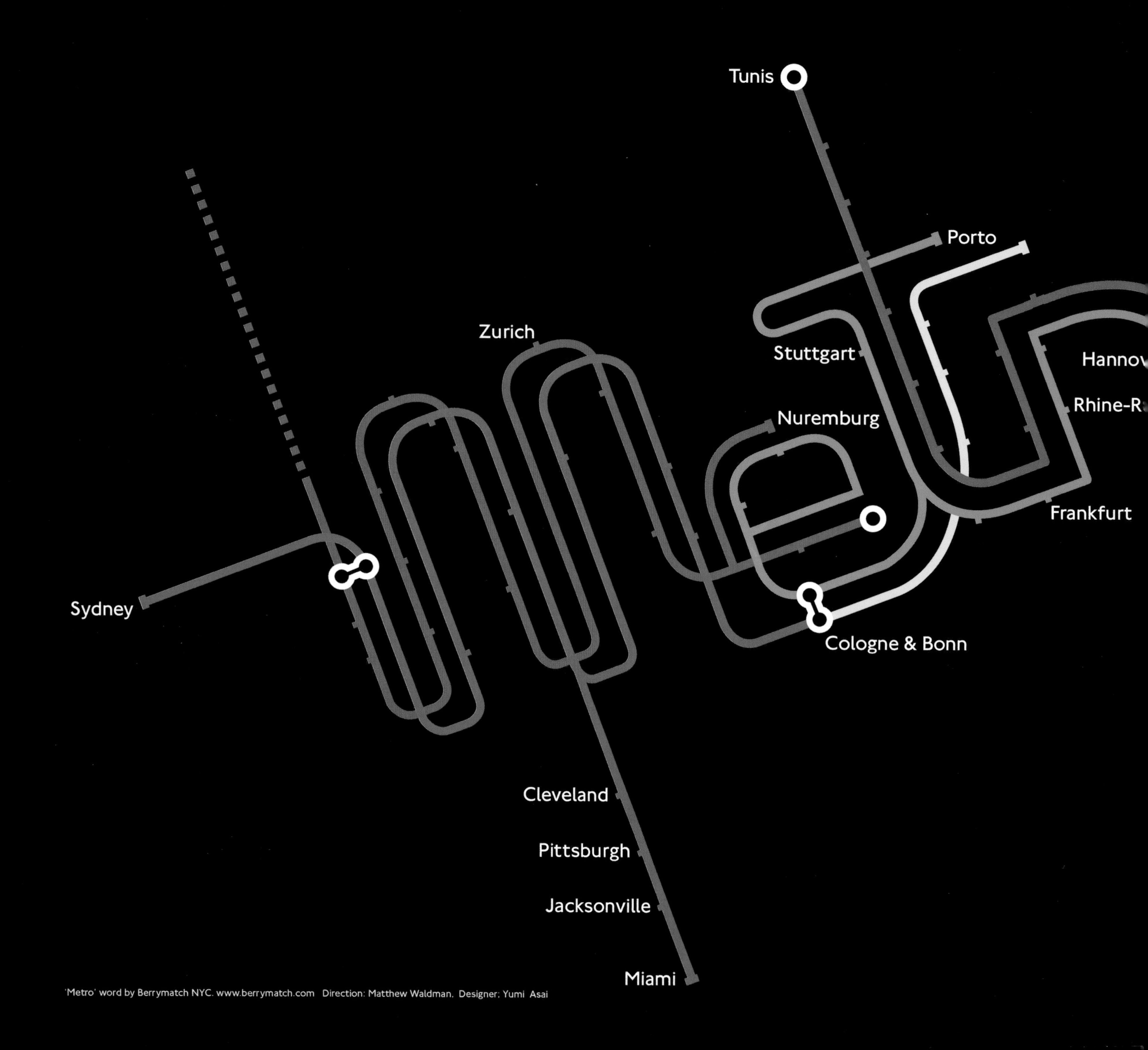

'Metro' word by Berrymatch NYC. www.berrymatch.com Direction: Matthew Waldman. Designer: Yumi Asai

These cities do not have "traditional" subway or metro systems; yet most have some underground sections or stations. Many would be classed as heavy-rail commuter or light-rail systems; Miami, for example, has one of each! They have been included in this book because they have successfully employed the cartographic style of the urban-transit map in their marketing and are generally perceived by their customers as "metro-like" services. More of these hybrid systems also appear in Zone 6.

Cologne & Bonn 118
Cleveland 118
Dublin 118
Frankfurt 119
Hannover 119
Jacksonville 119
Liverpool 120
Manchester 120
Melbourne 121
Miami 121
Nuremburg 121
Pittsburgh 122
Porto 122
Rhine-Ruhr 122
Stuttgart 123
Sydney 124
Tunis 125
Zurich 125

Zone 5

Cologne & Bonn

1 million. 1968. 54 stations. 28 miles underground.

The city region of Cologne, like most other German conurbations, boasts an impressive integrated mass-transit system. It includes links by the Verkehrsverbund Rhein-Sieg (VRS) to the smaller city and former West German capital of Bonn, just along the Rhein.

The network is mainly light rail, with many subsurface sections. It's a true hybrid; under large U signs escalators from the street descend into subterranean stations throughout Cologne and Bonn.

Most of the tunnels opened between 1968 and 1974. A 2.5-mile north–south tunnel is under construction in downtown Cologne.

Unless stated, all maps by kind permission of and copyright the system operator © 2007.

Cleveland

2.2 million. 1955. 17 miles. 41 stations.

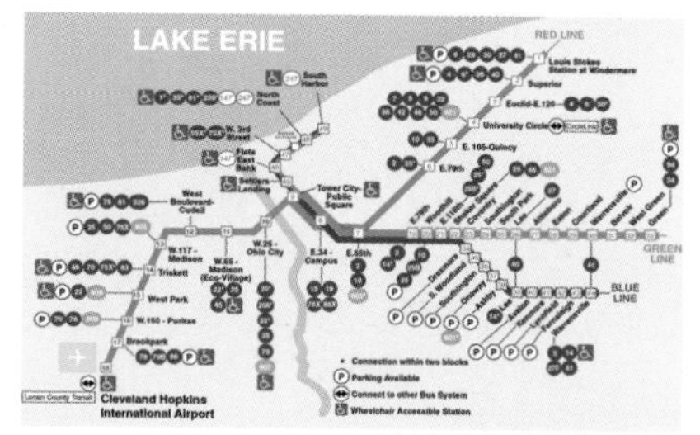

Less than a mile of the "Rapid" Red Line is underground, at the airport and by Tower City. Blue and Green light-rail vehicles share tracks with the Red Line between East 55 and Tower City. An extension is due from the airport to Berea.

Dublin

2 million. Due 2012. 43.5 miles and 72 stations planned.

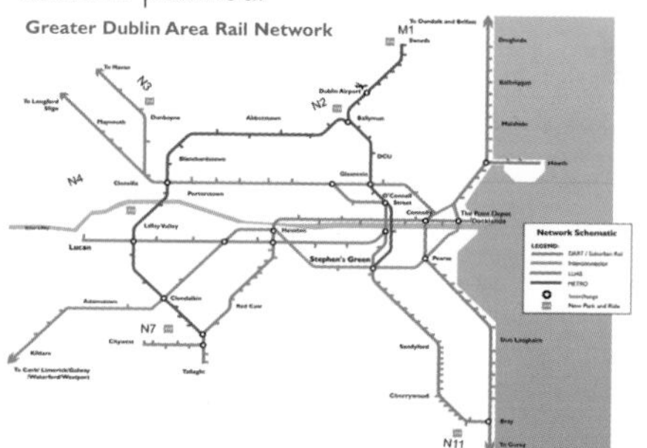

After the successful reintroduction of LUAS streetcars to Dublin, the Irish capital is now pressing ahead with long-awaited plans for a metro to complement its heavy-rail DART commuter lines. Metro North to the airport will be the first 10-mile section to be completed.

The full network proposed on this highly stylized and geographically distorted diagram combines the existing and extended DART, LUAS (green), and Metro (blue). The diagram is still relatively speculative at this stage.

Frankfurt

2 million. 1968. U-Bahn: 38 miles. 28 stations. 55 percent underground.

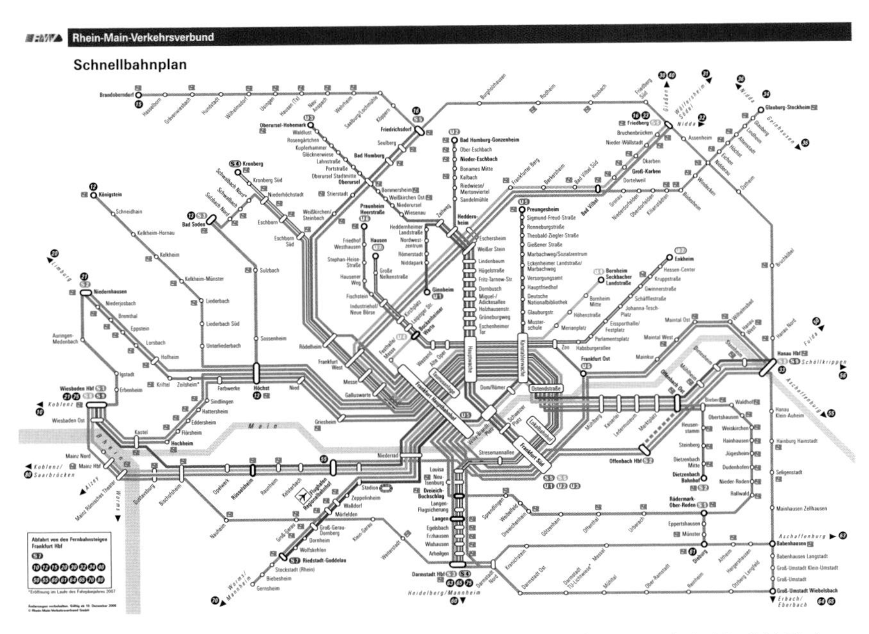

Proving again that Germans know how to do mass rapid transit properly, users of Frankfurt's urban-rail system hardly notice that it's a true hybrid of U-Bahn (lines U1 to U7), S-Bahn (S1 to S9), streetcar, and suburban commuter rail (numbered along with the regional rail lines and shown in gray as 10 to 80).

The diagram is therefore a triumph of design, depicting every component of this remarkable system. Hauptbahnhof is the true heart, allowing interchange between 25 U-Bahn and other rail routes, but the area appears incomprehensible at first glance. Is it perhaps the case that what computer-drawn maps add in terms of precision and neatness they sometimes lack in simplicity and legibility when looked at through the flawed medium of the human eye?

There are black circles with white centers used for station markers. Follow U6 from Frankfurt Ost through Zoo to Konstablewache, and it can be seen that even the major interchanges are simply rectangular elongated versions of the same graphic device.

The U-Bahn system started life with

trams that had underground tunnel sections in the city center.

Plans to expand U4 were recently shelved, but the northern section to Riedberg may be built as a branch off the existing A route, diverging from U1/U3 near Zeilweg station.

Hannover

1 million. 1975. 10.5 miles and 17 stations in tunnel.

Stadtbahnnetz
Hannover üstra

In creating another German hybrid network, Hannover converted its tram routes into modern light rail in the 1970s, using city-center tunnels with U-badged station entrances and known locally as the Üstra.

Although the diagram uses geographic distortion and 45-degree angles, notice how all the line branches have the same color, so as they converge at Aegidientorplatz, they appear as one thick line! Indentations, or full line breaks with white markers, are used instead of ticks for stations, and no station name crosses over a line.

A neat and effective diagram.

Jacksonville

1.3 million. 1989. 2.5 miles. 9 stations.

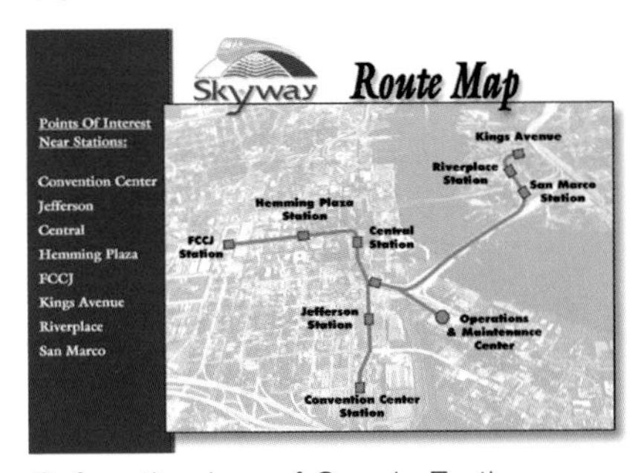

Before the days of Google Earth, Jacksonville's Skyway, an automated, elevated, people-mover monorail, produced this basic diagram laid over an aerial view of the city.

There are also proposals for a light-rail system running north–south from Gateway Mall to the airport.

Liverpool

1.4 million. 1977 (underground link and loop). 57 miles (4.2 in tunnel). 76 stations.

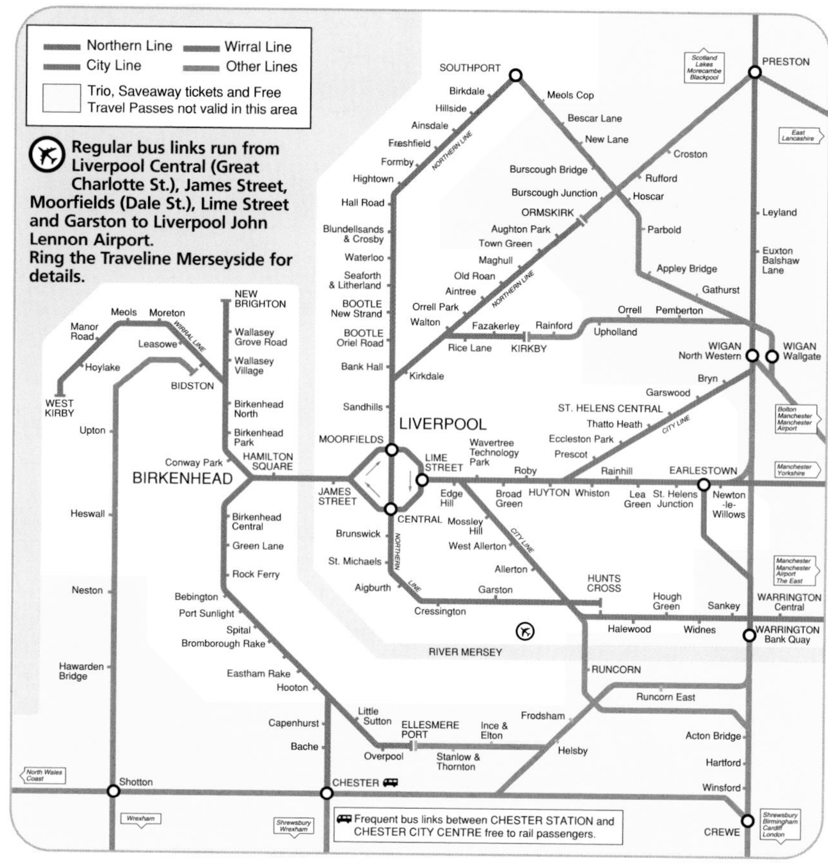

The first Mersey tunnel opened in 1886 and had three underground stations. It was converted from steam to electric in 1903. A new single-track loop linked it to all downtown stations in 1977. MerseyRail's suburban train service is frequent and on publicity its operator aims at a metro-like feel.

The current diagram, much like

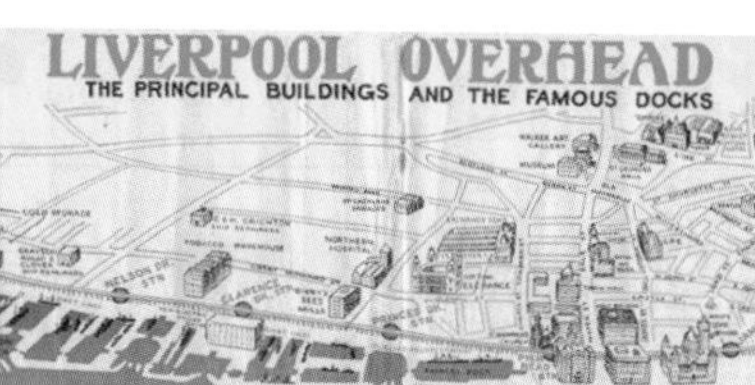

Glasgow's, tries to instill the notion of a unified, smoothly connected service despite its being standard heavy-rail suburban commuter lines.

The 6.2-mile Liverpool Overhead Railway opened in 1893 and was demolished in 1950, but ran along the docks with eighteen stations and an underground terminal at Dingle.

Plans for the return of light-rail streetcars have, to much local anger, been shelved.

Unless stated, all maps by kind permission of and copyright the system operator © 2007.

Manchester

2.6 million. 1992. 23 miles. 37 stations. 2 percent underground.

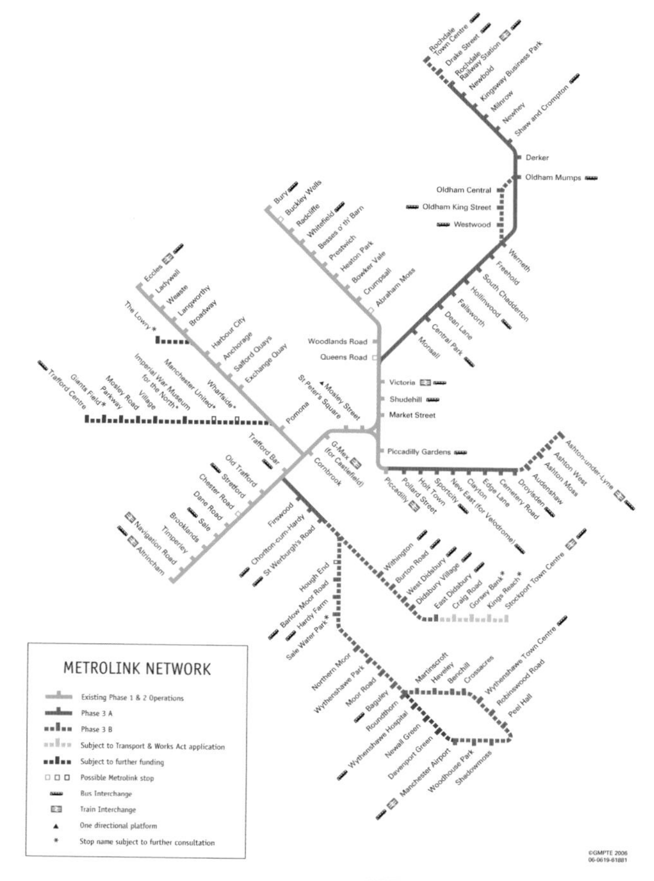

1.

Despite Manchester's numerous attempts to build a subway in the city, the United Kingdom's first modern light-rail system had no true underground stations apart from the hub below Piccadilly, which has a subway station feel (photo).

The wildly distorted, beautifully spaced diagram (1) of 45-degree diagonals, station ticks, and striking colors shows how the full Metrolink

network could look by 2012. The present extent of the system is shown on the in-car diagram (2).

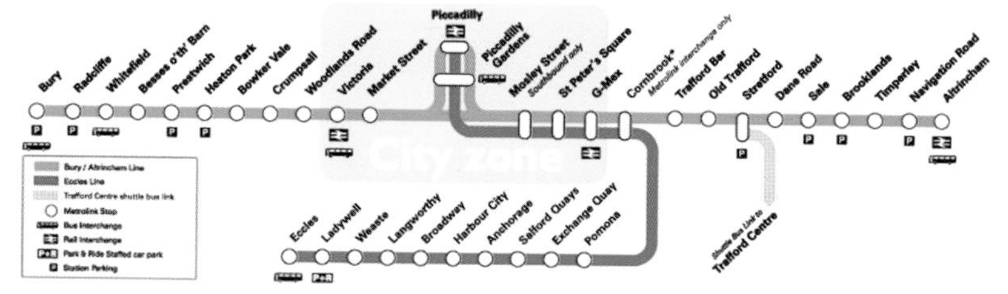

2.

Melbourne

3.6 million. 1984. 73.3 miles.
208 stations. 5 percent underground.

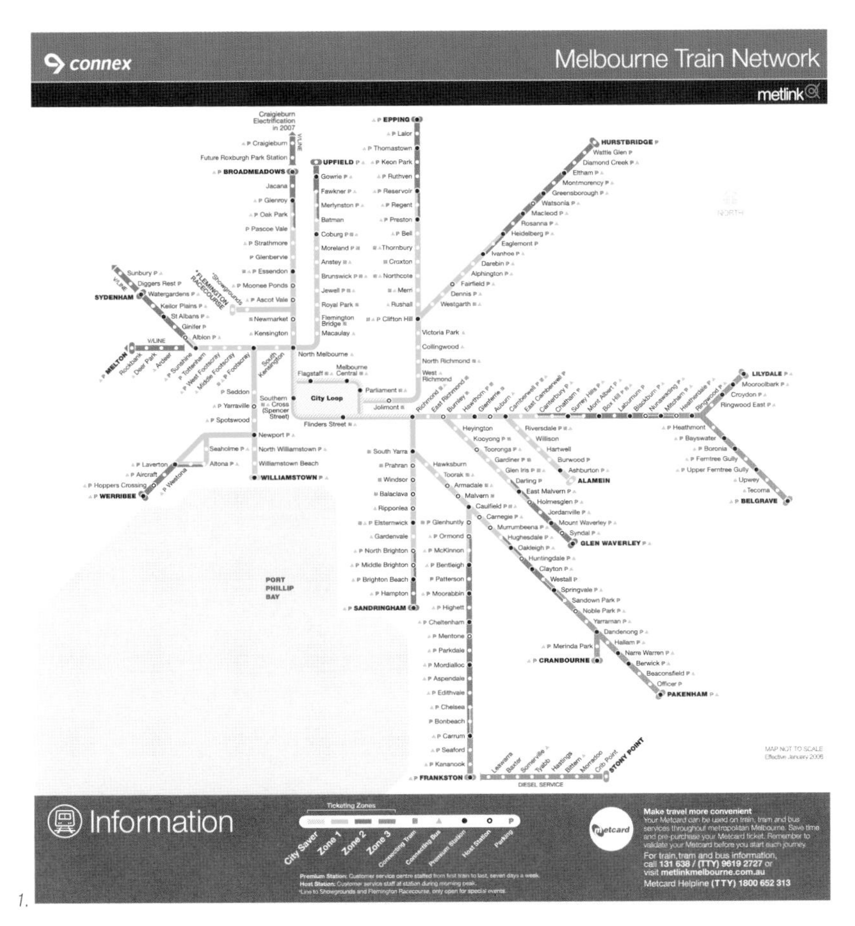

1.

Like those of Liverpool, Sydney, and some German cities, Melbourne's extensive train services (1) and vast streetcar network (2) are an integrated system knitted together by the construction of a fully underground loop in the city center, with five subsurface stations. Here is found the southern hemisphere's busiest railway station, Flinders Street.

The main diagram shows the commuter network of the heavy-rail suburban system. It crams in a lot of stations with ease, using 45-degree diagonals and aesthetic spacing, thus ironing out all those awkward geographic kinks in the lines.

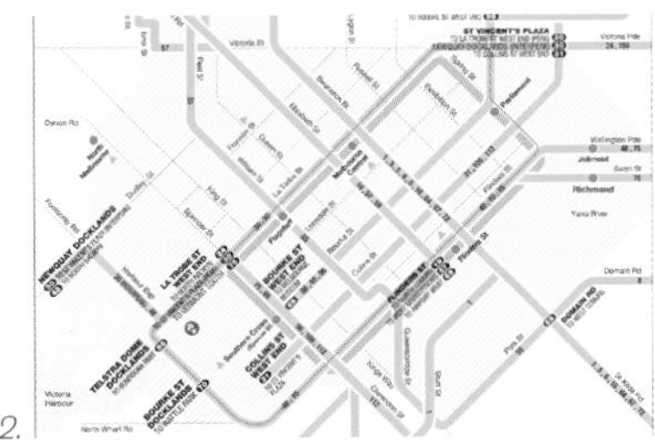

2.

Miami

5.4 million. 1984. 34 miles. 42 stations.

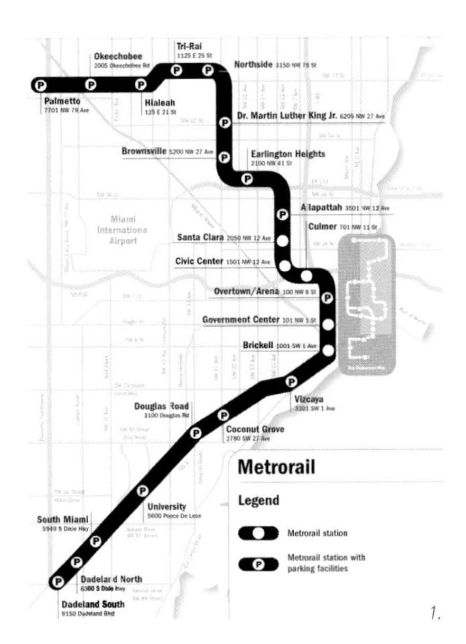

1.

Miami has the entirely elevated heavy MetroRail and the lighter elevated automatic MetroMover (2), which started as a 1.9-mile circle line and now has two branches. There are also plans for a streetcar called BayLink to connect South Beach with downtown.

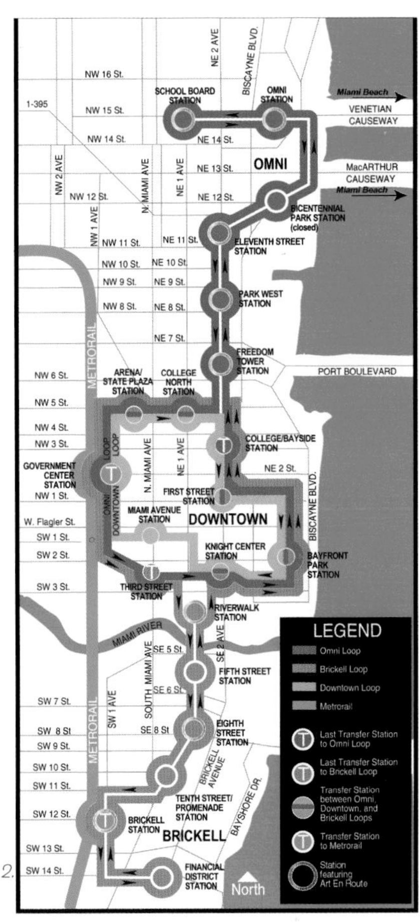

2.

Nuremburg

1.1 million. 1972. 24.8 miles. 40 stations. 70 percent underground.

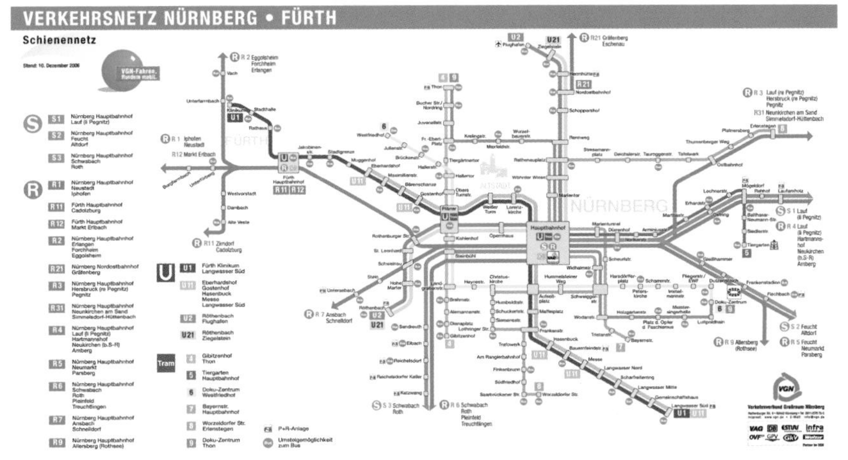

Unless stated, all maps by kind permission of and copyright the system operator © 2007.

Pittsburgh

2.3 million. 1987. 25 miles. 61 stops.

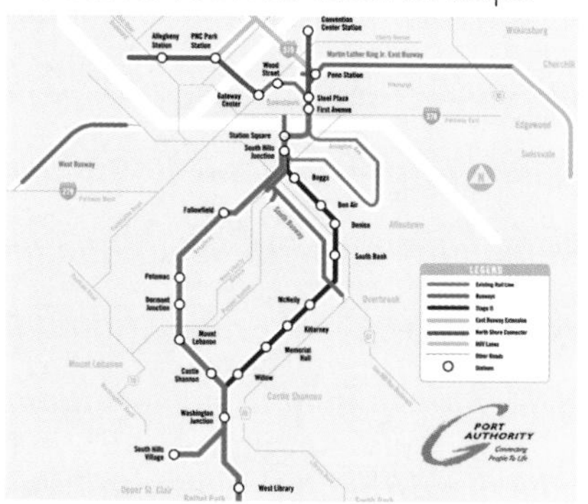

Pittsburgh's 125 years of streetcar operation were superseded by a modern light rail called the T, with a 2.5-mile downtown subway. An extension under the Allegheny River could connect the North Shore to downtown by 2011.

Porto

1.6 million. 2002. 36.8 miles. 60+ stops. 10 percent underground.

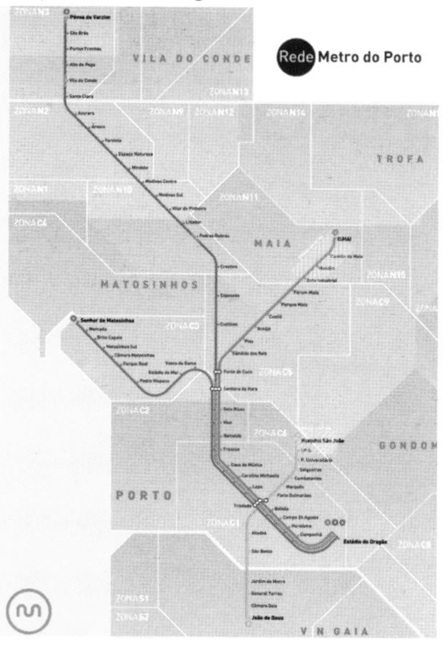

A light-rail network with 4.3 miles downtown in tunnel.

Rhine-Ruhr

9 million. 1954. U-Bahn/Schwebebahn: 77.7 miles. 125 stations.

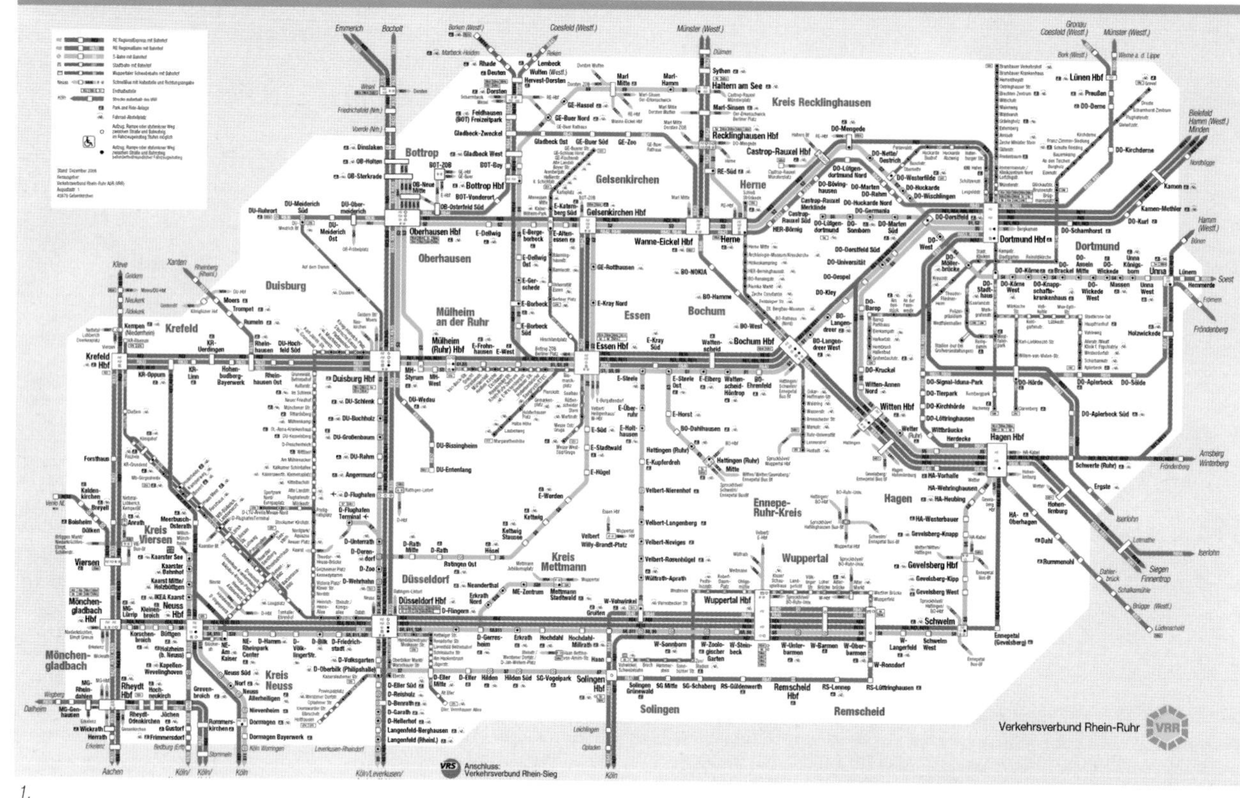

1.

Germany corrals its transport concerns under fare-zone agencies. The VRR oversees the largest conurbation of cities and towns in the country, collectively referred to as the Rhine-Ruhr, and publishes this first-class diagram (1).

However, if traveling within one of the centers, it may be useful to consult more detailed local maps for any of the cities VRR covers—for example, Bochum, Dortmund, Duisburg, Düsseldorf (3), Essen, Gelsenkirchen, Herne, Mülheim and Neuss—since there are multimodal options in each.

Wuppertal has the world's only suspended monorail (2), the Schwebebahn, opened in 1901.

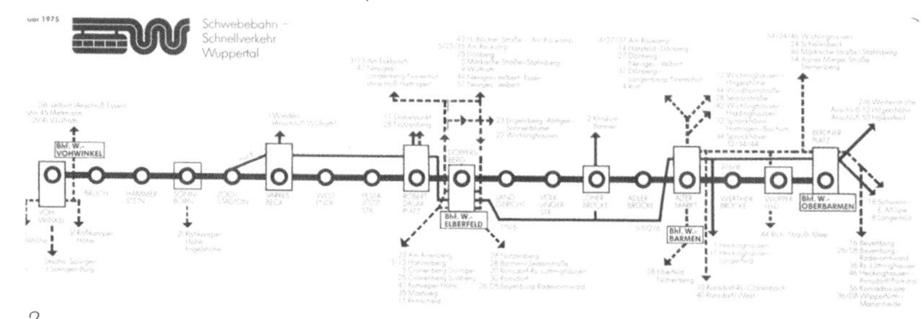

2.

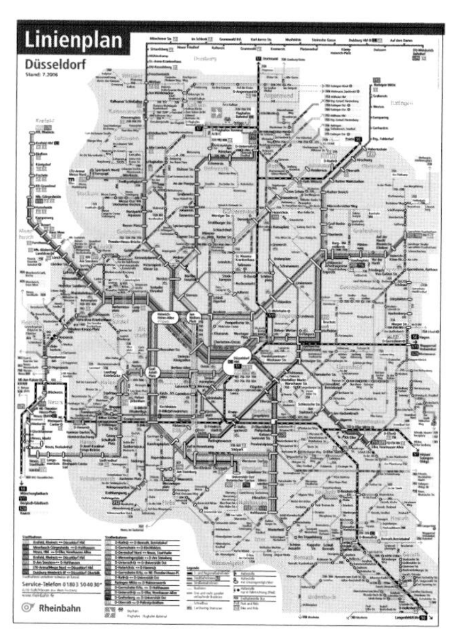

3.

Stuttgart

900,000. 1966. 70.2 miles. 13 stations in tunnel. 20 percent underground.

Verbund-Schienennetz

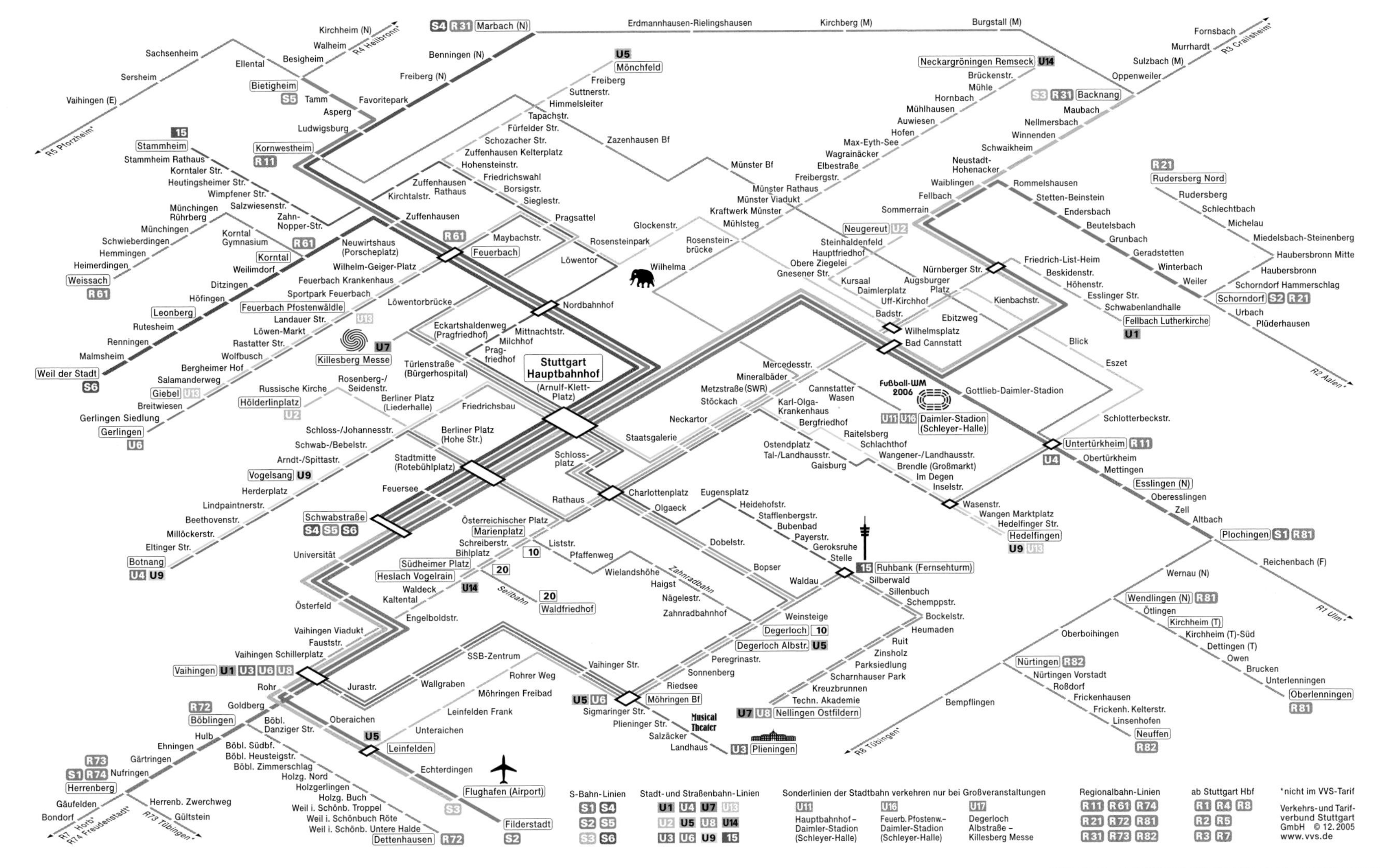

The Stuttgart system is tram based but with more than 15 miles running underground. Though its area is much smaller than that covered by the Rhine-Rhur, it is hoped that this schematic is considered worthy of inclusion at a larger size due to its exotic nature—a unique way of looking at an urban-rail map.

The diagonals are all at 30 degrees to the horizontal (this is not unique—see also Madrid among others), but here there is a complete absence of vertical or horizontal lines, thus giving the diagram an oddly and perhaps intentionally three-dimensional appearance.

This diagram is the only one of its type in the world, although Harry Beck did experiment briefly with a 60/120–degree variation of the London map in 1940. The rarity of this concept may be partially explained by the relative difficulty of following some routes.

Sydney

4.2 million. CityRail Sydney Suburban network: 1926. 220 miles. 122 stations. 4 percent underground.

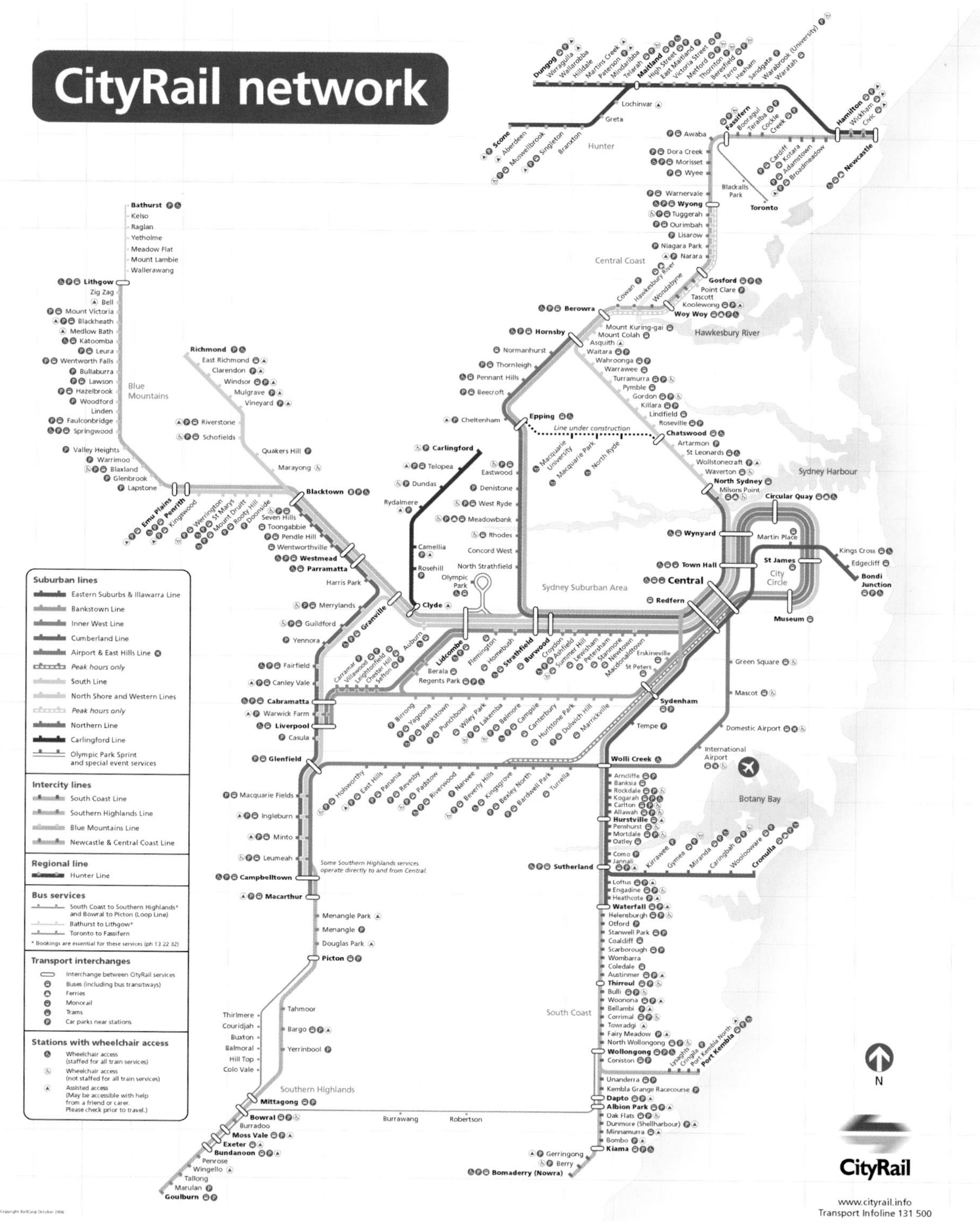

1.

Unless stated, all maps by kind permission of and copyright the system operator © 2007.

Sydney does not qualify as a true metro because it is a suburban rail system, spiraling miles out into the hinterland, and was never planned or built as a segregated, subterranean network. But its frequent service through the tunnels downtown, coupled with an excellent diagram (1) of 45-degree diagonals, bold primary colors, and evenly spaced stations denoted by ticks, proves how good design can lead to great expectations of fast and extensive rapid-transit-type services.

In the downtown area (2) there is also a light-rail streetcar (shown in blue) and a short monorail route (in red), neither of which is part of the CityRail network.

British influence saw much of the early design lean heavily on London Transport circa 1930. This plagiarism reached dizzying heights with the beautiful but brazen copies of the famous roundel at St. James and Museum stations and with the Sydney Railway Map of 1939 (3), a facsimilie of London's.

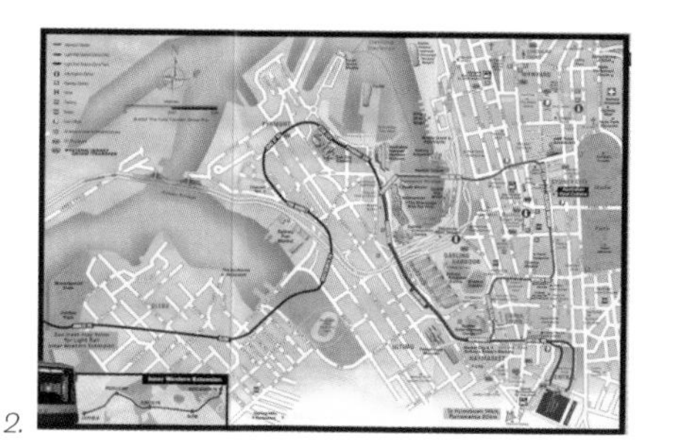

2.

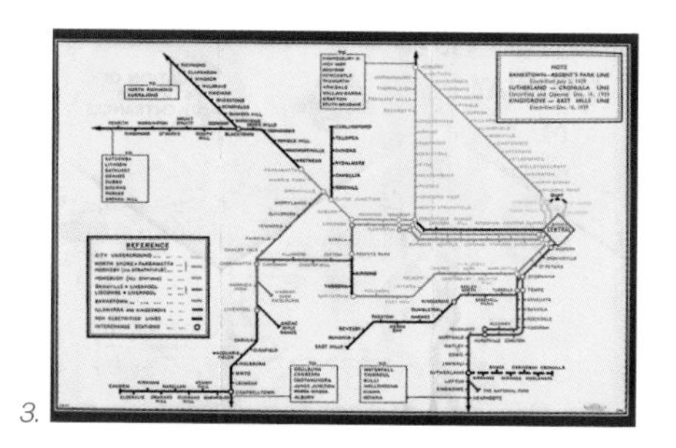

3.

Tunis 1.4 million. 1985. 19.8 miles. 47 stations.

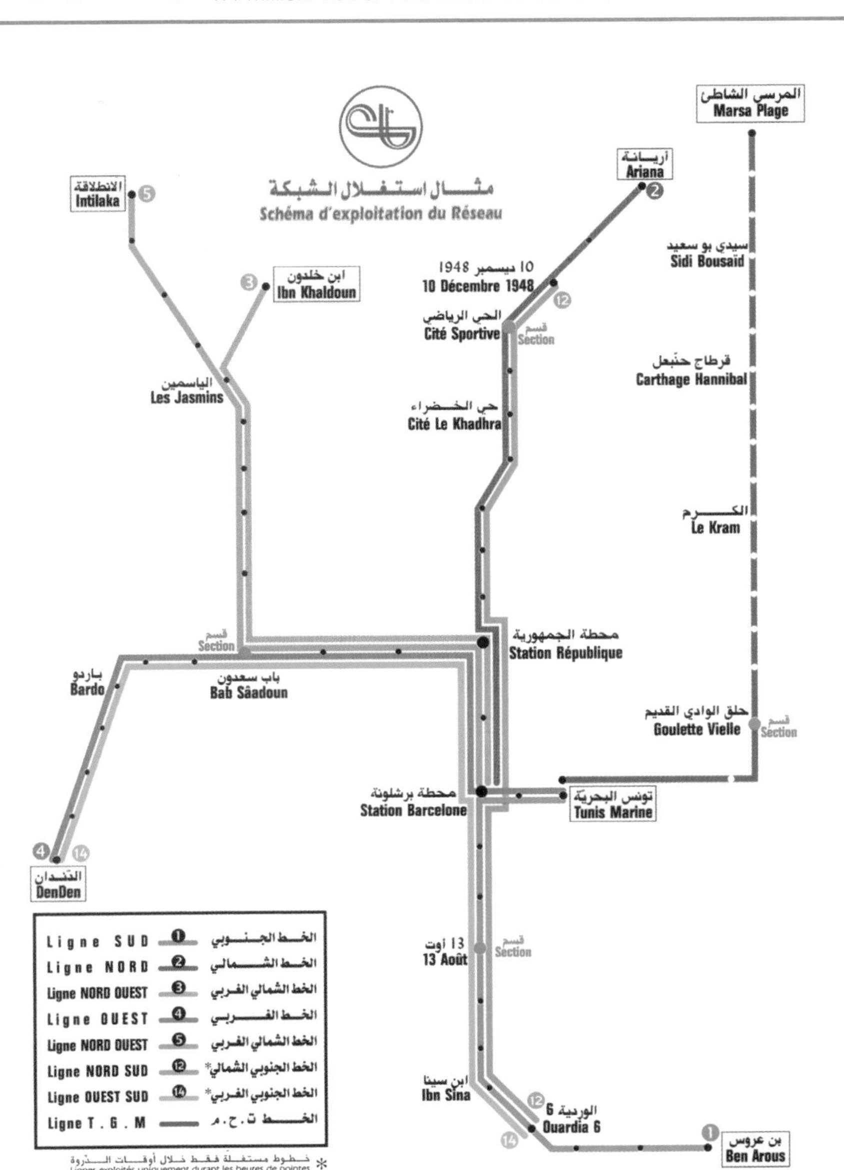

After many years of discussing plans to build Africa's first real underground, Tunis made the decision in the 1980s to cut costs and go for a metro-tram called Métro-Léger (literally: light metro), and the diagram is very metro-esque.

Zurich 1.3 million. 1999. 67.1 miles. 168 stations..

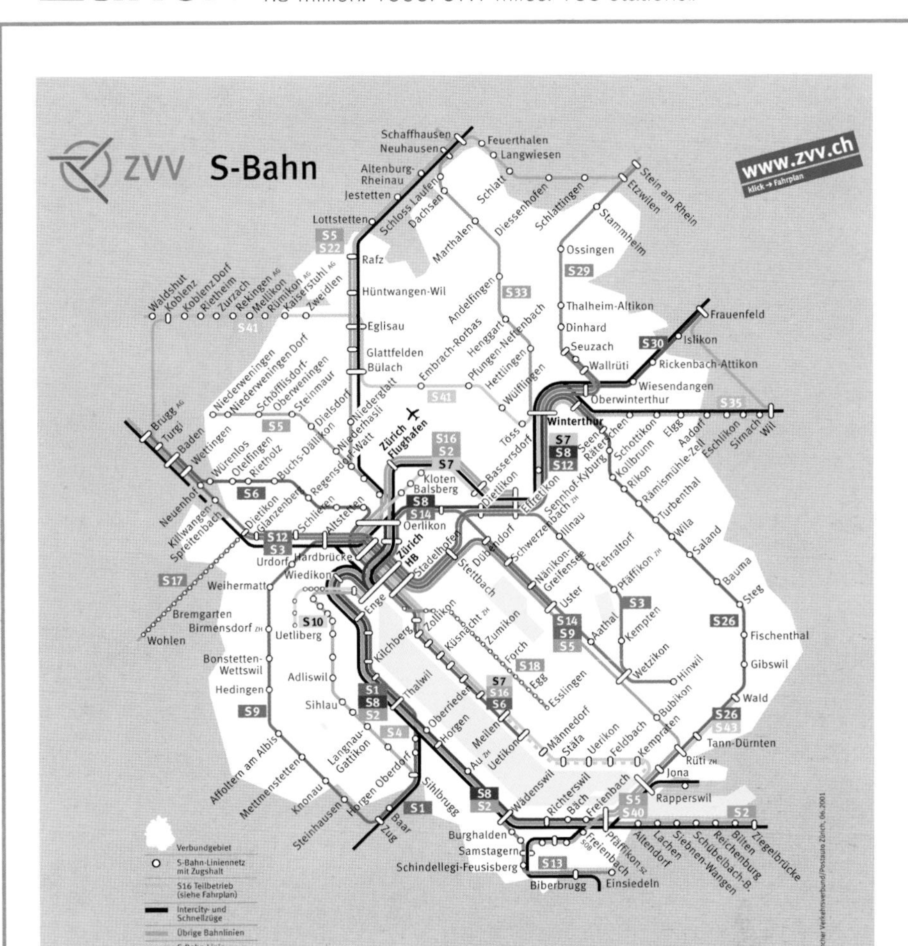

Zurich has eight stations underground but no official metro system. There was going to be a full mass rapid transit system, for which some of these subway stations were first constructed, but in the early 1970s the plan was dropped following a referendum.

However, the trams now use three underground stations originally dug for the subway. Other tunnels and subterranean stations that have opened since are used by the S-Bahn and the Sihltal-Zürich-Uetlibergbahn. This may partly explain why the diagram looks so much like an urban-rail map, with its 45-degree angles, geographic distortion of the center compared to the suburbs, even spacing of stations, bright, bold line colors, and copious use of the black circle with white-eye center symbol—proving once again that if an image gives a compelling enough impression of there being a full metro system, locals, visitors, and writers of books like this may actually believe that there is!

A huge variety of different vehicle types on rails now ply the cities listed in this zone. Most of the systems here are very recent, under construction, or planned. Or they are hybrid tram-trains, modern light-rail, or monorail services with metro-like maps or services. The section also includes a couple of more established, traditional urban-transit systems like the excellent networks of Helsinki, Caracas, Calgary, Busan, Minsk, Portland, Yokohama, Vancouver, Toulouse, Tashkent, and Sapporo, that did not make it into the previous zones simply because of their maps' size, quality, or orientation.

Adana–Belgrade 128
Belo Horizonte–Calgary 129
Caracas–Daejeon 130
Dallas–Fukuoka 131
Genoa–Houston 132
Incheon–Kobe 133
Kalkota–Manila 134
Maracaibo–Nottingham 135
Novosibirsk–Rennes 136
Rouen–San Juan 137
Santo Domingo–Tbilisi 138
Tehran–Yokohama 139

Zone 6

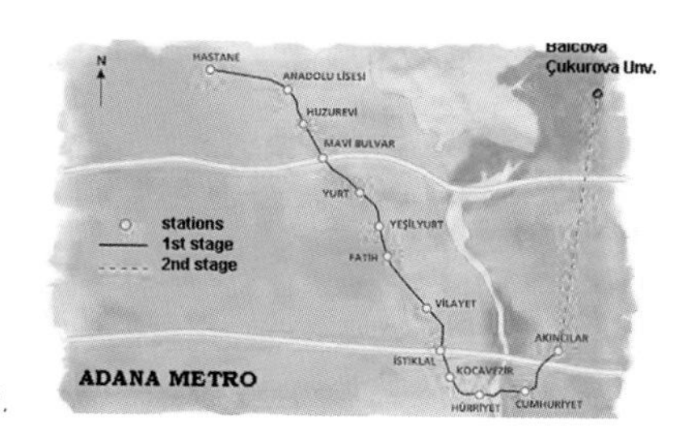

1.

Adana, Turkey (1)
1.5 million. Due 2010. 8 miles. 13 stations. A light metro line, similar to Izmir's or Bursa's, is being constructed.

Alexandria, Egypt
3.5 million. Plans for a 34-mile east–west urban-rail line seem to have been indefinitely postponed.

Algiers, Algeria (2)
2.1 million. Due 2010. 5.6 miles. Ten stations of a proposed three-line, 31-mile network under construction. Africa's most ambitious urban-rail project plans fifty-four stations. Progress is slow. A recent light-rail plan could beat the subway by some years.

Alicante, Spain (3)
725,000. 8.7 miles. 14 stops. An old rail line between Alicante and Denia via Benidorm is being upgraded to light rail. First section is open, and remaining 48 miles, with four subway stops in downtown Alicante, due soon.

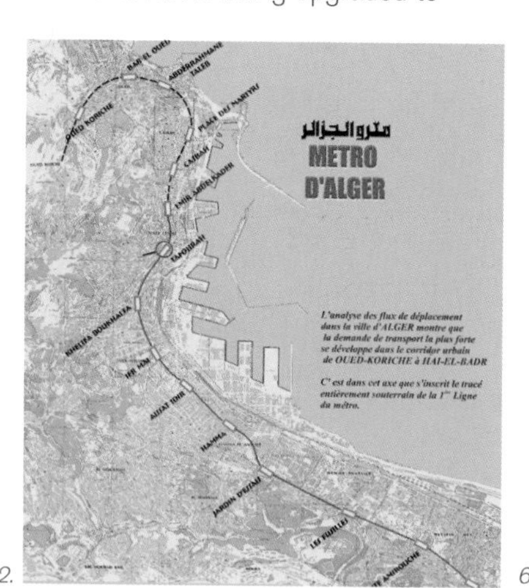

2.

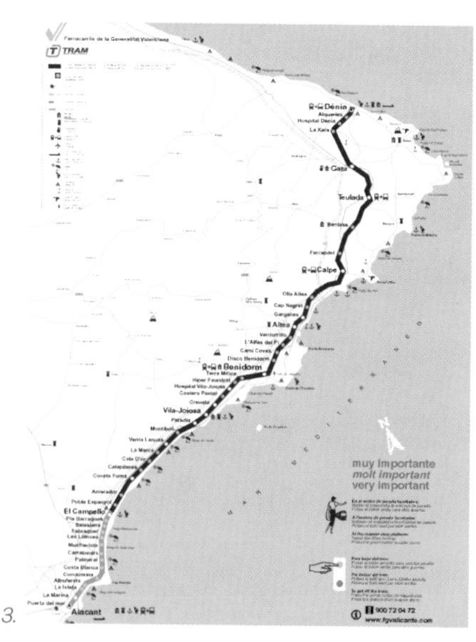

3.

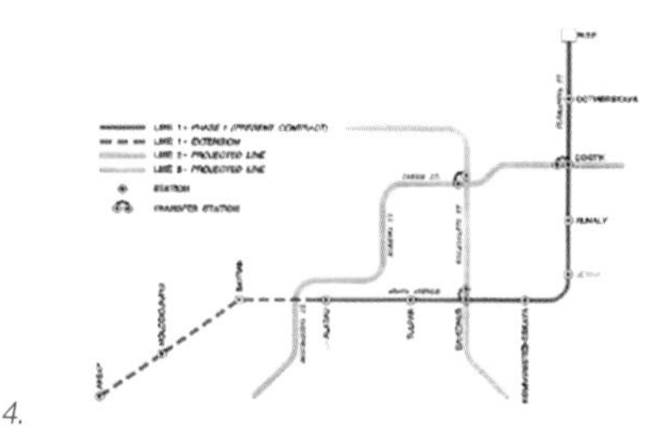
4.

Almaty, Kazakhstan (4)
1.1 million. 5.6 miles. One of the world's most landlocked cities, slap-bang in the center of the former Soviet Union, had built four miles of a planned 25-mile subway, but work stopped when money ran out, and now an American company is planning to build a monorail along a similar route with seven underground stations.

Ankara, Turkey (5)
4.3 million. 1997. 9 miles. 12 stations. 5.3-mile, eleven-station Ankaray light metro opened in 1996. With plans for many extensions, investment in heavy suburban commuter rail, and a monorail, Ankara might have an urban-transport network totalling 80 miles by 2015.

5.

Antwerp, Belgium (6)
529,000. 1975. 62 miles (4.7 in tunnel). 11 underground stations. A "pré-métro" system with streetcars running in tunnels under this, Europe's petrochemical capital. Other lines were partially constructed in the 1980s but never finished.

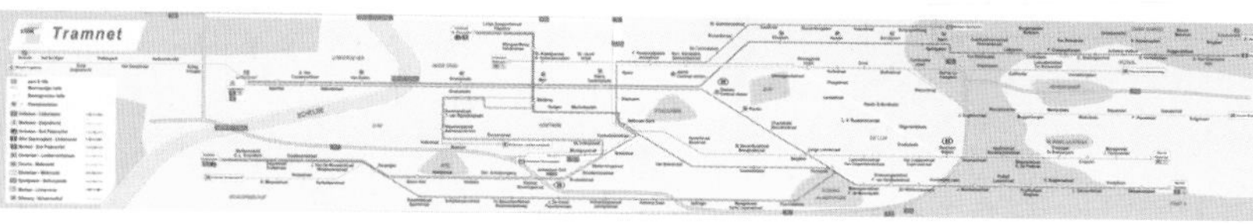

6.

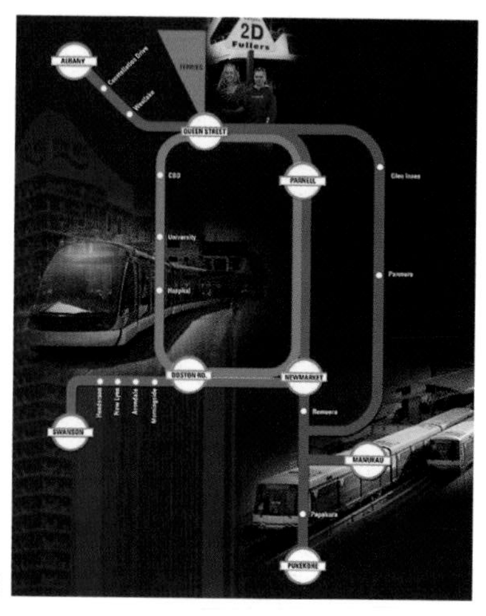

7.

Auckland, New Zealand (7)
1.2 million. Proposed for 2009. 14.3 miles. 20 stops. A plan to build mass transit here has been mothballed but the diagram was promising.

Baghdad, Iraq
5 million. Proposal to build a subway is unlikely to happen at the moment.

8.

Baku, Azerbaijan (8)
1.9 million. 1967. 18 miles. 20 stations. After a fire killed 286 people in 1995, there are plans for a 32-mile network. Many stations are in the typically ornate, grand Soviet style.

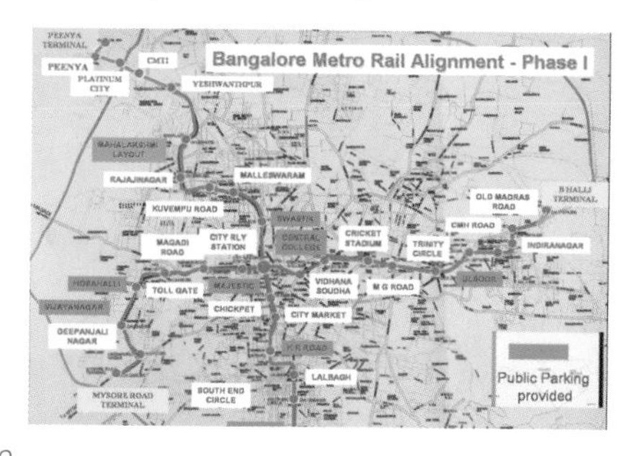

9.

Bangalore, India (9)
5.7 million. 18.6 miles of urban rail planned, including 4 miles of subway.

Belgrade, Serbia (10)
1.7 million. First section of an 18.6-mile light rail with subway sections is due soon to complement the existing tram and S-Bahn, which already have two underground stations.

Belo Horizonte, Brazil (11)
5.3 million. 1987. 18 miles. 19 stations.

10.

(***Belo Horizonte** cont.*) Brazil's third-largest city is building a second line.

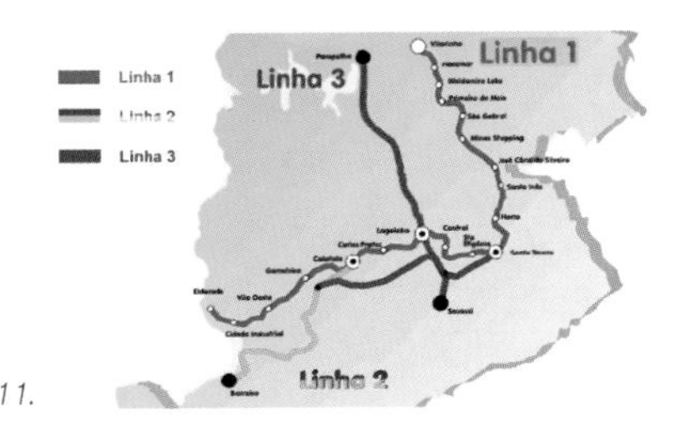

11.

Bielefeld, Germany (12)
325,000. 1971. 21 miles. 59 stations (7 underground). The 1970s oil crisis ended subway plans just as the first underground station was ready. The tunnel didn't open until 1991.

Birmingham, United Kingdom (13)
2.5 million. 1999. 12.4 miles. 23 stops. Typical UK modern light rail mostly on old alignment, with aspirations for many extensions as the diagram shows.

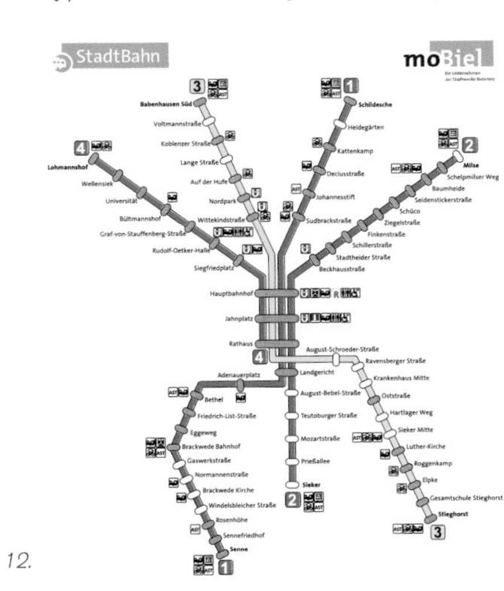

12.

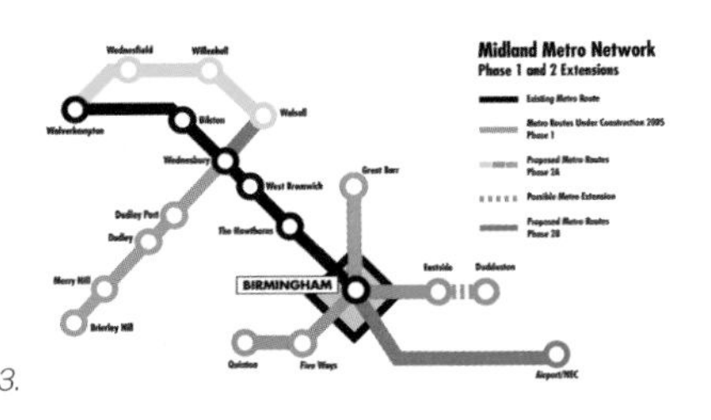

13.

Bologna, Italy
374,425. Plans revised in 2005 for a mini-metro and suburban rail upgrade.

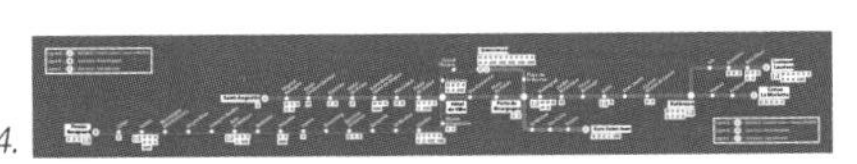

14.

Bordeaux, France (14)
925,253. 2003. 13.2 miles. Modern tram system with innovative ground-level power supply downtown.

Brasilia, Brazil (15)
1.7 million. 2001. 19.8 miles. 14 stations. Brazil's expanding capital will get more stations when the Ceilandia spur is finished.

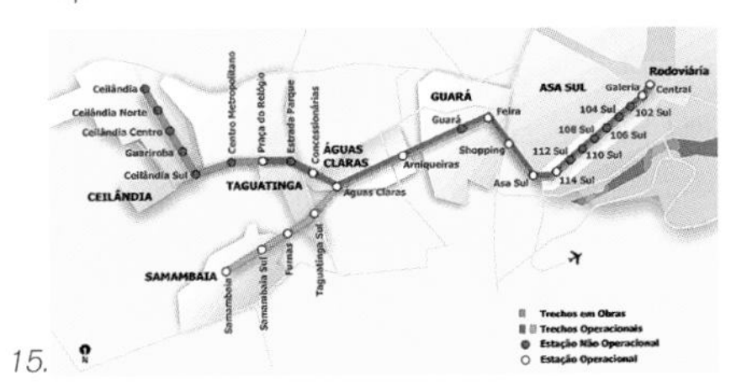

15.

Bratislava, Slovakia (16)
620,000. First 6.8 miles of a two-line, French VAL–type, automated metro could start construction soon.

16.

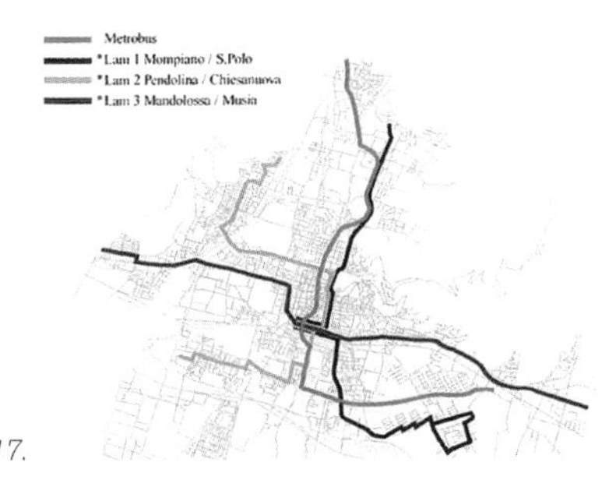

17.

Brescia, Italy (17)
1.1 million. Opens 2009. 11.1 miles. Twenty-three stations on a three-line system using automated cars.

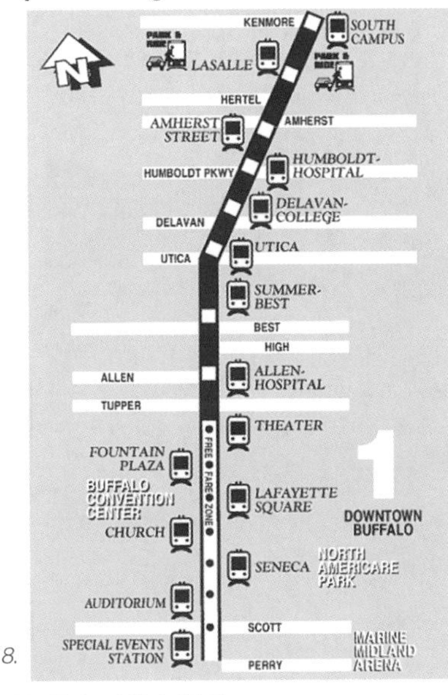

18.

Buffalo, USA (18)
1.1 million. 1984. 6.2 miles. 16 stations. Planned as an 11-mile link between north and south SUNY campuses via Main Street but never completed, giving a truncated air. Extension plans remain to Amherst and Niagara Falls. Unusually, tunnels are in the suburbs; cars run on the surface downtown.

Bursa, Turkey (19)
2.4 million. 2002. 10.5 miles. 17 stations. Fast-growing city has high hopes for its Bursaray; there are plans for up to 31 miles of lines.

19.

Busan, South Korea (20)
3.6 million. 1985. 45.4 miles. 86 stations. This seaport and second city is building a vast transit system.

20.

Calgary, Canada (21)
991,759. 1981. 26.2 miles. 36 stations. Light rail with metro-like high-level platforms, diagram, and frequency. System uses wind-generated clean energy to power the vehicles.

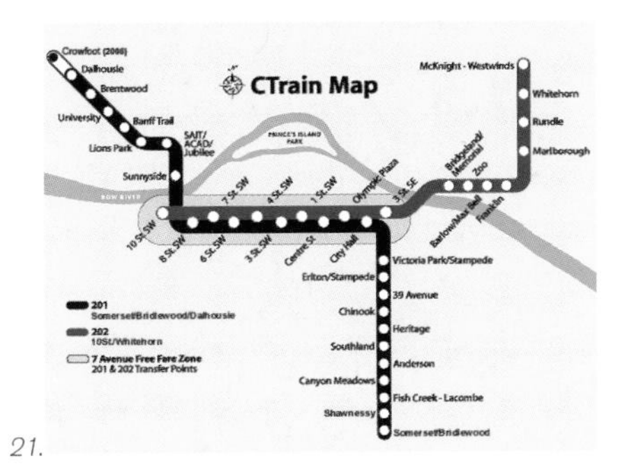

21.

Caracas, Venezuela (see map 1, p. 130)
4 million. 1983. 28 miles. 40 stations. Evenly spaced stations are distorted from their true geographical distances, resembling pieces of Meccano. The

Caracas–Daejeon

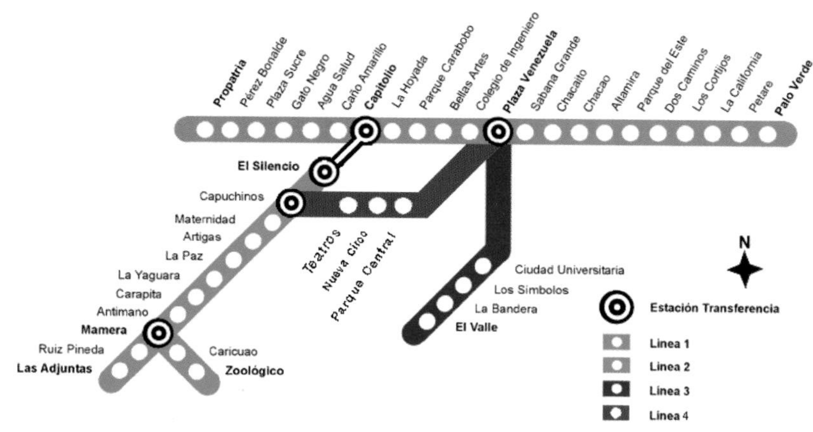

1.

(*Caracas cont.*) black circle with white center appears in another variant: as a two-ringed bull's-eye. Line 3 will grow by 8 miles and Line 4 will double to 7.45 by 2010.

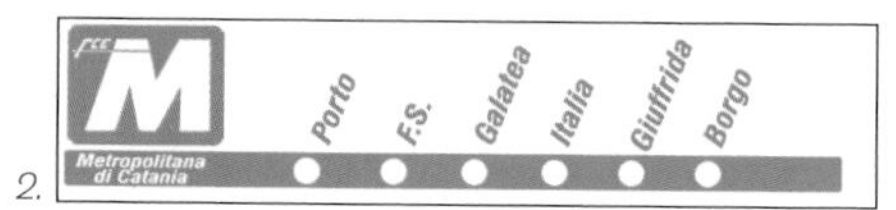

2.

Catania, Italy (2)
1 million. 1979. 2.5 miles. 6 stations. Extensions to this small line will link the center from Galetea to Giovanni XXIII by the main FGC rail station and downtown at Piazza Stesicoro. An existing line being upgraded may soon reach the airport.

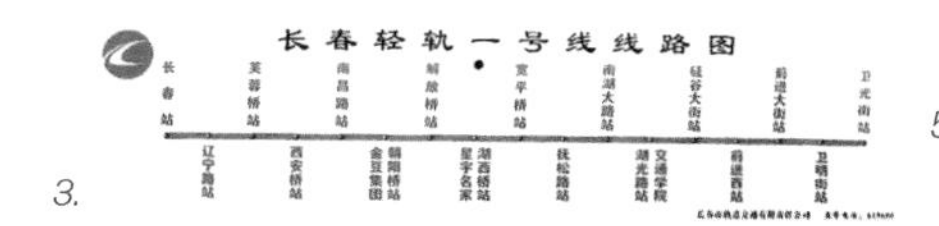

3.

Changchun, China (3)
6.8 million. 2006. 9 miles. 17 stations. Full loop of this 12-mile light rail still under construction. Elevated with a .75-mile underground section, it starts at the main railway station.

Charleroi, Belgium (4)
580,000. 1976. 10 miles. 20 stations. Total "pré-métro" network will be 15.5 miles when new, previously unopened sections are ready.

Chelyabinsk, Russia (5)
1.1 million. May open 2012. 18.6 miles. 28 stations. A stylish diagram, but the project is delayed.

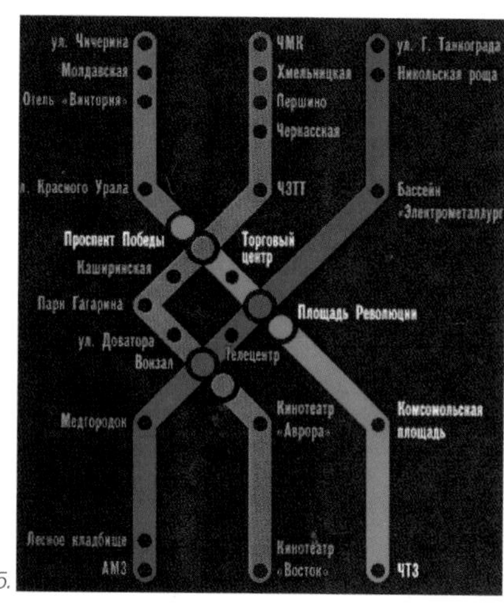

5.

Chengdu, China
9.8 million. Over 61 miles of subway are planned here on up to seven lines. The first 9.3 miles of Line 1 due 2010.

Chennai (Madras), India (6)
5.4 million. 1997. 9.3 miles. 14 stations. MRTS plans a 21.7-mile route, including 6.8 from Velachery to Mandavali and Tiruottiyur, which will form an inner circle, plus a spur from Anna Nagor to Villivakkam. Mostly elevated or at surface level.

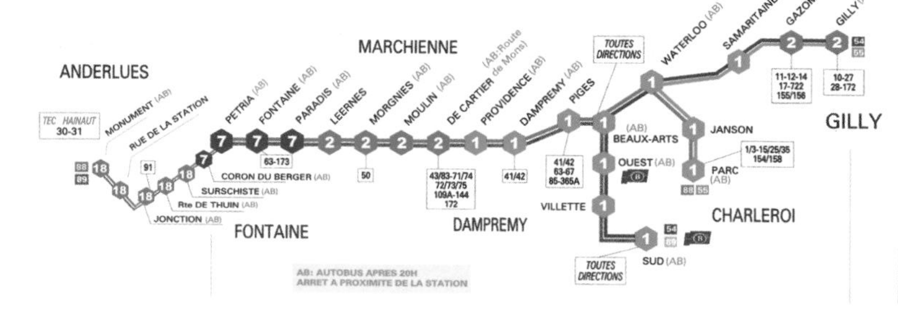

4.

Chiang Mai, Thailand (7)
700,000. Four-line elevated network with short tunnel section planned.

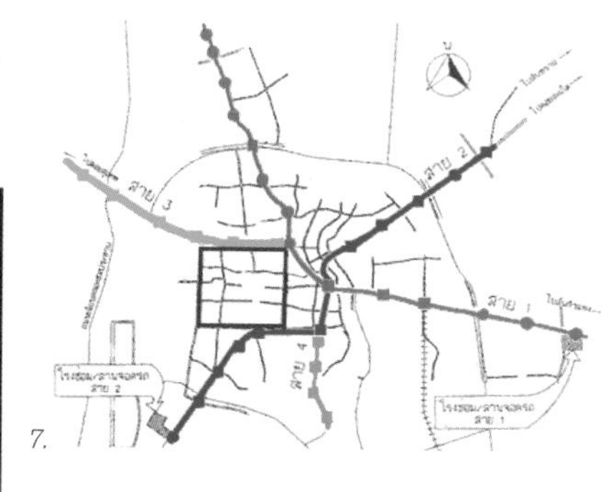
7.

Chongqing, China (8)
14 million. 2005. 12.1 miles. 18 stations. Streets are lined with elevated track pillars and huge holes descending into labyrinthine works while five lines and forty-five stations are built.

Cracow, Poland
1.2 million. A streetcar tunnel with two subway stations will form the central spine of two rapid tram lines, due 2008.

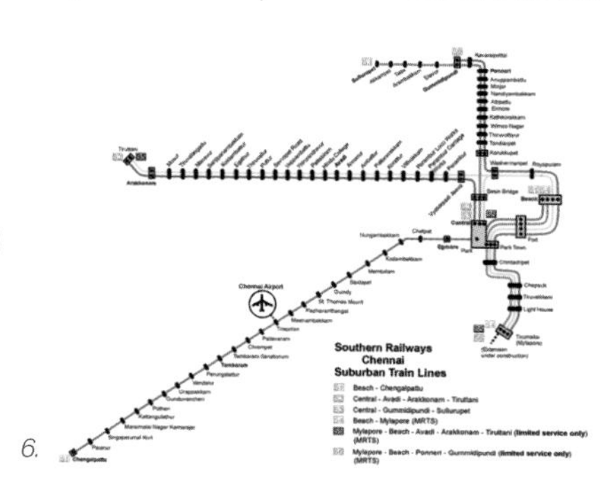

6.

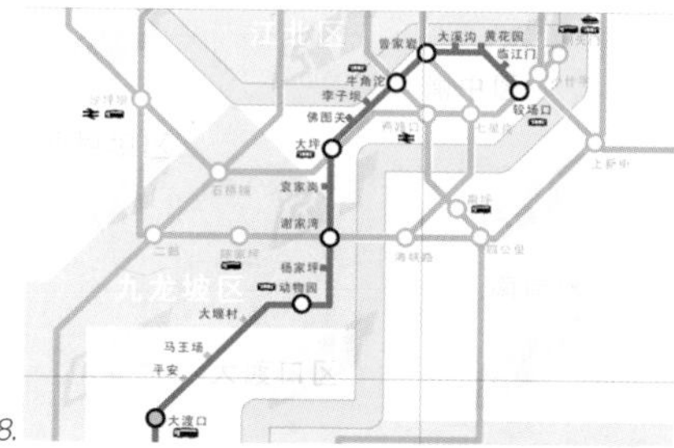
8.

Daegu (Taegu), South Korea (9)
2.5 million. 1997. 33 miles. 55 stations. The first two lines of the planned six-line, 99-mile system are now open, with Line 3 well on the way.

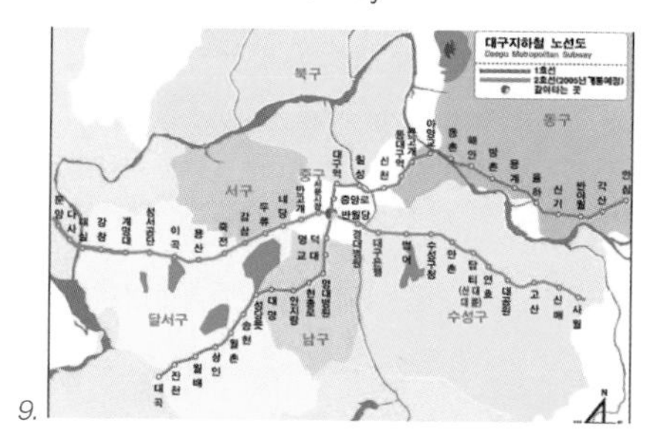

9.

Daejeon (Taejon), South Korea (10)
1.3 million. 2006. 7.7 miles. 12 stations. Part of a proposed five-line, 63.3-mile network has just opened.

10.

Unless stated, all maps by kind permission of and copyright the system operator © 2007.

Dallas, USA (11)
5.8 million. 1996. 45 miles (3.2 in tunnel). 35 stations. A varied collection of diagonals and key roads are shown on this schematic. Eventually, a 66-mile light-rail network is planned.

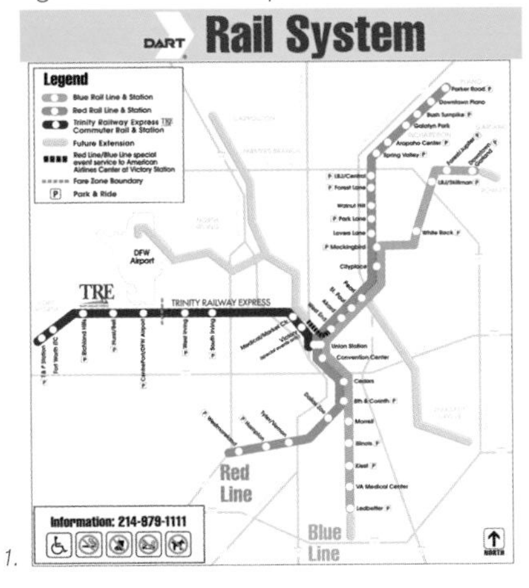

11.

Denver, USA (12)
2.3 million. 1994. 34.9 miles. 36 stations. Light-rail system called TheRide, with big expansion plans.

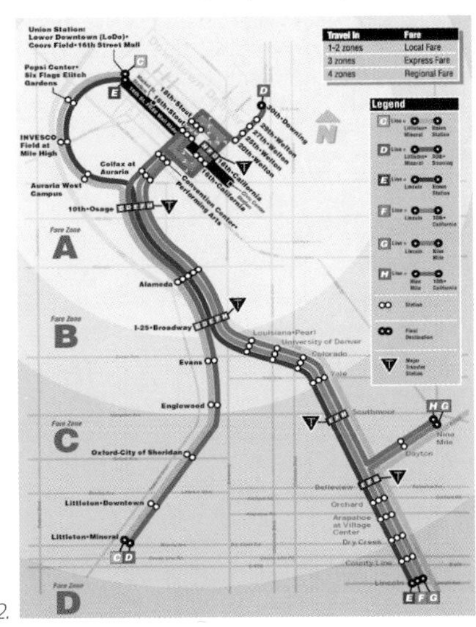

12.

Detroit, USA (13)
4.4 million. 1987. 3 miles. 13 stations.

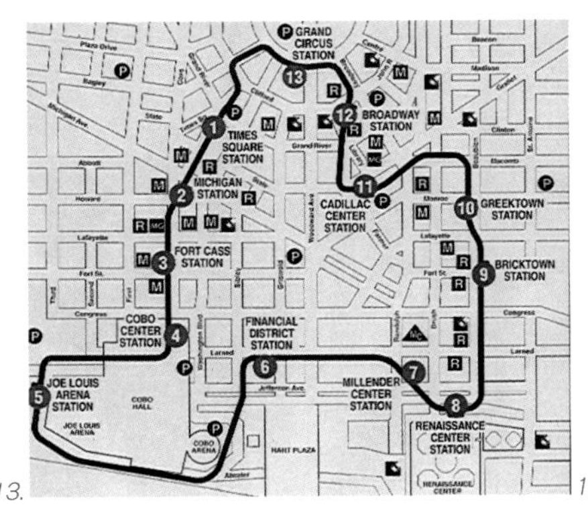

13.

Automated, elevated people-mover in downtown Detroit.

Dnipropetrovs'k, Ukraine (14)
1.3 million. 1995. 6.8 miles. 6 stations. Line 1 was planned from Komunarivs'ka in the west to Oktyabrskaya in the east, but has not quite made it and currently terminates at Vokzalnaya by the main rail station. The rest is being built and is due in 2009.

14.

Donets'k, Ukraine (15)
1.1 million. The first 3.3 miles and five stations of a projected 31-mile, forty-four station network in this former mining heartland are now finally under construction.

15.

Dubai, UAE (16)
1.3 million. A new two-line automated network of up to 31 miles is proposed. Thirty-seven stations, with elevated and subway sections, are expected by 2010.

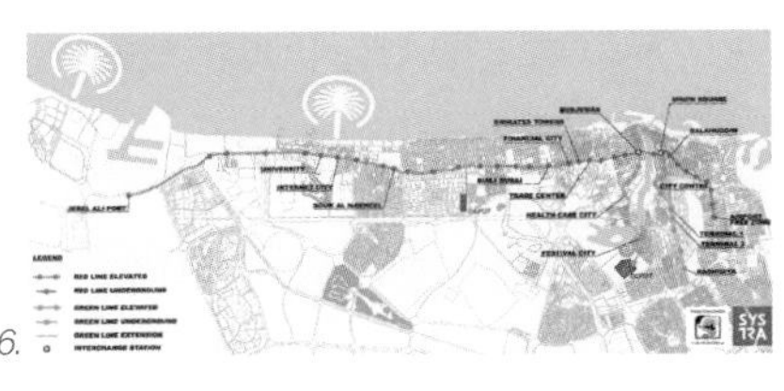

16.

Edinburgh, Scotland (17)
457,830. First 16-mile section of new light-rail tram expected 2010. Total network will be 24 miles.

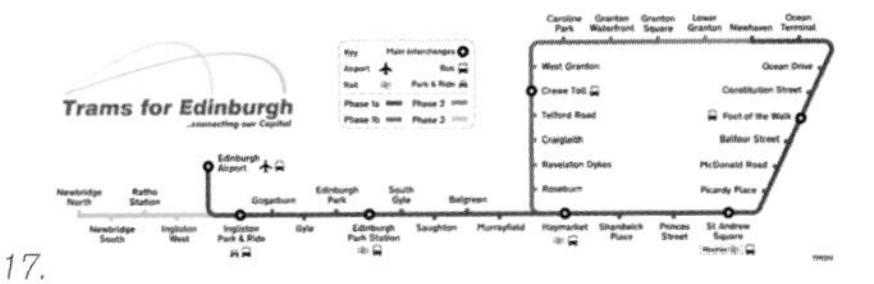

17.

Edmonton, Canada (18)
930,000. 1978. 8.1 miles. 10 stations. "Pré-métro"-style light rail with six central-area stations in subway section.

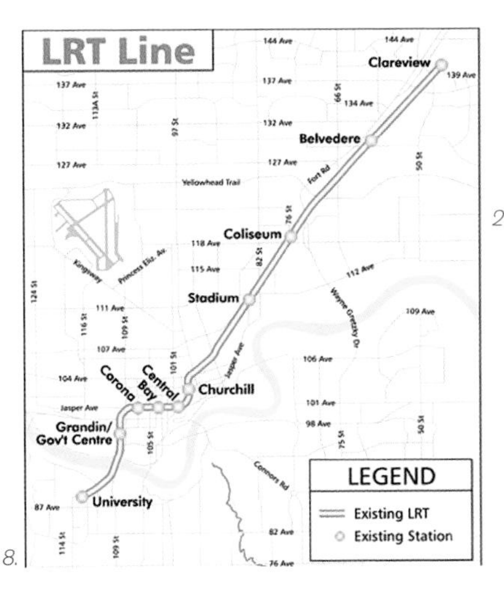

18.

Esfahan, Iran (19)
2 million. 7.7 miles. 15 stations under construction. The whole line will be underground.

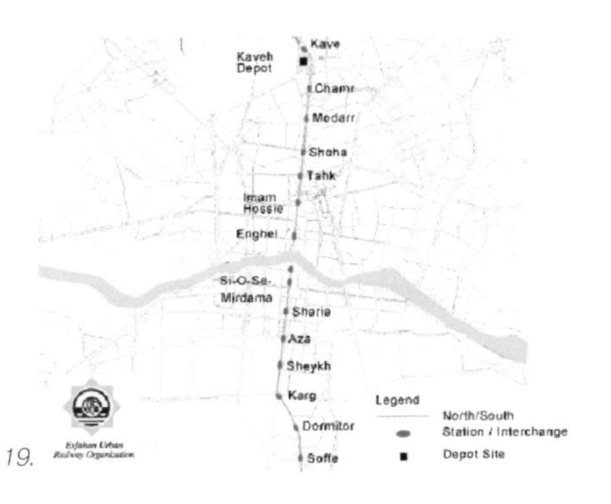

19.

Fortaleza, Brazil (20)
2.4 million. Opens 2008. 7.7 miles. 7 stations. Conversion of a freight line into a double-track metro is the backbone of a 26.7-mile system. Downtown, 2.5 miles will be underground.

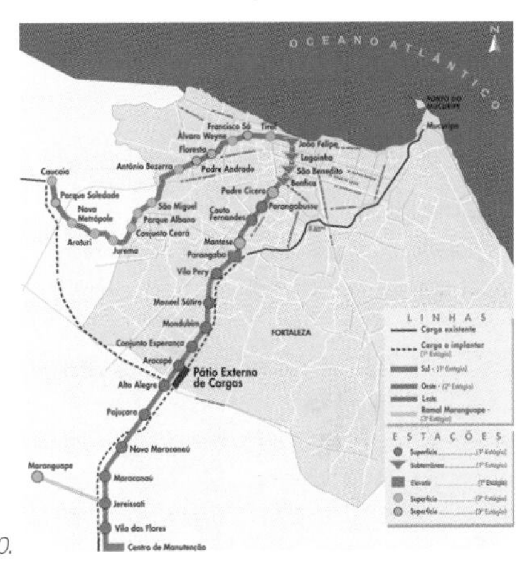

20.

Fukuoka, Japan (21)
1.3 million. 1981. 25 miles. 35 stations. All but two stations are underground. Unusually, the diagram uses pictograms of attractions nearby as station markers.

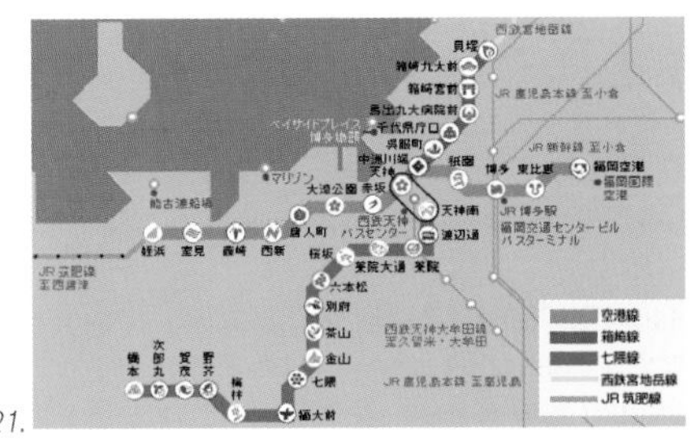

21.

Genoa, Italy (1)
800,000. 1990. 3.4 miles. 7 stations. Dinegro to Brin in the former Certosa tram tunnel opened just in time for the World Cup.

1.

Goiânia, Brazil (2)
1.2 million. 7.2 miles. North–south elevated line with some subway. Due 2012. The first part being built will run between Terminal do Cruzeiro do Sul and Terminal Rodoviário.

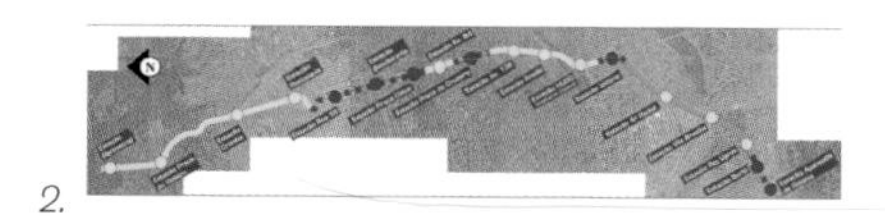
2.

Gothenburg, Sweden (3)
879,000. 72.7 miles. 131 stops (2 underground). Europe's largest tram system in Sweden's second city has plans for more extensions.

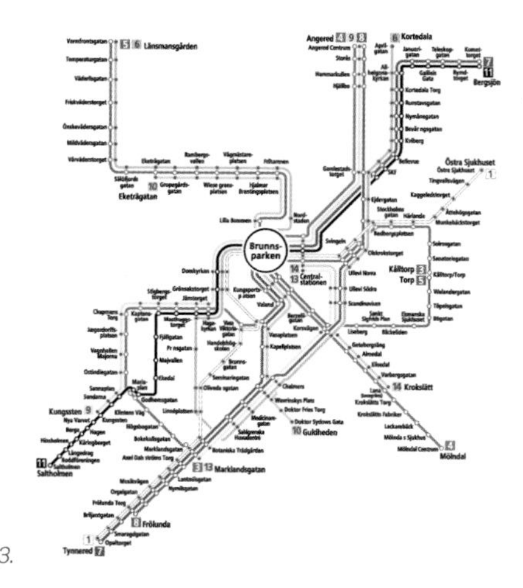

3.

Grenoble, France
362,792. 1987. 20-mile, three-line tram.

Guadalajara, Mexico (4)
4 million. 1989. 15 miles. 30 stations. Line 2 and central area of Line 1 are underground. Both are light rail. Three lines planned for 2010.

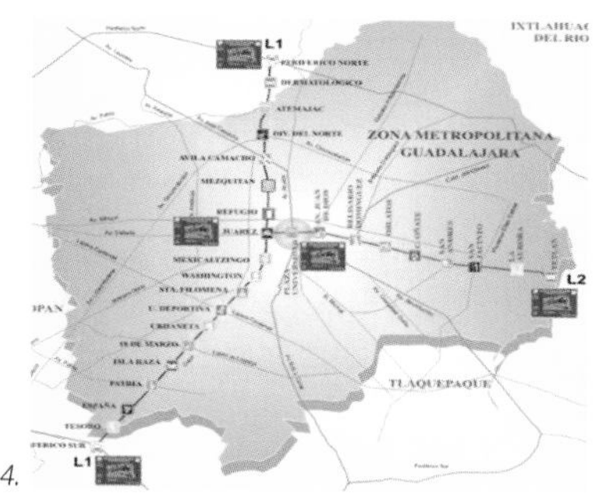

4.

Gwangju, South Korea (5)
1.4 million. 2004. 7 miles. 14 stations. The first part of Line 1 recently opened in what will eventually be a five-line network. Most maps are rich and geographically based, but this diagram mainly shows Line 1.

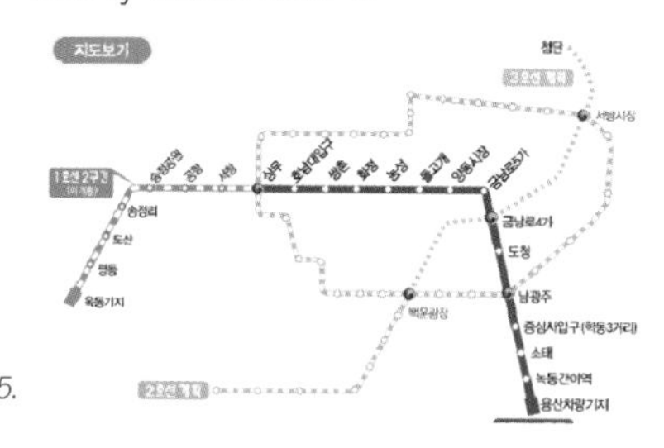

5.

The Hague, Netherlands
975,000. 2004. 0.8 miles. Tunnel added to 71-mile tram network to form part of RandstadRail. Zoetermeer Line converted from heavy to light rail with new stops and 1.5-mile branch to Oosterheem. Rotterdam–Den Haag Line converted from heavy rail to metro, and after completion of 1.8-mile tunnel in 2008, RandstadRail will be directly linked to the Rotterdam Metro.

Haifa, Israel (6)
830,000. 1959. 1.2 miles. 6 stations. The commercial area is on a narrow coastal strip, industry is halfway up a steep rise, and residential areas are higher up the cliffs, so funicular Carmelit Subway was built inside the mountain to link the three tiers.

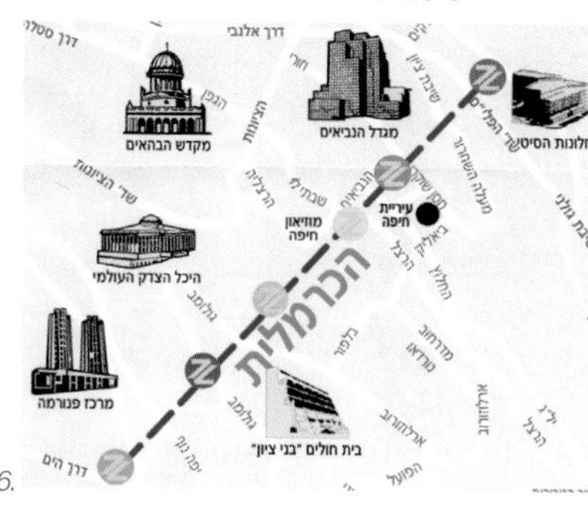

6.

Hangzhou, China
6.4 million. Planning 172 miles on eight lines of subway. First is due by 2010.

Hanoi, Vietnam
3.1 million. Plans for a 9-mile light-rapid system with 1.5 in tunnel by 2010.

Helsinki, Finland (7)
850,000. 1982. 13 miles (5 in tunnel). 16 stations. Plans for a long westerly extension are now approved. The full network could be more than 50 miles by 2025.

Hiroshima, Japan (8)
2.8 million. 1997. 9.9 miles. 21 stops. The Miyajima Line is a large horseshoe-shaped monorail, mostly elevated but with three underground stations in the center.

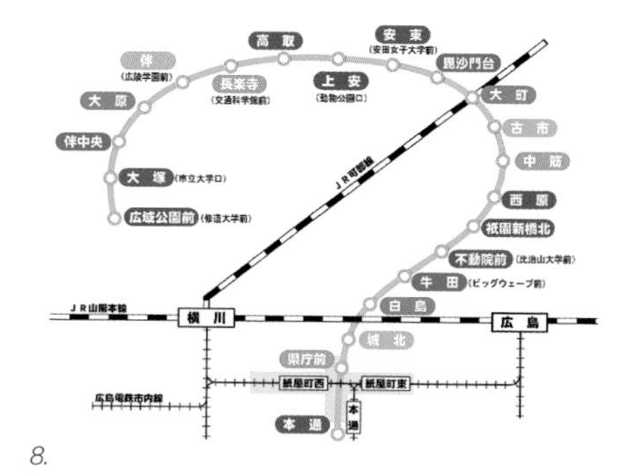

8.

Ho Chi Minh City, Vietnam
3.3 million. Planning a citywide network of 124 miles by 2020.

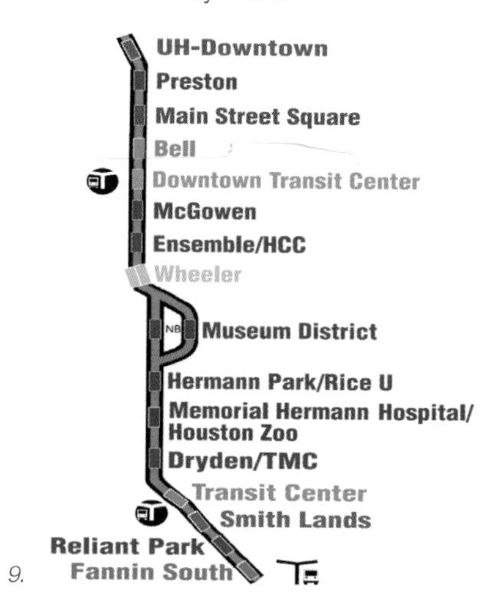

9.

Houston, USA (9)
4.2 million. 2004. 7.4 miles. 16 stops. For an urban area with such a huge

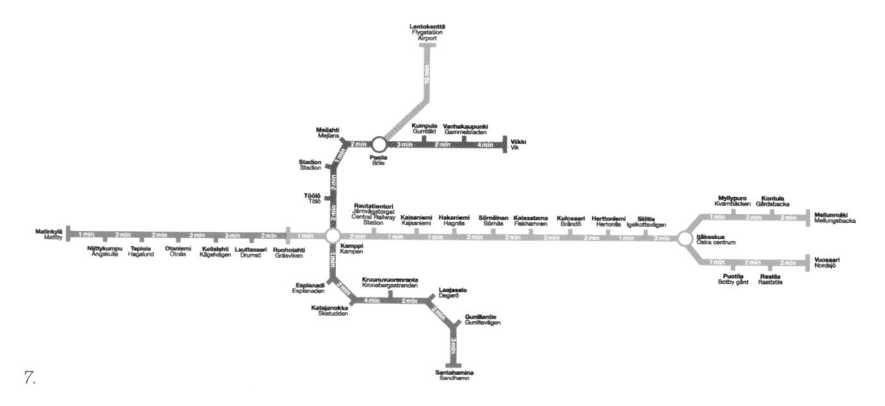
7.

Unless stated, all maps by kind permission of and copyright the system operator © 2007.

population, it lagged behind most in public transport, but MetroRail should be 72 miles by 2025.

Incheon, South Korea (10)
3.8 million. 1999. 15.5 miles. 23 stations. Growing so fast it has its own 18.6-mile branch from the Seoul Subway and now needs a separate network, which will eventually be 77.7 miles on five lines.

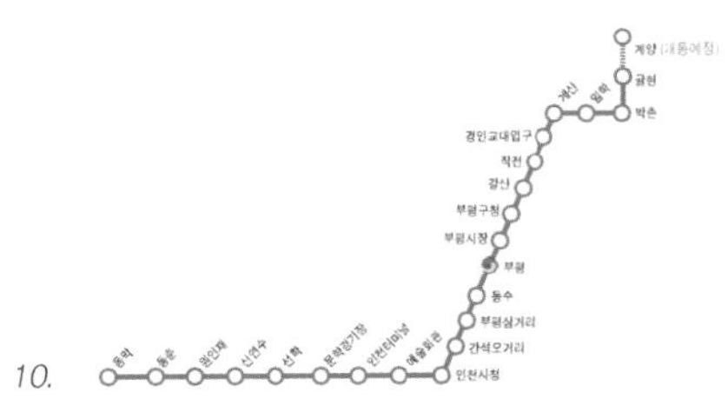

10.

Istanbul, Turkey (11)
11.5 million. 2000. 4.3 miles. 6 stations. Tunneling under violent earthquake zones meant the first full metro from 4 Levent to Taksim didn't open until 2000. The next stretch, over the Golden Horn bridge and underground through the old city, is under way. The 11-mile light-rail Aksaray opened in 1989 and has an underground section. A funicular subway called Tünel opened in 1875.

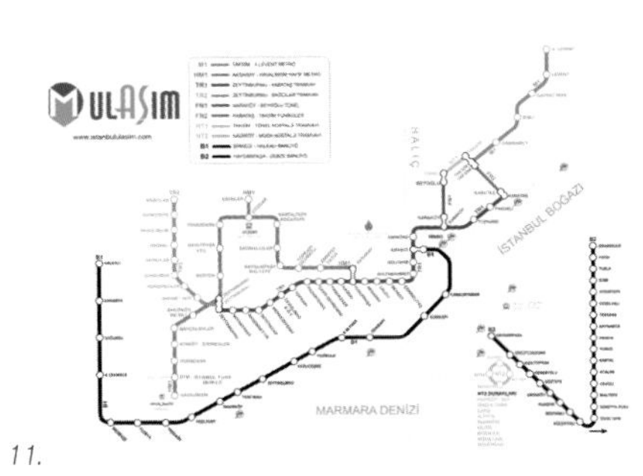

11.

Izmir, Turkey (12)
3.5 million. 2000. 6.8 miles. 10 stations. The first section runs between Uçyol and Bornova, but the full network will be 35 miles. Three stations are underground.

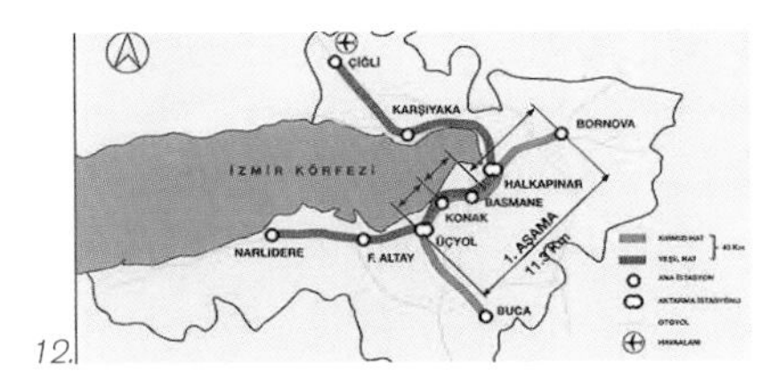

12.

Jakarta, Indonesia (13)
11 million. Due to open soon. 16.7 miles. 29 stations. Building an elevated monorail due to the high water table and flood risks.

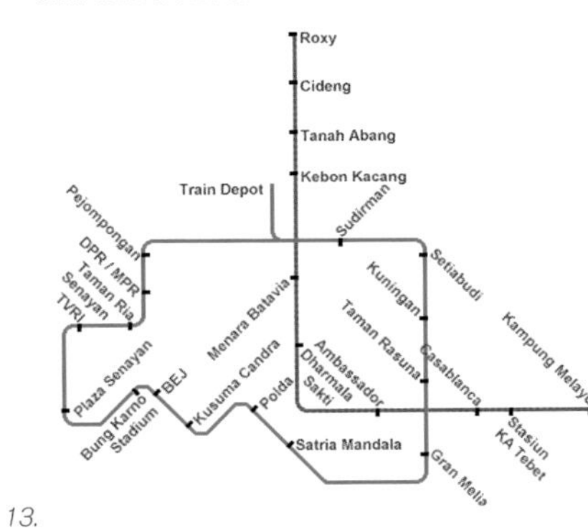

13.

Jerusalem, Israel (14)
650,000. Plans a light-rail system with subway sections and street running.

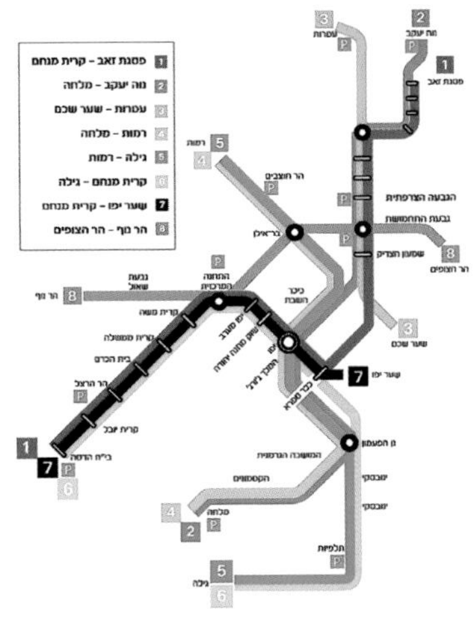

14.

Kao-hsiung, Taiwan (15)
3 million. 2007. 17.6 miles. 24 stations. A network stretching to almost 49.7 miles with more than seventy stations is proposed. Following construction problems on the underground section of the Orange Line, the Red Line will now open first.

15.

Karlsruhe, Germany (16)
283,959. Pioneer in radical urban-rail transit policies, with a fully integrated light- and heavy-rail network, now has plans to build a tram/road tunnel.

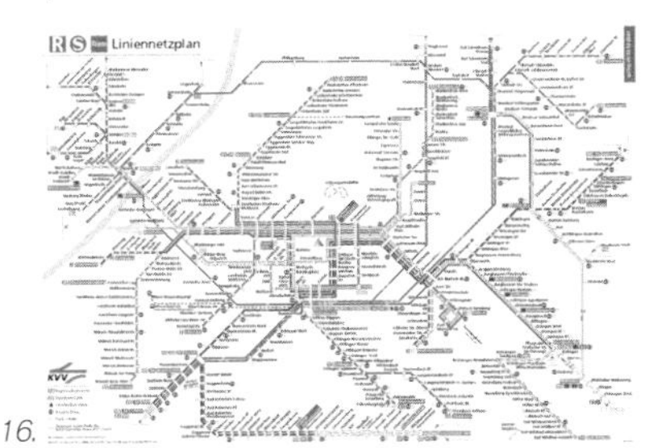

16.

Kawasaki, Japan
1.2 million. Construction of a 9.3-mile route has now been shelved.

Kazan, Russia (17)
1.1 million. 2006. 6.8 miles. 5 stations. First part of a master plan to give the Tartarstan capital at least three lines and 28 miles of routes. Initial line was a long time coming to fruition due to finances.

17.

Kitakyushu, Japan (18)
1.1 million. 1985. 5.6 miles. 13 stations. A simple, busy, and effective one-line monorail.

18.

Kobe, Japan (19)
1.5 million. 1977. 19.9 miles (9.3 in tunnel). 25 stations. In addition there is a private metro-like service on the Sanyo Electric Railway along the coastal strip. Kobe uses a black background diagram with two schematic curves to represent the system like a human smiley face.

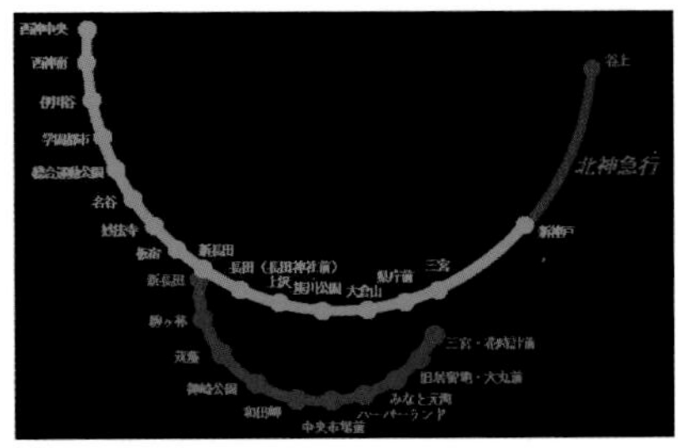

19.

Unless stated, all maps by kind permission of and copyright the system operator © 2007.

Kolkata–Manila

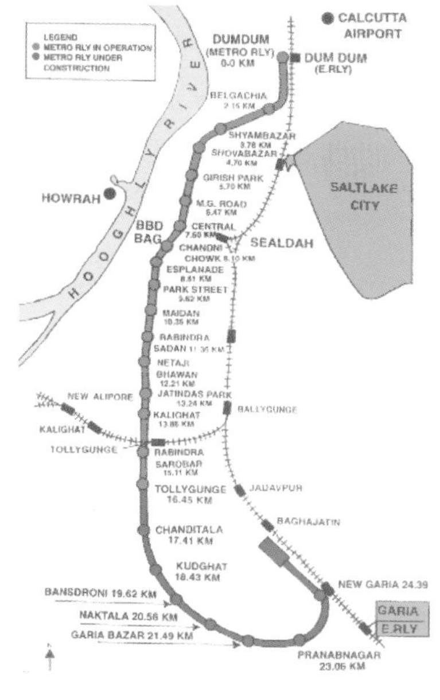

1.

Kolkata (Calcutta), India (1)
10.9 million. 1984. 10 miles. 17 stations. India's first subway. The full network will be 62 miles on five lines. Current project is to extend existing Line 1 by 5.6 miles from Tollygunge to New Garia.

Krasnoyarsk, Russia
917,200. Opening delayed. 3.1 miles. 5 stations. Like most former Soviet cities, Krasnoyarsk has a plan; if all goes well (and so far it has not) Siberia's second-largest city will have a three-line system by 2025.

Kryvy Rih, Ukraine (2)
700,000. 1986. 11.1 miles (4.2 in tunnel). 12 stations. A rapid tram in Volgograd style.

Kyoto, Japan (3)
1.5 million. 1981. 17.9 miles. 29 stations. An 18.6-mile extension is planned between Nagaoka and Rokujizo.

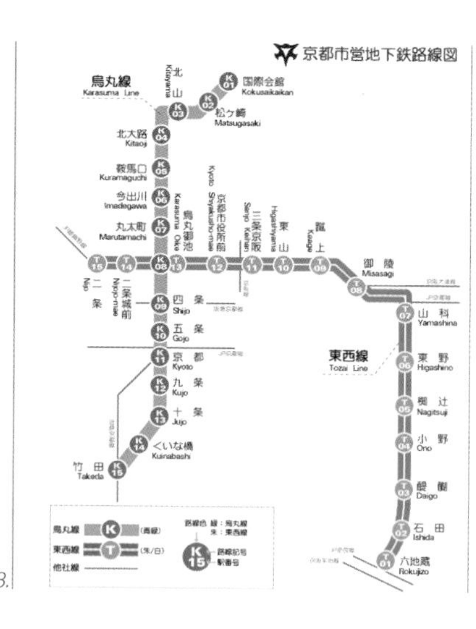

3.

Lagos, Nigeria
17 million. Plans for two lines in what is set to be the world's fifth most populous city are now being revived.

Laon, France
26,265. 1989. 0.9 mile. 3 stations. Mini-metro in small town using cable power.

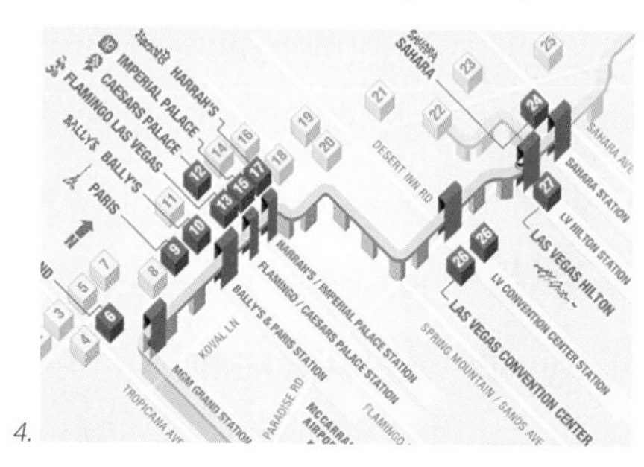

4.

Las Vegas, USA (4)
1.4 million. 1995. 4 miles. 7 stations. A driverless monorail in the Nevada desert boomtown of Las Vegas. Nice 3-D style attempted on map.

Lausanne, Switzerland (5)
244,000. 1991 (1877). 8.8 miles. 15 stations. Fully automated, rubber-tired, light-rail lines, with ten subway stations.

2.

Unless stated, all maps by kind permission of and copyright the system operator © 2007.

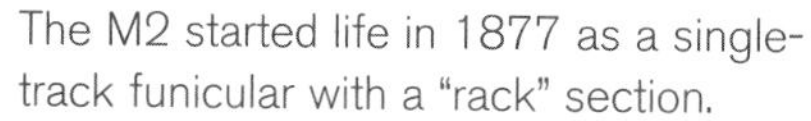

The M2 started life in 1877 as a single-track funicular with a "rack" section.

5.

Lima, Peru (6)
6.3 million. 2003. 6.2 miles. 6 stations. The Peruvian capital has had subway ambitions since at least 1972. The first stage of a modern five-line system has just opened.

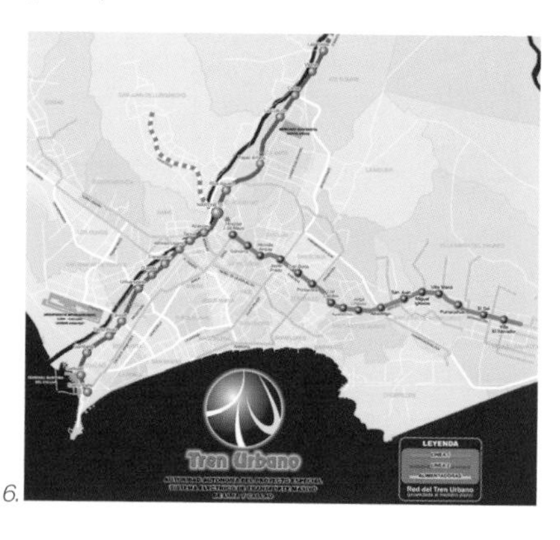

6.

Linz, Austria (7)
203,000. 21 miles. 85 stops. A 1.2-mile tunnel with two underground stations has been added (in 2004) to the extensive Linz tram system. Also has Europe's steepest rack ascension rails.

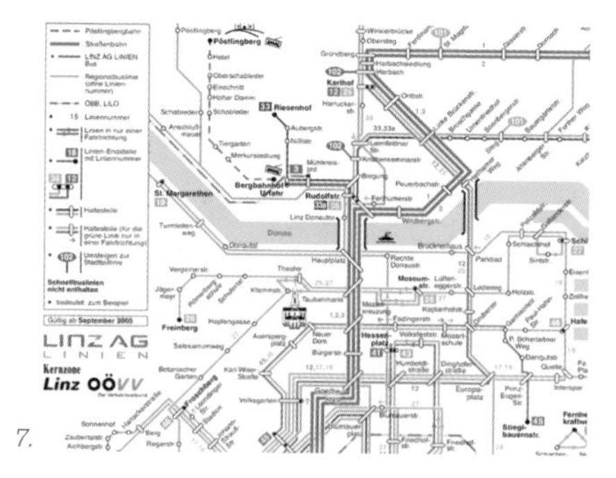

7.

Ludwigshafen-Mannheim, Germany (8) 500,000. 1969. 2.5 miles and 11 stations in tunnel. Together with Heidelberg these cities form the Rhein-Neckar tariff union. Like those in Rhine-Ruhr, Stuttgart, and Frankfurt, this is a light-rail system with many tunnels and underground stations. Some sections were built to "pre-metro" standards, but the full plan was never realized.

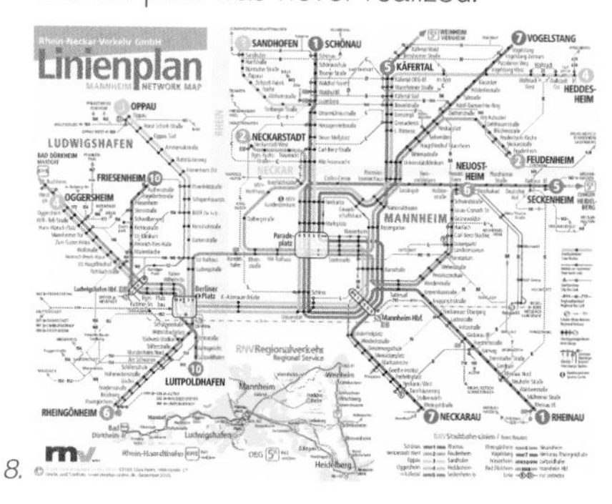

8.

Malaga, Spain (9)
748,000. Up to 18.6 miles. 40 stations. Ambitious plans on the Andalusian coast to furnish Spain's sixth-largest city with modern light rail by 2012 have the green light for the first phase of construction. Up to 90 percent of the first two lines will be underground.

9.

Manila, Philippines (10)
11.8 million. 1984. 28.4 miles. 39 stations. A transport solution to tackle the rapid urban growth was born in the 1970s. The first elevated light-rail line, opened in 1984, was Metrostar. Two more lines have been added, with more due soon.

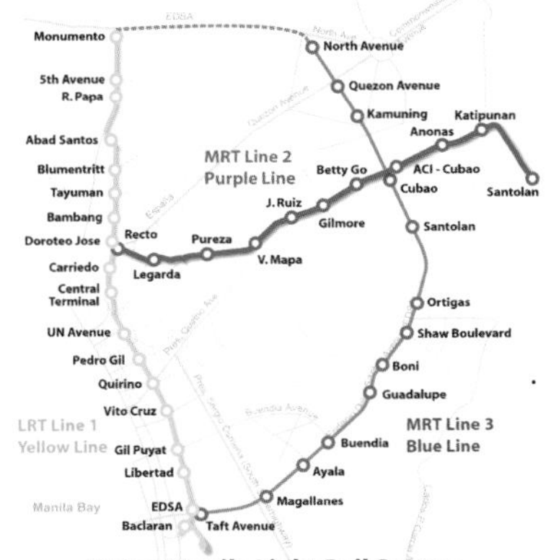

10. **Metro Manila Light Rail System**

Maracaibo, Venezuela (11)
2.7 million. Opens 2008. 3.7 miles. 6 stations. Will be a four-line, 37.2-mile network. Line 1 will have underground sections later.

11.

Mashhad, Iran (12)
3 million. Opens 2008. 11.2 miles. 22 stations.

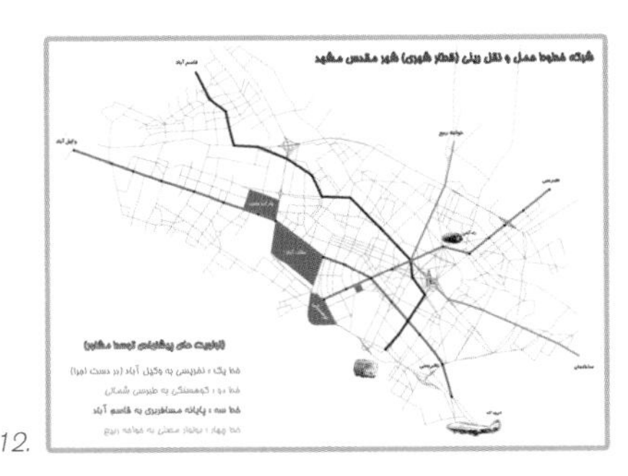

12.

Medellín, Colombia (13)
2.4 million. 1995. 18.1 miles. 25 stations. Extension due.

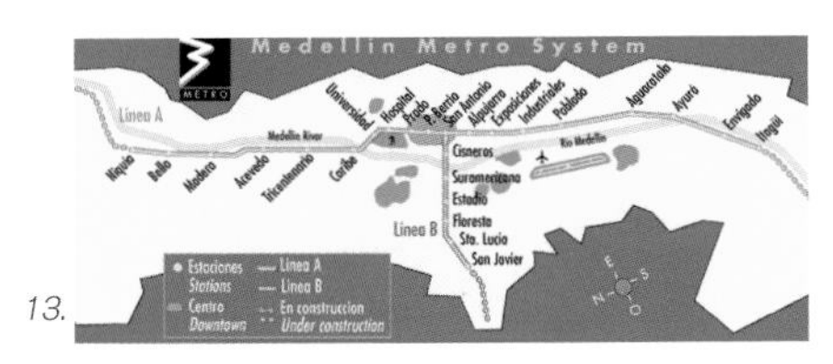

13.

Minneapolis/St. Paul, USA (14)
2.9 million. 2004. 11.9 miles. 17 stops. Light rail called Metro Transit, with an underground station at the airport.

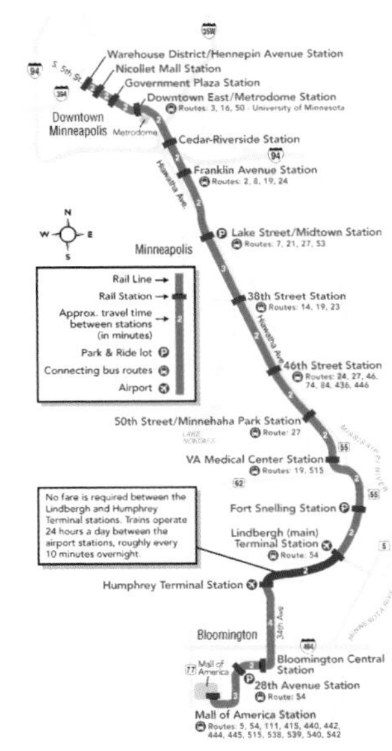

14.

Minsk, Belarus (15)
1.7 million. 1984. 17.1 miles. 22 stations. The first two lines of what could be a four-line system traversing this ancient city are already up and running, and work has started on at least two extensions.

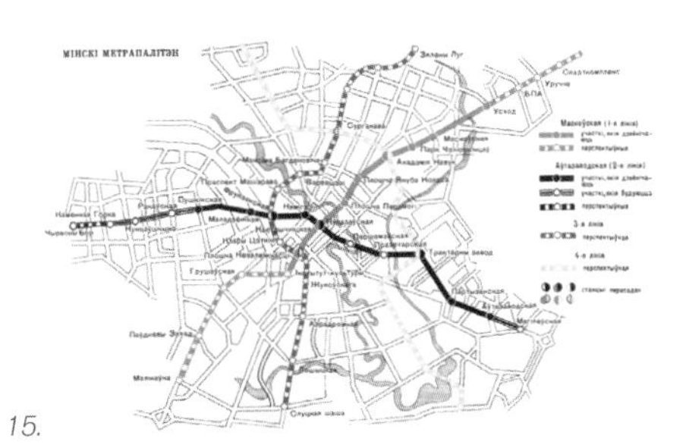

15.

Monterrey, Mexico (16)
2.8 million. 1991. 14.3 miles. 24 stations. Partly elevated, partly underground, a small system set to expand to 49.7 miles serving Mexico's third-largest city.

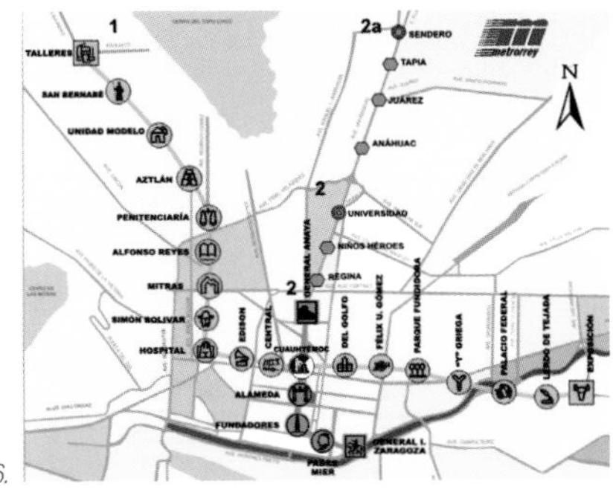

16.

Montevideo, Uruguay (17)
1.5 million. A proposed 28.5-mile system has not yet materialized.

17.

Montpellier, France
225,392. 2000. 21-mile light-rail with two lines and two more planned.

Mumbai (Bombay), India (18)
12.6 million. 1857. Heavy suburban commuter-rail network with metro-like frequency and plans for subway, LRT, and a monorail.

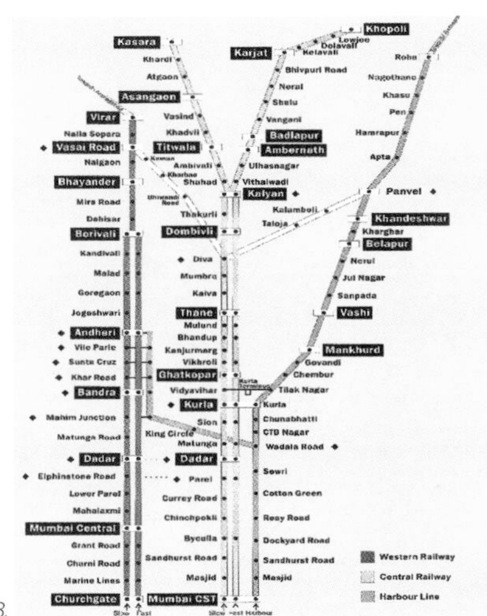

18.

Naha, Japan (19)
300,000. 2003. 8 miles. 15 stations. Totally elevated monorail serving downtown and the airport, proving again Japan's commitment to public transit even in smaller cities.

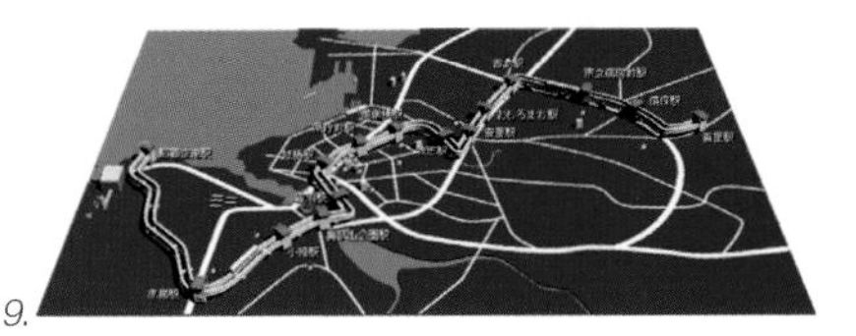

19.

Nanjing, China (20)
5.3 million. 2005. 13.4 miles. 16 stations. Investment on a grand scale will give this massive city up to six new mass-transit lines by 2025.

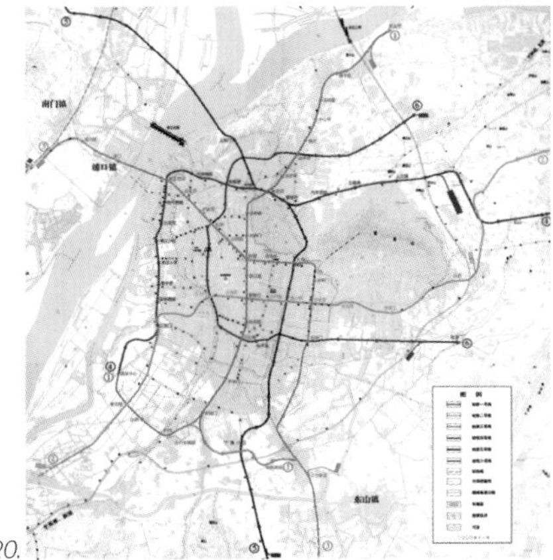

20.

Nantes, France
280,600. 1985. 24-mile, three-line tram.

Nottingham, United Kingdom (21)
640,000. 2004. 8.7 miles. 23 stops. Like most UK light rail, the NET utilizes old rail alignments. There are plans for two more lines to take the total network to over 24.8 miles.

21.

Unless stated, all maps by kind permission of and copyright the system operator © 2007.

Novosibirsk–Rennes

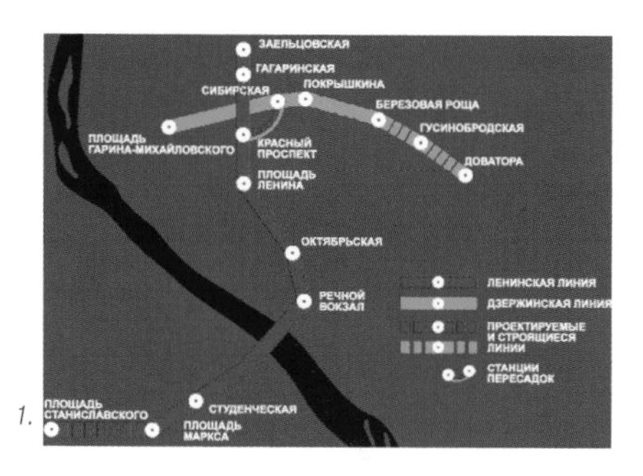

1.

Novosibirsk, Russia (1)
1.7 million. 1985. 8.8 miles. 11 stations. Siberia's biggest city (Russia's third in after St. Petersburg and Moscow) has a harsh environment, with winters in the minus-40s and occasional earthquakes. Another 37.2 miles are planned, on four or possibly five lines, but progress is painfully slow and cash in short supply.

Odessa, Ukraine
1.2 million. An initial line with eight underground stations is planned.

Omsk, Russia (2)
2.1 million. Supposedly due to open 2016 with an initial route length of 5 miles and five stations. Uncertainty hovers over its future: one report says less than a mile of tunnel exists, another that there's construction across the city! Either way it will be a long, cold wait for a train in Omsk.

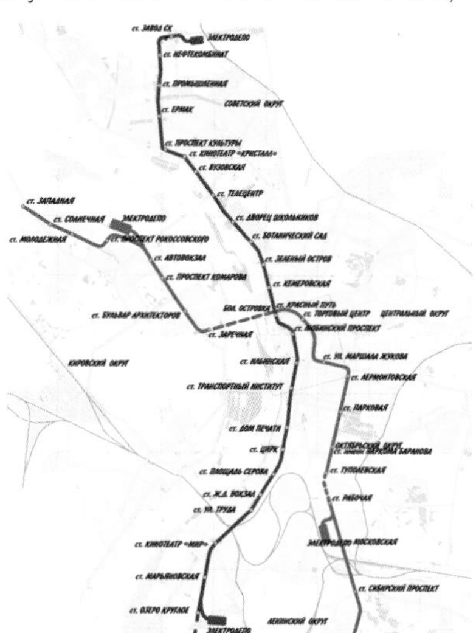
2.

Orléans, France
274,000. 2000. 11.2 miles. 24 stations. Modern light-rail system.

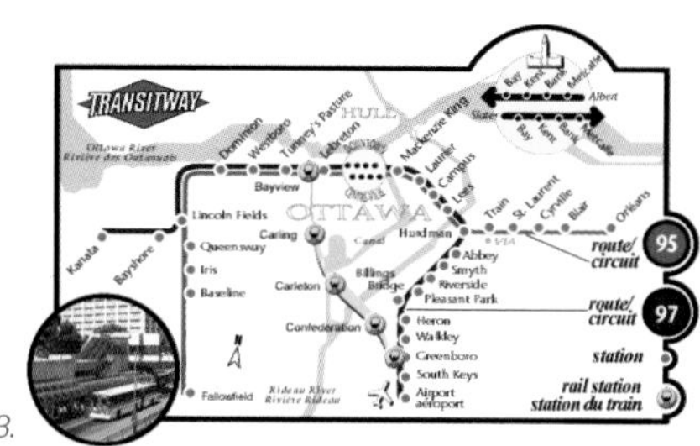

3.

Ottawa, Canada (3)
808,391. 2001. 5 miles. 7 stations. The diesel powered O-Train is a pilot for a full-scale light-rail system that may include some subway sections. A new north-south line is due in 2009.

Palermo, Italy (4)
1 million. 6.2 miles and 11 stops of a light metro system are under construction and due to open by 2009.

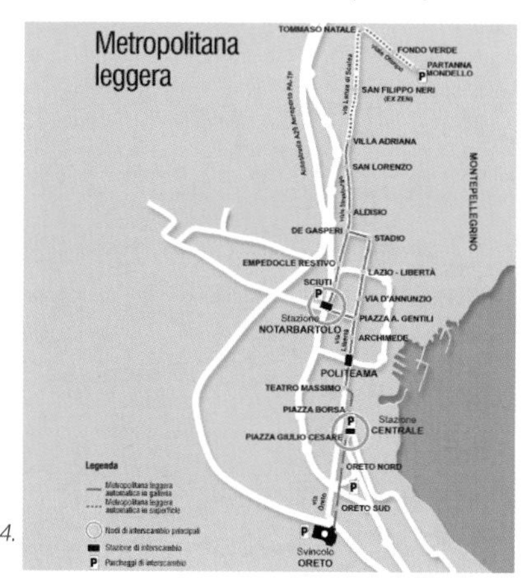

4.

Parma, Italy
175,789. A 7-mile light metro with fourteen stations, half of them underground, due to open around 2010.

Perm, Russia
2.8 million. Work started in 2005 on a three-line system to be ready for 2015.

Phoenix, USA (5)
1.3 million. Light rail opening 2008.

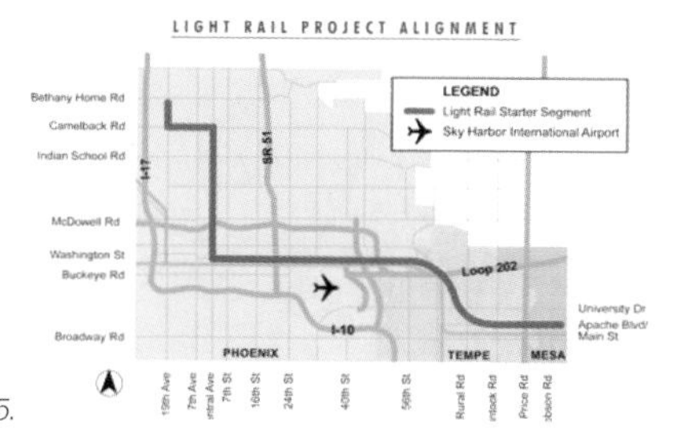

5.

Portland, USA (6)
1.3 million. 1986. 38.5 miles. 56 stations. Born from public objections to a new highway, the MAX (Metropolitan Area Express) is more a light rail than a metro but has a long tunnel under Washington Park, with the deepest U.S. station, 262 feet below the zoo.

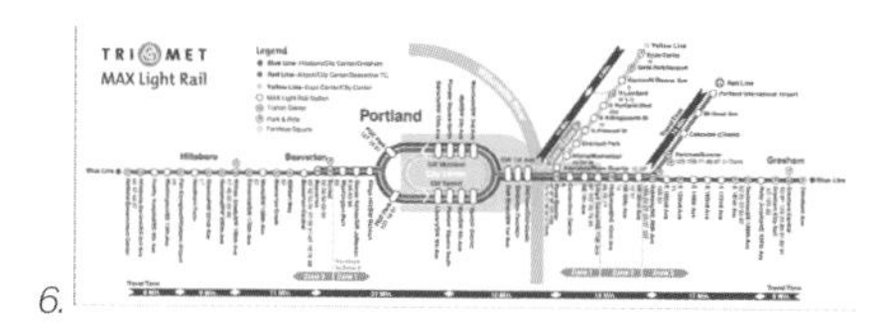

6.

Porto Alegre, Brazil (7)
3.2 million. 1985. 21 miles. 17 stations. Line 1 stretches up the coast to link outlying towns to the center of this sprawling Brazilian metropolis. A second line of 12.4 miles will be built in the center and toward the eastern suburbs.

Poznan, Poland
600,000. A 3.7-mile, six-stop fast tram. City crisscrossed with old trams has recently built an express line.

Pyongyang, North Korea (8)
2.5 million. 1973. 14 miles. 15 stations. Shrouded in secrecy, this is rumored to be one of the most beautiful subways in the world, with some cathedralesque stations of exquisite decor named after phrases from the socialist revolution.

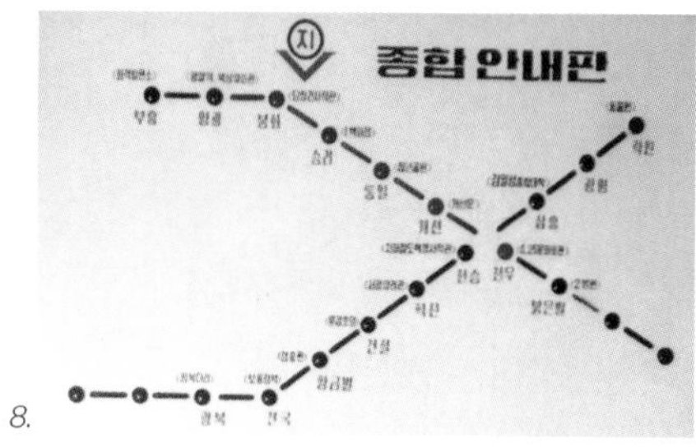

8.

Rennes, France (9)
364,000. 2002. 5.6 miles. 15 stations. Proving to smaller cities across the world that you do not have to be a metropolis to build a metro, Rennes has France's third automatic VAL system. A further east–west line is also on the drawing board.

Rouen, France (10)
720,000. 1994. 9.3 miles. 31 stations. Only a small section of this light-rail

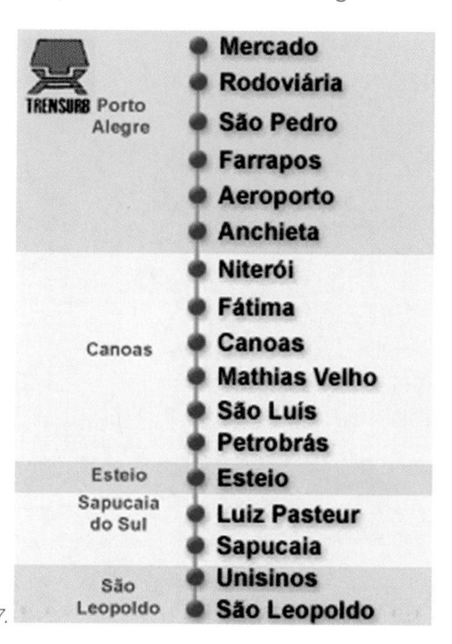

7.

Unless stated, all maps by kind permission of and copyright the system operator © 2007.

9.

system is underground, but it's officially called Métro by the locals and the operator. The map is typically French, with nice reversed-out blue blocks at termini to help identify direction.

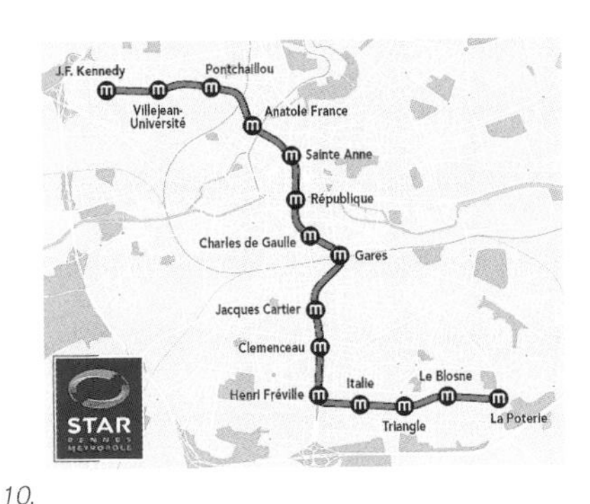

10.

Sacramento, USA (11)
1.2 million. 1987. 37.3 miles. Forty stops of modern light-rail system for California's state capital.

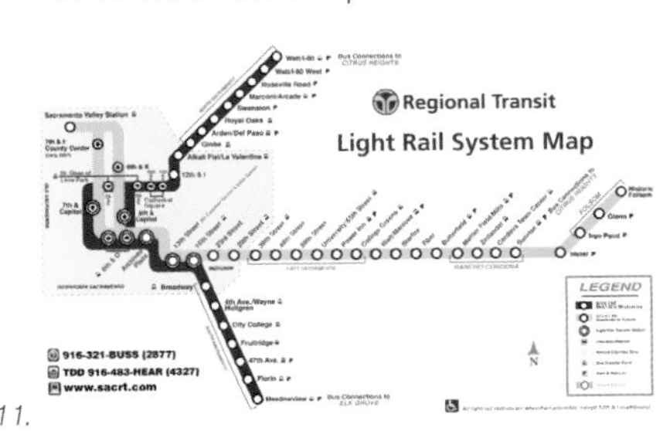

11.

St. Louis, USA (12)
2.7 million. 1993. 37.6 miles. 37 stations (6 underground). More of a light rail, the MetroLink system is built on old rail alignments, tunnels and bridges. Several extensions are now being planned.

12.

Salerno, Italy
150,000. Opens 2008. 4.8 miles. 8 stations. This short route will eventually link directly to the airport.

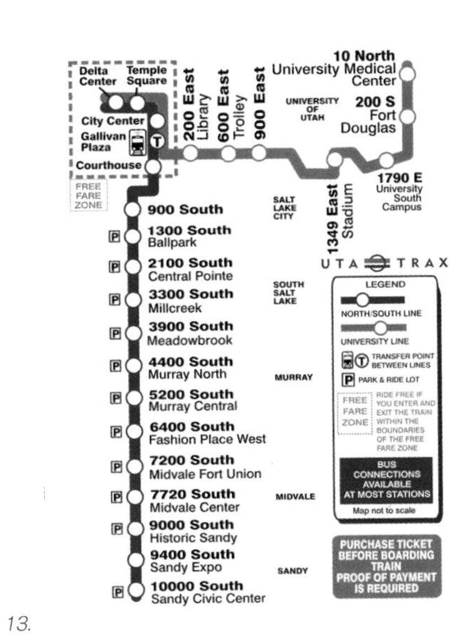

13.

Salt Lake City, USA (13)
1 million. 1999. 18.7 miles. 23 stops.

Salvador, Brazil (14)
2.25 million. 2007. 7.4 miles. 8 stations. Bahia state capital gets mass transit.

Samara, Russia (15)
1.2 million. 1987. 5 miles. 7 stations.

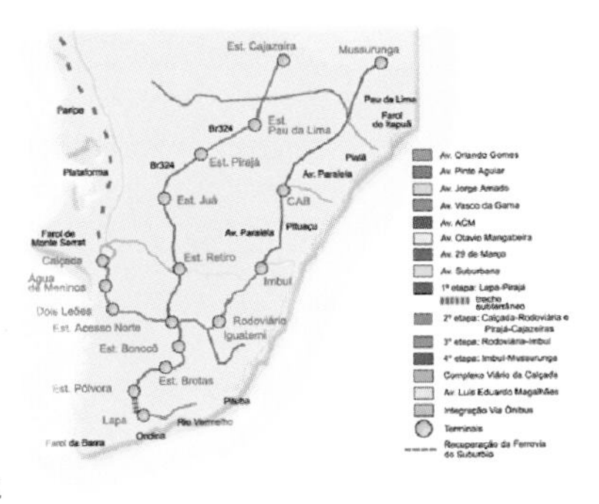

14.

Just one line of a planned four-line network is currently operating.

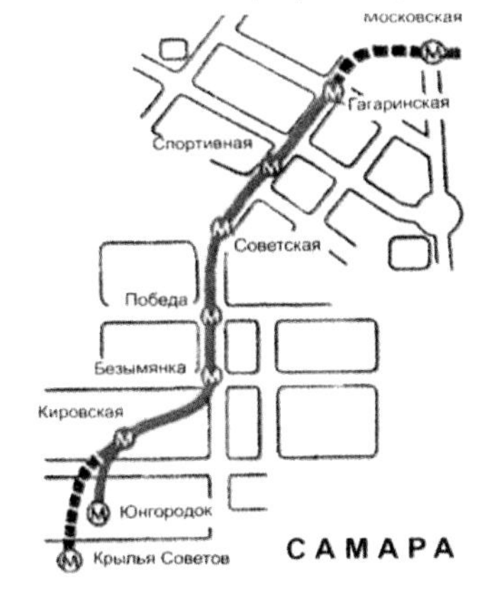

15.

San Diego, USA (16)
2.7 million. 1995. 22.4 miles. 49 stations. Not a true metro but the diagram is very strong, with its black circles and white centers, white-line connectors, and ironing-out of geographic kinks. The smoothed-out Orange Line has the sun for its logo, and the ironed-out Blue Line a wave.

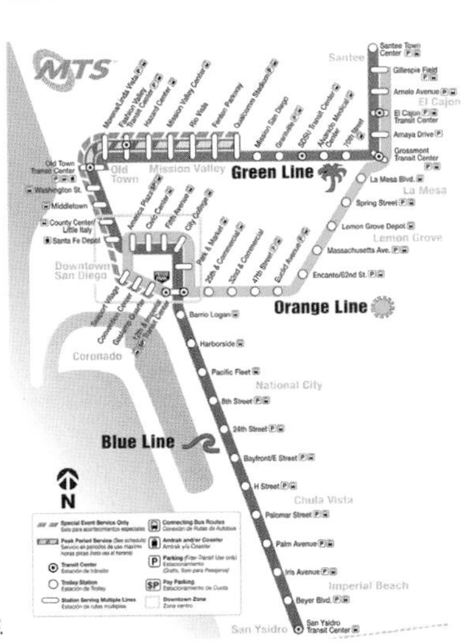

16.

San Jose/Santa Clara, USA (17)
1.7 million. 1987. 29.8 miles. 46 stations. Silicon Valley's older urban centers have a light rapid transit system that is starting 8.4 miles of extensions, at least one section in tunnel. The extensions will increase it to 43.5 miles, which may connect to the southern end of BART (see p. 68) at Fremont.

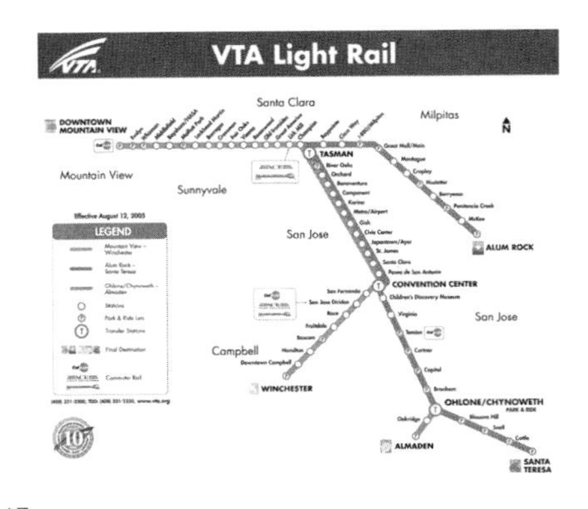

17.

San Juan, Puerto Rico (18)
1.25 million. 2003. 10.7 miles. 16 stations. The biggest construction project in the Caribbean. Extensions to the airport are planned.

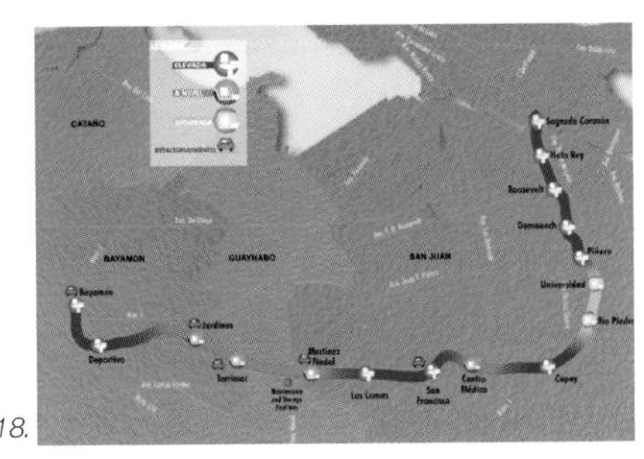

18.

Santo Domingo–Tbilisi

Santo Domingo, Dominican Republic
2.2 million. 10 miles. 15 stations. Apparently due by 2008.

Sapporo, Japan (1)
1.8 million. 1971. 29.8 miles. 46 stations. Japan's fifth-largest city, and the fourth after Tokyo, Osaka, and Nagoya to build a mass-transit system.

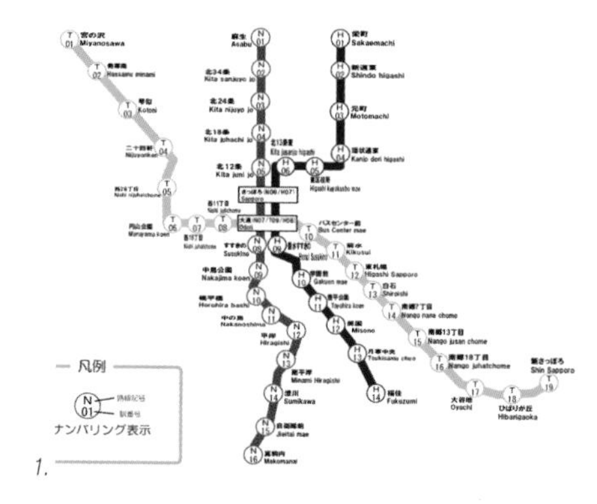

1.

Seattle, USA (2)
570,000. An ambitious plan, with a promising diagram, to build a new monorail network has recently collapsed, and light rail is proposed.

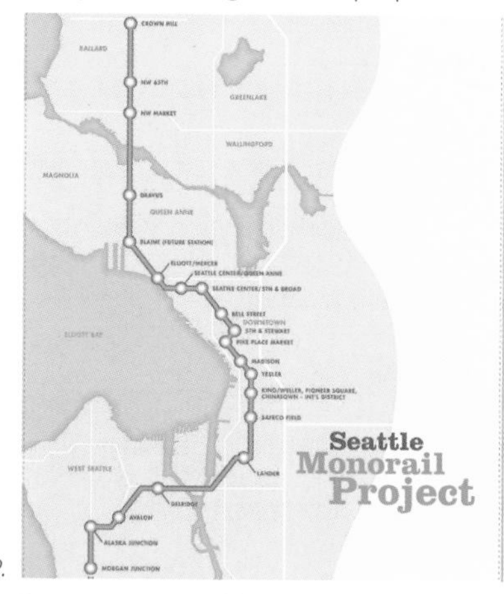

2.

Sendai, Japan (3)
1.4 million. 1987. 9.3 miles. 17 stations. Also planned is an east–west Line 2.

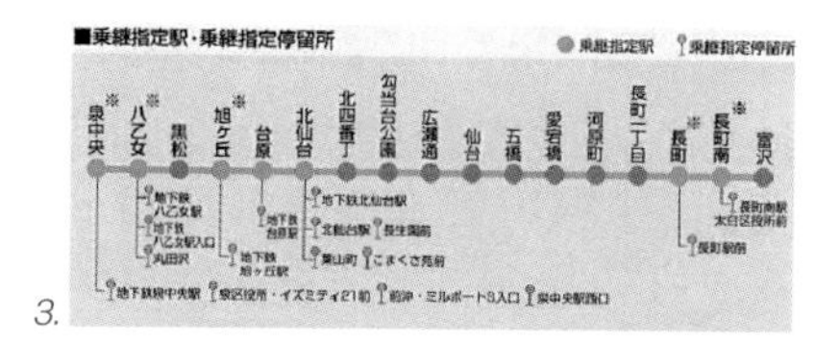

3.

Seville, Spain (4)
1.1 million. Opens 2008. Total planned network length of 31 miles and sixty stations. Half of the stations on Line 1 will be underground.

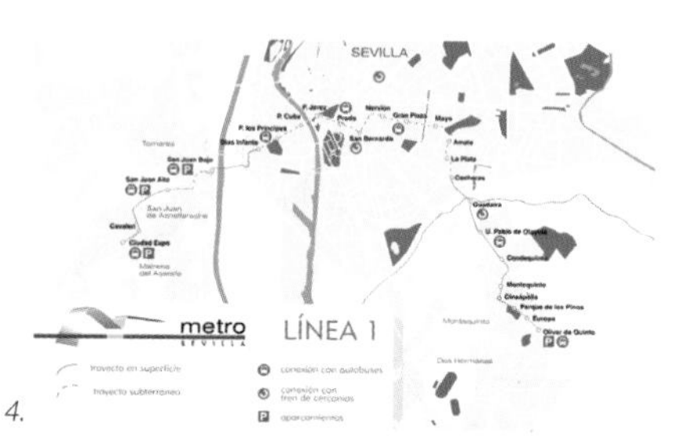

4.

Sheffield, United Kingdom (5)
875,000. 1994. 18 miles. 48 stops. Modern light rail. More than 50 percent runs on street level, but much of the rest is on old rail alignments.

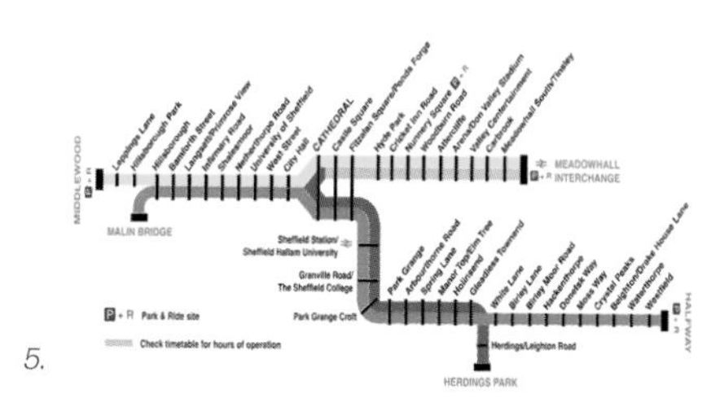

5.

Shenyang, China
7.2 million. 39-mile network planned.

6.

Shenzhen, China (6)
7 million. 2003. 13.4 miles. 19 stations. This new city has grown by 5 million people in the past decade alone. The Shenzhen Metro should eventually become a network of over 100 miles.

Shiraz, Iran
1.3 million. Due by 2010. 16.6 miles.

7.

Sofia, Bulgaria (7)
1.1 million. 1998. 6.2 miles. 8 stations. Light, wide, spacious, and heavily influenced by Russian and Eastern European systems. The downward *M* logo is used as station marker on this map showing all planned lines.

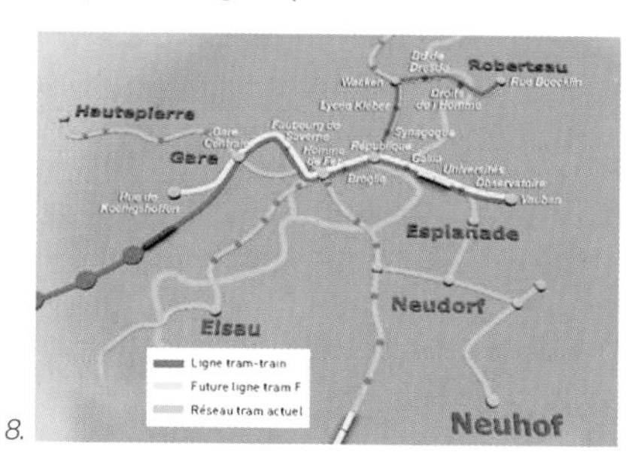

8.

Strasbourg, France (8)
650,000. 1994. 33.4 miles (by 2008.) Light rail with one subway station.

Tashkent, Uzbekistan (9)
2.3 million. 1977. 22.4 miles. 26

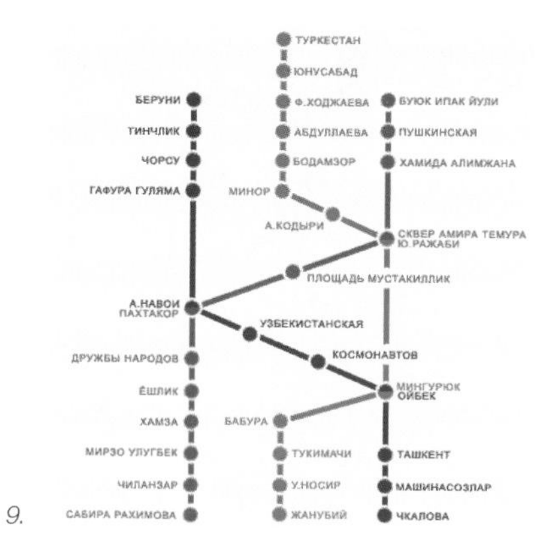

9.

stations. Typically Russian-style diagram, with dots for stations and dashes for lines. The stylization heavily distorts the as-yet-mainly-unbuilt Yunosobod Line, but the effect is dynamic and looks great.

10.

Tbilisi, Georgia (10)
1.2 million. 1966. 12 miles. 21 stations. Two lines in the Georgian capital span the long, thin city between the Ural Mountains.

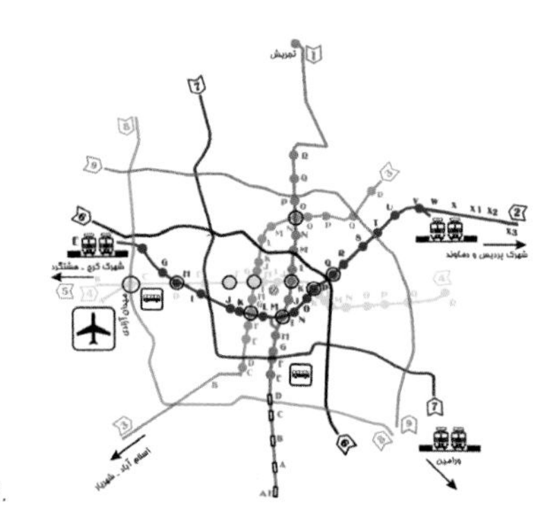

11.

Unless stated, all maps by kind permission of and copyright the system operator © 2007.

Tehran, Iran (11)
10.7 million. 1999. 59.7 miles. 54 stations. Much-delayed mass transit for Iran's vast capital. Line 1 and Line 2 through downtown and suburbs are predominantly underground.

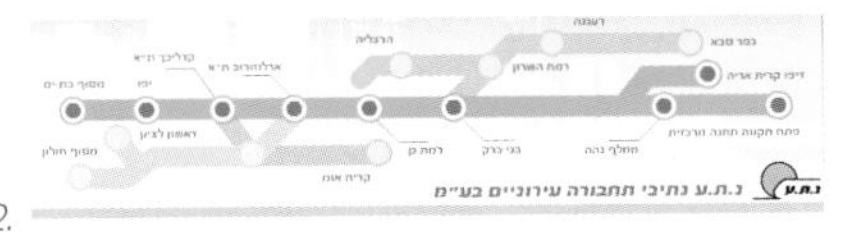

12.

Tel Aviv, Israel (12)
350,000. Proposed light-rail scheme with a possible subway section.

Thessaloníki, Greece
1 million. Opens 2013. 6.2 miles. 14 stations. The second city of Greece has started building a much-needed metro system. Given Athens's achievements, hopes are high.

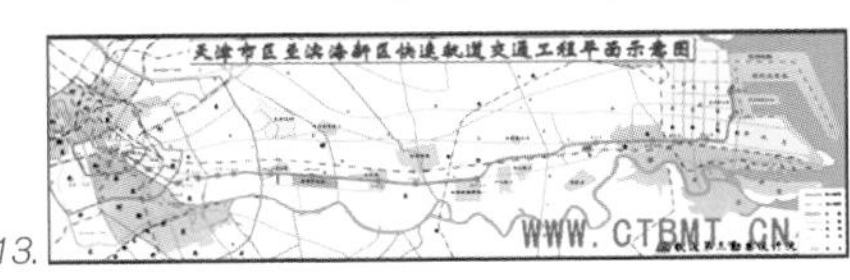

13.

Tianjin, China (13)
9 million. 1980. 4.3 miles. 10 stations. Existing lines built in a canal bed were closed to allow extensions at either end. This huge metropolis is planning seven metro lines of over 93 miles by 2010.

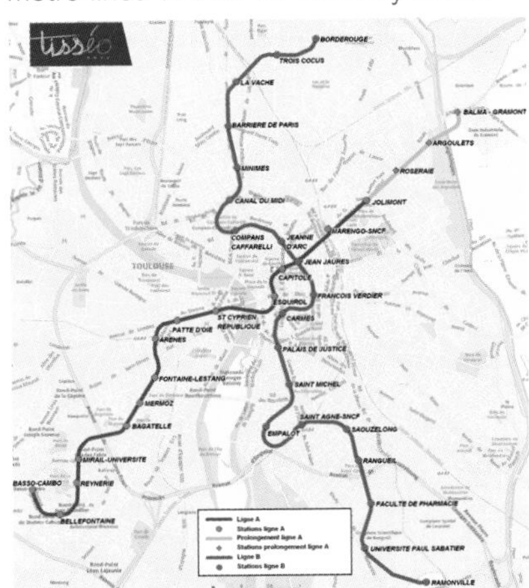

14.

Toulouse, France (14)
750,000. 1993. 7.6 miles. 18 stations. An underground-automated VAL system with new southeast-to-northwest line.

Turin, Italy (15)
1 million. 2006. 9.3 miles. 11 stations. This automated, driverless metro was just ready for the 2006 Winter Olympics.

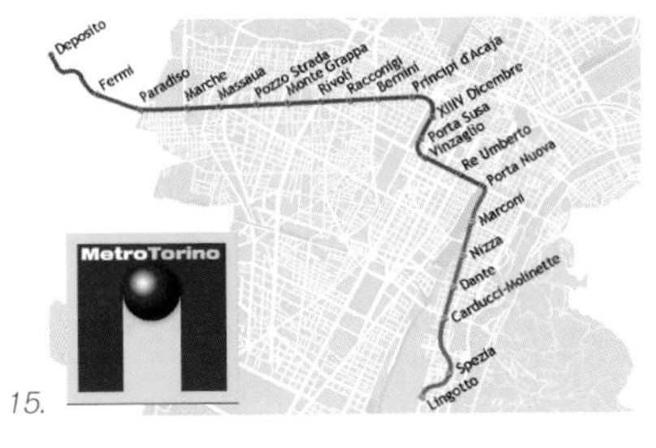

15.

Ufa, Russia
1.1 million. Opens 2010. The first wholly new line under construction in Russia since 1997 will have six stations in 5.6 miles.

Valencia, Venezuela
1.1 million. 2000. 3.8 miles. 7 stations. This is the first stretch of what will be a two-line metro tram system. The full network may not be seen until 2020.

Valparaíso, Chile
900,000. 2005. 26.7 miles. 20 stations (4 in tunnel).

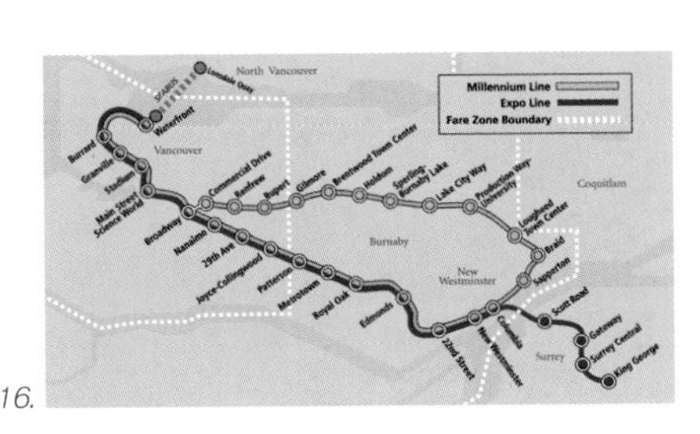

16.

Vancouver, Canada (16)
2.2 million. 1986. 31 miles. 32 stations. Mainly elevated, automated Sky Train. The Expo Line opened first, followed by the Millennium Line.

Volgograd, Russia (17)
1.1 million. 8.4 miles. 1984. 18 stops. A rapid "express-tram" with metro-like service and two subway stations in a 2-mile tunnel. A 2.6-mile underground extension is being built, and a second line is planned.

17.

Wuhan, China
8 million. 2004. 6.3 miles. 10 stations. This new elevated system is the first part of a proposed seven-line network totaling 136 miles.

Xi'an, China
7.8 million. Due 2011. 16.4 miles. Two lines, with 90 percent underground.

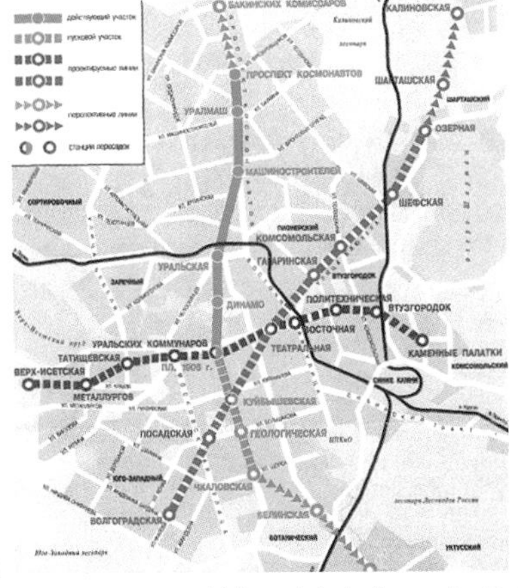

18.

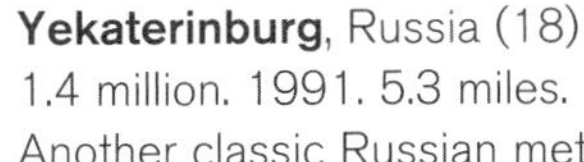

Yekaterinburg, Russia (18)
1.4 million. 1991. 5.3 miles. 7 stations. Another classic Russian metro with equally impressive palatial stations.

Yerevan, Armenia (19)
1.2 million. 1981. 8 miles. 10 stations. New line planned from downtown to the northeast and a possible third line later.

19.

Yokohama, Japan (20)
3.5 million. 1972. 25.6 miles. 37 stations. For a city in a country with such an excellent rail infrastructure, Yokohama, which claims to be Japan's second- largest city, has relatively low transit track mileage.

20.

Unless stated, all maps by kind permission of and copyright the system operator © 2007.

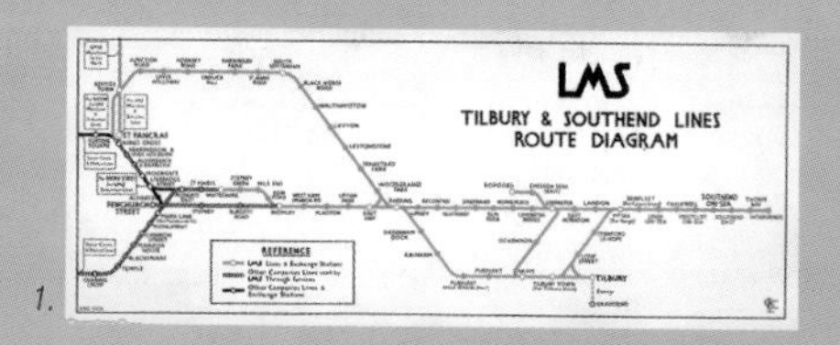

1.

Easy guide to Inter-City from the West Midlands

Inter-City makes the going easy

2.

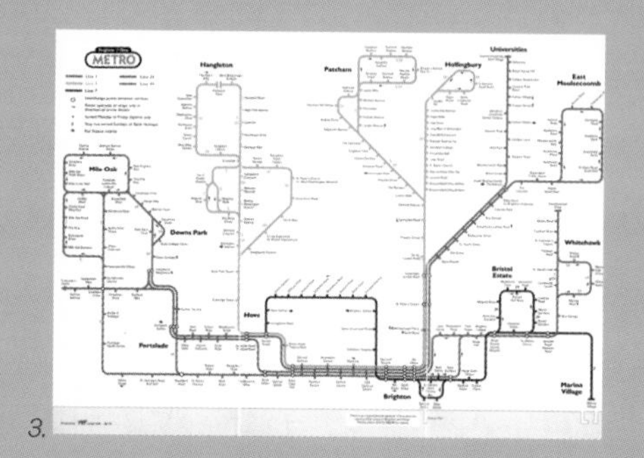

3.

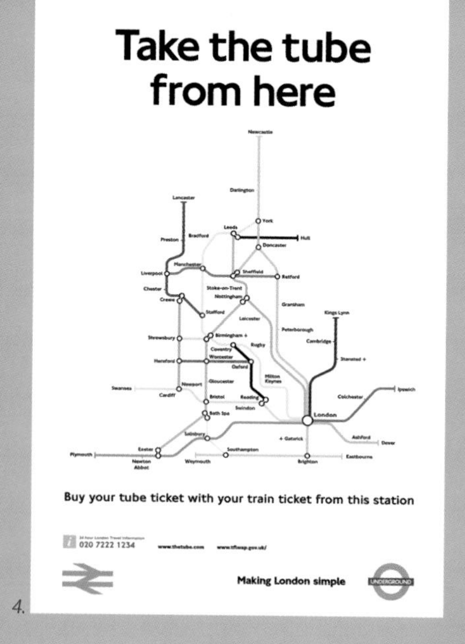

4.

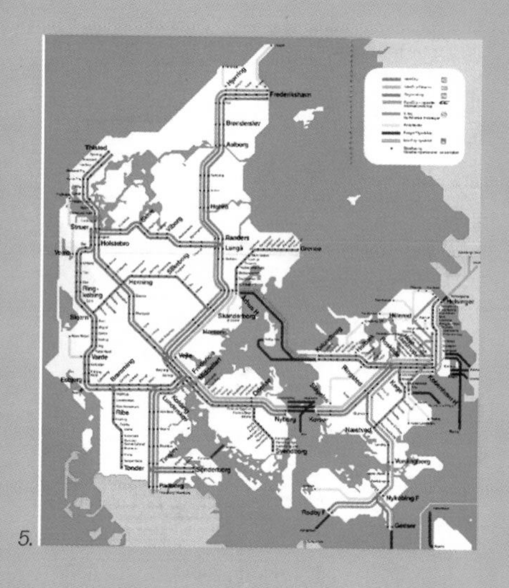

5.

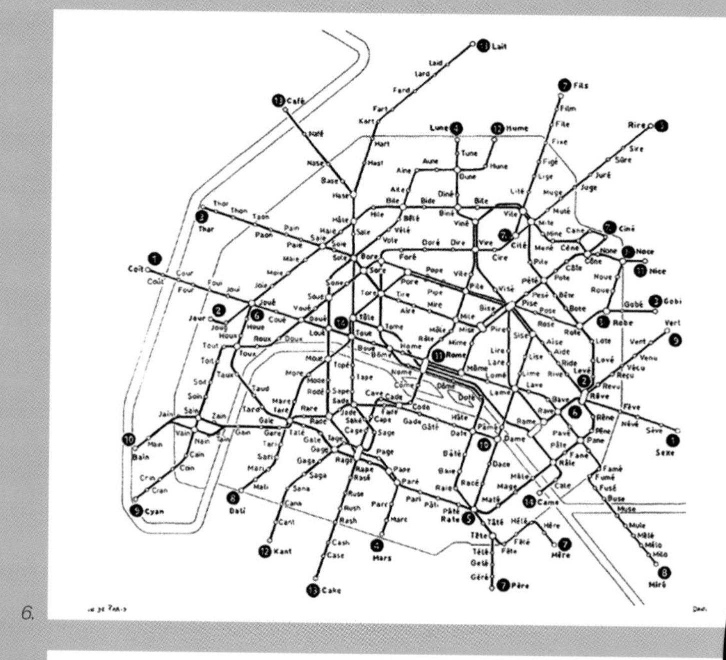

6.

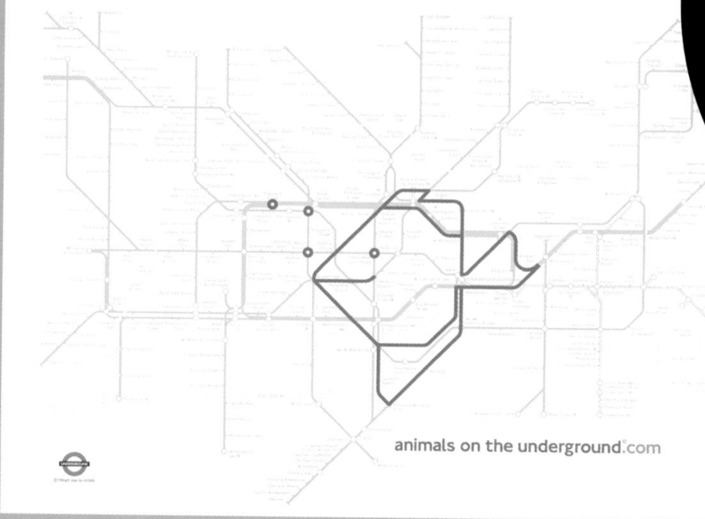

7.

8.

From George Dow's LMS in-car panel map of 1935 (1) right through to the fantasy Gallifrey Underground (14), the transit-maps concept of 45-degree diagonals remains much copied. West Midlands Inter-City in 1978 (2), TfL in 2001 (4), and Danish Railways (5) in 2003 have all used the style to promote heavy rail. Even bus operators have used it—e.g. Brighton and Hove in 1997 (3). Wit inspired Daniel Lehman's four-letter Paris map (6), and endless commuting got Paul Middlewick spotting animals on the London Underground diagram (7)—he now has an officially sanctioned Web site (www.animalsontheunderground.com) where the fish and others can be downloaded. Ralph Gray has produced all manner of geographic variations, including globes (8) and even a stylish solar system (12), using the urban-rail diagram style. The Goethe Institute turned their metro-map-style branch locator into a mouse mat (9). Scott Rosenbaum's mind-blowing piece from 2003 based on the New York City subway system is titled *Service Change* (10). Computer giant Interoute shows its fiber-optic network on this mass-transit style diagram (11). Science fiction is ideal for phantom networks, as the in-car map prepared for the sinister *Dark City* movie shows (13). Fans of U.K. sci-fi TV series *Dr Who* drew up an imaginary guide to the Time Lord's domain (14). Fantasy networks are invented worldwide; here, an idea for a subway in Manchester to complement the light-rail Metrolink (15).

9.

10.

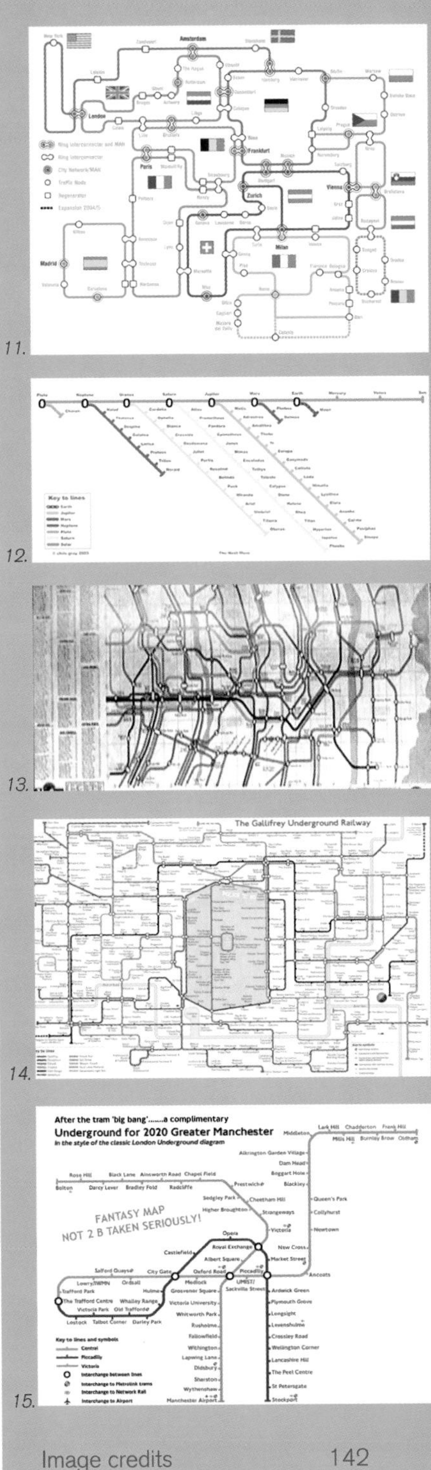

11.

12.

13.

14.

15.

Apendix

Image credits 142
Bibliography 142
Index 143

Image credits

Key to main collections and resources

A = Mike Ashworth Collection. **C** = Capital Transport Collection. **D** = David Pirrman of New York City Subway Resources, www.nycsubway.org. **L** = London's Transport Museum, +44 20 7379 6344. **N** = New York City Transit Museum, +1 718 694 1867. **O** = Author's collection, mark@metromapsoftheworld.com, +33 1 4636 3140. **P** = Peter Lloyd Collection, http://easyweb.easynet.co.uk/~ursa/. **R** = Robert Schwandl Collection, www.urbanrail.net. **U** = UITP Library, Thierry Maréchal, + 32 2 673 66 43.

Image sources page by page

Page 1 world diagram by: Mark Ovenden, Alan Foale, Transport for London. **p. 4** Author photo: by Darren Tossell. 1, 2: O. **p. 5** Hans Rat photo: U. Mike Ashworth photo: by Darren Tossell. **p. 6** 1: A. 2: O. 3: Library of Congress. **p. 7** 4: L. 5: N. **p. 8** 1: A. 2: Library of Congress. 3, 7,12: O. 4: P. **p. 9** 5: L. 6: Alfred B Gottwaldt. 8, 11,13: C. 9,10: A. 14: L. **p. 10** Chicago, New York (Union Dime), Paris: A. Tokyo, New York (1959), London, Madrid: O. Moscow: Artemy Lebedev. Berlin: U. **p. 11** New York, Chicago, London: O. Berlin S-Bahn, Paris: A. Madrid, Tokyo: U. **pp. 12–15** 1, 5, 6, 9, 13: Alfred B Gottwaldt. 2, 3, 4, 8, 10, 12,14: www.netzspinnen-berlin.de. 7: O. 11: U. 15: BVG, Berlin. Photos: C. **pp. 16–19** 1, 2, 3, 4, 5, 6, 7: A. Photo: Brian Patton. 8: O. 9: CTA. **pp. 20–23** 1, 2, 3, 4, 5, 8: L. 6, 7, 9, 10: O. 11: Bureau Mijksenaar. 12: QuickMap. 13: TfL. **pp. 24–27** 1: P. 2, 3, 4: U. Monochrome photo: Metro de Madrid. Color photo: D. 5: A. 6, 7: O. Turnstiles photo: C. Entrance photo: O. 8: CTM. 9: Metro de Madrid. **pp. 28–31** 1, 2, 5, 7, 8, 9, 10, 11,12, 13, 14,15,16,17,18,19, 20: Artemy Lebedev. 3: O. Photo: Moscow Metro. **pp. 32–35** 1, 3, 4, 6: N. 7, 8, 9, 10: A. 2, 5: www.bmt-lines.com. 11: O. D train photo: O. A. 8th Ave photo: C. 12: Joseph Brennan. 13: Kick Design Inc. 14: MTA. Photo: O. **pp. 36–39** Photo: D. 2, 3, 4, 5, 7: P. 1937 cover: P. 6, 8: A. In-car, 14: J Pepinster. 9: C. 10,11,13, 16: O. 12: C Spandonide. 17: RATP. **pp. 40–43** 1, 2, 3, 4: U. 5, 7: A. 6, 8, 9, 10: O. 11: Tokyo Metro/TOEI. **p. 44** Mexico pole: Ciudad de Mexico. Subte sign: Diseño Shakespear. U-Bahn sign: O. Barcelona pole: TMB. **p. 45** Budapest pole: O. **p. 46** 1, 2: U. 3: A. Photo: R. **p. 47** 4: TMB. **p. 48** 1, 2, 4: A. 3: O. **p. 49** 5, 6, photo: MBTA. **p. 50** 1, 4: O. 2, 3, 5, 6, 7: BKV. **p. 51** 8, 9 & photos: BKV. **p. 52** 1, 2: U. 3: A. 4, 5: Diseño Shakespear. **p. 53** 6: Diseño Shakespear. **p. 54** 1, 2, 3: U. Photo: R. **p. 55** 4: HVV. **p. 56** 1, Sign photo: Peter Olsen. Entrance photo: MTR. 2: U. 3, 4: O. **p. 57** 6: MTR. **p. 58** 1, 2, 3: Metropolitan de Lisboa. 4: U. **p. 59** 5, main map: Metropolitano de Lisboa. Photo: by David Pirmann of O. **p. 60** 1, 2: U. 3: O. 4: STC. **p. 61** 5: STC. **p. 62** 1: Benoît Clairoux. 2, 3: STCUM. **p. 63** Photo: by Jean Rene Archambault. 4, 5: STCUM. **p. 64** 1, 2: O. Photo: MVG. **p. 65** 2: MVV. **p. 66** 1, 2: U. 3: O. **p. 67** 4: U. 5: OKK. Photo: U. **p. 68** 1, 2: U. 3: Josh Lehan. 5: Steve Boland **p. 69** 4: SF Muni Metro. Photo: C. 6: BART. **p. 70** 1: L. 2, 3: SMRT **p. 71** 4: 5,6, 7: Seoul Metropolitan Government. Photos: Patrick Debrisson. **p. 72** 1, 2 3: U. **p. 73** 4, 5, photo: Kommet/St. Petersburg Metro. **p. 74** 1, 2: WMATA. **p. 75** Photo upper: U. Photo lower: C. 3: WMATA. **p. 76** Copenhagen: Ørestadsselskabet. Oslo: from Trafikanten. Singapore photos: Roman Hackelsberger. Prague photo: U. São Paulo: Companhia Do Metropolitano De São Paulo. Lyon photos: O. Rio: Metro Rio. Philadelphia: SEPTA. **p. 77** Bilbao: Metro Bilbao. Rome photos: U. Vienna: Wiener Linien. Rotterdam diagram: RET. **p. 78** 1: GVB. 2, 3, cover: U. **p. 79** 1: O 2, 3: Attiko Metro. Photo: C **p. 80** 1: Beijing Publishing House. 2, 3: Beijing Top Result. 4: Beijing Subway. Photo: Wikiverse. **p. 81** 1: Foster and Partners. 2, 3: Metro Bilbao. **p. 82** 1: STIP. 2: A. Photo: O. **p. 83** 1, 2: U. Photo: Metrorex. **p. 84** 1: DSB S-Tog. 2, 3, 4 photo: Ørestadsselskabet. **p. 85** 1, 2, 3, photo: Delhi Metro Rail Corporation. **p. 86** 1, 3, photo: SPT. 2: A. 4: Adam Gordon. **p. 87** 1, 2, 4: Metro Kyyvskiy Metropoliten. 3: Hans-Ulrich Riedel. 5, 6, 7: www.metropoliten.kiev.ua. Photos: U. **p. 88** Photo by Neil Madhvani. 1: STAR. 2: Stesen Sentral. 3: KTM. 4: Integrated Rail Project. **p. 89** 1 & top photo: MTA. Lower photo: O. **p. 90** map & photo: TCL. **p. 91** 1, 2, 3: City of Nagoya Transportation Bureau. Photo: Mark Kavanagh. **p. 92** 1, 2: NEXUS. Photo: O. **p. 93** 1, 2: Trafikanten. 3: U. **p. 94** 1: SEPTA. 2: PATCO. Photo by Richard Brome. **p. 95** 1, 2: U. 3, 4: Dopravni Podnik Photo by Mark Thomas. **p. 96** 1, 3: U. 2, photo: Metro Rio. **p. 97** 1, U. 2, : ATAC & Steer Davies Gleave. Photo: Wikiverse. **p. 98** 1: RET. 2, 3: U. Photo: Eddy Konijnendijk. **p. 99** 1, 2: Metro São Paulo. 3: U. **p. 100** 1, photo: SMRT. 2, 3: O. **p. 101** 1: U. 2, 3: L. 4, photo: O. 5: SL. **p. 102** 1, 2: TRTS. Photo: Chaffee Yiu. **p. 103** 1: TTC. 2, 3: U. Photo: O. **p. 104** 1, 2, 3: FGV. Photo: C. **p. 105** 1: U. 2, 3: Wiener Linien. Lower left photo: R. Middle left photo: C. **pp. 106–107** All logos system operators (top line left to right): Warsaw, Milan, Lille, Shanghai; (bottom line left to right): Bangkok Skytrain, Guangzhou, Nizhniy Novgorod, Santiago. **p. 108** Atlanta photo: Robert Ferreira. Atlanta map: MARTA. Baltimore maps: MTA Maryland. **p. 109** Bangkok maps: BTS and BMCL. Cairo map: NAT. Cairo photo: U **p. 110** Guangzhou maps: GZMTR . Guangzhou photo: www.gakei.com. All Kharkiv maps from Metro Kharkiv. Kharkiv postcard from: U. **p. 111** Lille maps and photo: Transpole. Marseille map and photo: RTM. **p. 112** Milan 1: U. 2: L. 3: ATM. Naples 1: ACAM. 2: Metro Napoli. Photo by Werner Huber. **p. 113** Newark 1, 3: PATH. 2: NJ Transit. Nizhniy Novgorod 1, 2, 3, photo: meta.metro.ru. **p. 114** Recife map and photos: MetroRec. Santiago map: Metro de Santiago. Photo by Raúl Moroni. **p. 115** Shanghai 1 and photo: Shanghai Metro. 2: Shanghai Daily. Warsaw map from Warsaw Metro. Photo: Witold Urbanowicz. Line map photo: Maria Holubowicz **pp. 116–117** Metro typo-cartography: Berrymatch NYC, www.berrymatch.com. **p. 118** Cologne & Bonn map: VRS. Photo: R. Cleveland map: RTA. Dublin map: Railway Procurement Agency. **p. 119** Frankfurt map: RMV. Photo: R. Hanover map: GVH. Jacksonville map: JTA. **p. 120** Liverpool map and photo: Merseytravel. Overhead map: A. Manchester maps from: GMPTE. Photo: Geoff Edwards. **p. 121** Melbourne maps: MetLink. Photo: Connex Melbourne. Miami maps from MDTA. Nuremburg map: VAG. **p. 122** Pittsburgh map from PAT. Porto map: Metro Porto. Rhine-Ruhr main map: VRR. Wuppertal map: A. Dusseldorf map: Rheinbahn. **p. 123** Map and photo: VVS. **p. 124** 1: CityRail. 2: MetroLightRail 3: L. **p. 125** Tunis map from: U. Photo by Dmitriy Dmitriadi. Zurich map: ZVV. **p. 126** (L–R) Istanbul: Istanbul Ulasim. Calgary: Calgary Transit. Vancouver: Skytrain. **p. 127** (L–R) Helsinki: HKL. Tehran: Tehran Metro. Ankara: Ankara Metrosu. Saporro: Saporro City Transportation Bureau. Dnepropetrovsk: Dnepropetrovsk Metro. **p. 128** 1: Adana Metro Project. 2: Entreprise Metro d'Alger. 3: FGV Alicante. 4: Almaty Monorail. 5: Ankara Metrosu. 6: DeLijn 7: Auckland City Rail Project. 8: Baku Metro. 9: Bangalore MRT Ltd. **p. 129** 10: Beovoz. 11: Metro Belo Horizonte. 12: MoBiel. 13: Centro. 14: TBC. 15: Metro DF. 16: Vladimir Simko. 17: Brescia Mobilita. 18: NFTA metro. 19: Bursaray. 20: Busan urban Transit Authority. 21: Calgary Transit. **p. 130** 1: Metro de Caracas. 2: Metropolitana Catania. 3: Changchun Railway Company 4: TEC Charleroi. 5: www.chelmetro.ru. 6: Hannes Neugebauer. 7: 2BangkokTravel.com. **p. 131** 8: CQ Metro. 9: Daegu Metropolitain Subway. 10: Daejeon Subway Construction. 11: DART. 12: RTD. 13: Detroit People Mover. 14: http://gorod.dp.ua/metro. 15: Donets'k Metro. 16: Dubai Metro. 17: Trams For Edinburgh. 18: Edmonton SLRT. 19: Esfahan Urban Railway Organization. 20: Metrofor. 21: Fukuoka City Subway. **p. 132** 1: Genova Metro. 2: Metro de Goiania. 3: Sparvagen Goteborg. 4: SITEUR. 5: Gwangju Metropolitan Rapid Transit Corporation. 6: The Carmelit Subway. 7: HKL. 8: Hiroshima Astram Line. 9: Metropolitan Transit Authority of Harris Country, Texas. **p. 133** 10: Incheon Rapid Transit Corporation. 11: Istanbul Metro. 12: Izmir Metro. 13: PT Jakarta Monorail. 14: Rakevetkala-Jerusalem. 15: Kaohsiung Rapid Transit Corporation. 16: VBK. 17: Kazan Transport. 18: Kitakyushu Urban Monorail. 19: Kobe City Subway. **p. 134** 1: Metro Railway Kolkata. 2: MetroTram. 3: Kyoto Municipal Transportation Bureau. 4: Las Vegas Monorail. 5: T-L. 6: Metro de Lima. 7: Linz Linien. 8: RNV. 9: Metro Malaga. **p. 135** 10: Manila Light Rail Transit Authority. 11: Metro de Maracaibo. 12: Mashhad Urban Railway 13: Metro de Medellín. 14: MetroTransit. 15: Minsk Metro. 16: Metrorrey. 17: Fundacion Metro Montevideo. 18: MumbaiRail. 19: Yui-Rail. 20: Nanjing Metro. 21: NET. **p. 136** 1: Novosibirsk Metro. 2: Mostovik. 3: OC Transpo. 4: Commune Palermo. 5: Valley MetroRail. 6: Portland MAX. 7: Trensurb. 8: Carl Yui. **p. 137** 9: TCAR. 10: STAR Rennes Metropole. 11: Sacramento RTD. 12: MetroLink. 13: UTA. 14: Metro de Salvador. 15: Metro Samara. 16: San Diego Trolley. 17: VTA. 18: ATI. **p. 138** 1: Sapporo Transportation Bureau. 2: Seattle Monorail. 3: Sendai City Transport. 4: Metro de Sevilla. 5: Stagecoach Supertram. 6: Shenzhen Metro. 7: Sofia Metro. 8: Compagnie des Transports Strasbourgeois. 9: Tashkent Metro. 10: Metro Tbilisi. 11: Metro Tehran. **p. 139** 12: Tel Aviv LRT. 13: TJDT. 14: SEMVAT. 15: Metrotorino. 16: Translink. 17: Volgograd.metro.ru. 18: Yekaterinburg Metro. 19: R. 20: Yokohama Transportation Bureau. **pp. 140–141** 1, 3: O. 2: A. 4: TfL. 5: DSB. 6: Daniel Lehman. 7: Paul Middlewick. **p. 141** 8, 12: R Gray. 9: Benjamin Oliver. 10: Scott Rosenbaum. 11: Interoute.com. 13: New Line Cinema. 14: Shaun Lyon. 15: O. **p. 144** Image: r-town.

Bibliography

Useful Books

Andreu, Marc, et al. *La ciutat transportada; Dos segles de transport collection al servei de Barcelona.* Barcelona: Transports Metropolitans de Barcelona, 1997.

Bennett, David. *Metro: The Story of the Underground Railway.* London: Mitchell Beazley, 2004.

Berton, Claude, and Alexandre Ossadzow. *Fulgence Bienvenue et la construction de Métropolitain de Paris.* Paris: Presses, Ponts et Chaussées, 1998.

Blackwell, Lewis. *20th Century Type.* London: Laurence King, 2004.

British Railway Maps of Yesteryear. Shepperton: Ian Allan, 1991.

Cercanias de RENFE y Fundacion de los Ferrocarriles Espanoles. *Madrid en sus Cercanias; un recorrido por la Metropoli u su ferrocarril.* Madrid: Editec@Red SL, 2003.

Childs, Nick, et al. *Railway Maps and the Railway Clearing House: The David Garnett Collection in the Brunel University Library.* Uxbridge: Brunel University Library, 1986.

Clairoux, Benoît. *Le Métro de Montréal.* Montreal: Hurtubise HMH, 2001.

Clarke, Bradley H. *Boston Rapid Transit Album.* Cambridge, MA: Boston Street Railway Association, 1981.

Cudahy, B. J. *Under the Sidewalks of New York: The Story of the Greatest Subway System in the World.* 2d. ed. rev. New York: Fordham University Press, 1995.

Diseño Shakespear Catalogue. Buenos Aires, 2002.

Dow, Andrew. *Telling the Passenger Where to Get Off.* London: Capital Transport, 2005.

Fischler, Stan. *Subways of the World.* Osceola, FL: MBI, 2000.

_______. *Uptown, Downtown, A Trip through Time on New York's Subways.* New York: Hawthorn Books, 1976.

Garbutt, Paul. *World Metro Systems.* London: Capital Transport, 1997.

Garland, Ken. *Mr Beck's Underground Map.* London: Capital Transport, 1994.

Gottwaldt, Alfred B. *Das Berliner U-und S-Bahnnetz; Eine Geschichte in Streckenplanen.* Stuttgart: Transpress, 1994.

Green, Oliver, and Jeremy Rewse-Davies. *Designed for London.* London: Laurence King, 1995.

Hackelsberger, Christoph. *U-Bahn Architektur in München.* Munich: Prestel, 1997.

Halliday, Stephen. *Underground to Everywhere.* London: Sutton, 2001.

Hardy, Brian. *Paris Métro Handbook.* London: Capital Transport, 1993.

_______. *The Berlin U-Bahn.* London: Capital Transport, 1996.

Haresnape, Brain. *British Rail1948–78, A Journey by Design.* Shepperton: Ian Allan, 1979.

Hinkel et al. *Underground Railways Yesterday-Today-Tomorrow from 1863–2010.* Vienna: Schmid Verlag, 2004.

Hollis, Richard. *Graphic Design: A Concise History.* London: Thames and Hudson, 1994.

Howes, Justin. *Johnston's Underground Type.* London: Capital Transport, 2000.

Krambles, George, and Arthur H. Peterson. *CTA at 45.* Chicago: Transit Scholarship Fund, 1993.

Kuhlmann, Bernd. *Stadt-schnellbahnen der Sowjetunion.* Vienna: Verlag Josef Otto Slezak, 1981.

Landers, John. *Twelve Assorted Historical New York City Street and Transit Maps, Volume II 1847–1939.* New York: H&M Productions, 2000.

Lawrence, David. *A Logo for London.* London: Capital Transport, 2000.

Le Patrimonie de la RATP. Paris: Flohic Editions, 1998.

Leboff, David, and Tim Demuth. *No Need to Ask—Underground Maps 1863–1933.* London: Capital Transport, 1999.

Maryland Department of Transportation. *Baltimore Region Rail System Plan.* Committee Report. Baltimore, 2002.

McDermott, Catherine. *20th Century Design.* London: The Design Museum/Carlton Books, 1999.

Moffat, Bruce G. *The "L," The Development of Chicago's Rapid Transit System, 1888–1932, Bulletin 131.* Chicago: Central Electric Railfans Association, 1995.

Moscow Metro. Official souvenir book. Moscow: Moscow Metro, 1997.

The Moscow Underground. Official brochure. Moscow: Metro Moscow, circa 1994.

New York Transit Museum. *Subway Style.* New York: Stewart Tabori & Chang, 2004.

Nock, O. S. *World Atlas of Railways.* London: Victoria House, 1978.

Pattison, Tony, ed. *Jane's Urban Transport Systems.* Coulsdon: Jane's Information Group, annual.

Picture History of the Liverpool—Manchester Railway. Liverpool: Scouse Press, 1970.

Richens, Daniel. *Metros in Europa.* Stuttgart: Transpress, 1996.

Roberts, Maxwell J. *Underground Maps After Beck.* London: Capital Transport, 2006.

Schwandl, Robert. *Berlin U-Bahn Album.* Berlin: metroPlanet, 2002.

_______. *Hamburg U-Bahn and S-Bahn Album.* Berlin: Robert Schwandl Verlag, 2004.

_______. *Metros in Spain.* London: Capital Transport, 2001.

_______. *U-Bahnen in Skandinavien.* Berlin: Robert Schwandl Verlag, 2004.

SELNEC PTE. *Public Transport Plan for the Future.* Manchester: SELNEC PTE, 1973.

Ström, Marianne. *Metro-Art in the Metro-polis.* Paris: ACR Edition Internationale, 1994.

Taylor, Sheila. *The Moving Metropolis.* London: Laurence King, 2001.

The Times Concise Atlas of the World. London: Guild, 1986.

Tyneside PTE. *Public Transport on Tyneside, A Plan for the People.* Newcastle: Tyneside PTE, 1973.

Walmar, Christian. *The Subterranean Railway.* London: Atlantic Books, 2004.

Wildbur, Peter, and Michael Burke. *Information Graphics, Innovative Solutions in Contemporary Design.* London: Thames and Hudson, 1998.

Wright, John. *Circles under the Clyde: History of the Glasgow Underground.* London: Capital Transport, 1997.

Useful Web sites

www.uitp.com
www.urbanrail.net
www.reed.edu/~reyn/transport.html
www.nycsubway.org
www.subwaynavigator.com
www.subways.net
www.mic-ro.com/metro/

Museums

New York: www.mta.nyc.ny.us/mta/museum/
London: www.ltmuseum.co.uk
Budapest: www.bkv.hu/angol/muzeum
Stockholm: www.sparvagsmuseet.sl.se/
Glasgow: www.glasgowmuseums.com

Index

Geographic Index

Adana 128
Africa 8, 107, 125, 128
Alexandria 128
Algeria 128
Algiers 128
Alicante 128
Almaty 128
America, North 6–9, 14, 16, 28, 48, 52, 60, 62, 68, 74, 94, 103, 128
America, South 52, 60, 98
Amsterdam 78, 98
Ankara 128
Antwerp 128
Armenia 138
Asia, east/southeast 41, 64, 70, 80, 91
Athens 77, 79, 139
Atlanta 108
Auckland 128
Austria 6, 134
Azerbaijan 128
Baghdad 128
Baku 128,
Baltimore 6, 74, 107,
Bangalore 128
Bangkok 109
Barcelona 24, 25, 26, 27, 45–47
Bavaria 64
Beijing 77, 80
Belarus 135
Belgium 6, 82, 111, 128, 130
Belgrade 128
Belo Horizonte 128, 129
Berlin 7–9, 12–15, 38, 54
Bielefeld 129
Bilbao 26, 81
Birmingham 6, 129
Bochum 122
Bologna 129
Bombay, see Mumbai
Bonn 118
Bordeaux 129
Boston 8, 9, 48, 49
Brasilia 129
Bratislava 129
Brazil 99, 114, 128, 129, 131, 132, 137
Brescia 129
Brighton & Hove 140
Britain 4, 6, 8, 9, 16, 38, 52, 56, 76, 86, 92, 103, 125
Brussels 7, 77, 86
Bucharest 83
Budapest 7, 45, 50, 51
Buenos Aries 45, 52
Buffalo 129
Bulgaria 138
Bursa 128, 129
Busan 126, 129
Cairo 109
Calcutta, see Kolkata
Calgary 126, 129
Canada 129, 131, 136, 139
Canton 110
Caracas 126, 129–130
Catania 130
Changchun 130
Charleroi 130
Charleston 8
Chelyabinsk 52, 130, 138
Chengdu 130
Chennai 85, 130
Chiang Mai 130
Chicago 7, 8, 16–19, 46
Chile 139
China 56, 80, 114, 130, 132, 135, 138, 139
Chongqing 130
Cleveland 130
Cologne 118
Copenhagen 84
Cracow 130
Daegu 130
Daejeon 130
Dallas 131
Delhi 85
Denver 131
Detroit 131
Dnipropetrovs'k 131
Dominican Republic 138
Donets'k 131
Dortmund 122
Dubai 131
Dublin 7, 118
Duisburg 122
Düsseldorf 122
Edinburgh 131
Edmonton 131
Egypt 109, 128
Esfahan 131
Essen 122
Europe 6, 8, 12, 27, 36, 47, 48, 50, 63, 79, 95, 98, 103, 111, 128, 132, 134, 138
Finland 132
Fortaleza 131
Foshan 110
France 6, 8, 39, 62, 83, 90, 103, 111, 129, 132, 134–139
Frankfurt 54, 119, 135
Fukuoka 60, 131
Gelsenkirchen 122
Genoa 132
Georgia 138
Germany 6, 7, 12, 15, 55, 64, 118, 119, 121–123, 128, 133, 135
Glasgow 7, 86, 120
Goiania 132
Gorky 113
Gothenburg 132
Grenoble 132
Guadalajara 132
Guangzhou 110
Gwangju 132
Hague, The 132
Haifa 132
Hamburg 45, 54, 55
Hangzhou 132
Hannover 119
Hanoi 132
Helsinki 126
Hiroshima 132
Ho Chi Minh City 132
Holland 7
Hong Kong 7, 56, 57, 110
Houston 132, 133
Incheon 71, 133
Iran 131, 135, 138
Iraq 128
Israel 132, 133, 138
Istanbul 133
Italy 6, 97, 112, 129, 130, 132, 136, 137, 139
Izmir 128, 133
Jacksonville 119
Jakarta 133
Japan 7, 40, 41, 60, 66, 131–135, 138, 139
Jerusalem 133
Kao-hsiung 133
Karlsruhe 133
Kawasaki 133
Kazakhstan 128
Kazan 133
Kent 8
Kharkiv 110
Kiev 87, 131
Kitakyushu 133
Kobe 133
Kolkata 85, 134
Krasnoyarsk 134
Kryvy Rih 134
Kuala Lumpur 88
Kyoto 134
Lagos 134
Lantau Island 56
Laon 134
Las Vegas 134
Lausanne 134
Lille 111
Lima 134
Linz 134
Lisbon 58, 59
Liverpool 4, 6–8, 120, 121
London 1, 4, 6–9, 16, 20–24, 26, 29, 35, 38, 39, 56, 78, 86, 98, 100, 123, 124, 144
Long Island 7, 32,
Los Angeles 9, 77, 89
Luidwigshafen 134
Lyon 90
Madras, see Chennai
Madrid 7, 24–27, 46, 123
Malaga 134
Manchester 4, 6–8, 92, 120, 141
Manhattan 7, 8, 32, 110
Manila 134
Mannheim 134
Maracaibo 135
Marseille 111
Mashhad 135
Mataró 46
Medellín 135
Melbourne 121
Mexico 60, 132, 135,
Mexico City 7, 45, 60, 61
Miami 117, 121
Milan 112
Milton Keynes 4
Minneapolis/St. Paul 135
Minsk 126, 135
Monterrey 135
Montevideo 135
Montpellier 135
Montreal 7, 52, 62, 63, 103
Moscow 13, 25, 27–31, 34, 63, 72, 136,
Mumbai 135
Munich 64, 65
Nagoya 91, 138
Naha 135
Nanjing 135
Nantes 135
Naples 112
New England 8, 48
New Jersey 94
New York 6–8, 11, 16, 32–35, 68, 140
New Zealand 128
Newark 7, 113
Newcastle 7, 92
Nizhniy Novgorod 113
North Korea 136
Nottingham 6, 135
Novosibirsk 136
Nuremburg 121
Odessa 136
Omsk 136
Orléans 136
Osaka 41, 45, 66, 67, 138
Oslo 93
Ottawa 136
Palermo 136
Paris 4, 7, 8, 21, 26, 27, 34, 36–39, 60, 62, 140, 144
Parma 136
Perm 136
Peru 134
Philadelphia 94
Philippines 134
Phoenix 136
Pittsburgh 130
Poland 130, 136
Portland 126, 136
Porto 130
Porto Alegre 136
Poznan 136
Prague 95
Promontory Summit 8
Puerto Rico 137
Pyongyang 136
Quebec 62
Recife 114
Rennes 136
Rhine-Ruhr 122, 135
Richmond 48
Rio de Janeiro 96
Rome 97
Rotterdam 98, 132
Rouen 136, 137
Russia 6, 72, 113, 128, 130, 133, 134, 136–139
Sacramento 137
Salerno 137
Salt Lake City 137
Salvador 137
Samara 137
San Diego 137
San Francisco 7, 68, 69
San Jose/Santa Clara 137
San Juan 137
Santiago 114
Santo Domingo 138
São Paulo 77, 99
Sapporo 126, 138
Scotland 8, 86, 131
Seattle 138
Sendai 138
Seoul 45, 70, 71, 133
Serbia 128
Seville 27, 138
Shanghai 115
Sheffield 138
Shenyang 138
Shenzhen 138
Shiraz 138
Siberia 134
Singapore 77, 100
Slovakia 129
Sofia 138
South Korea 70, 129, 130, 132, 133
Southport 9
Spain 7, 24, 26, 27, 46, 81, 83, 104, 114, 128, 134, 138
St. Louis 7, 137
St. Petersburg 72, 73, 136
Stockholm 101
Strasbourg 138
Stuttgart 123, 135
Swansea 6
Switzerland 134
Sydney 7, 124
Taegu, see Daegu
Taejon, see Daejeon
Taipei 102
Taiwan 133
Tashkent 126, 138
Tbilisi 138
Tehran 139
Tel Aviv 139
Thailand 109, 130
Thessaloníki 139
Tianjin 139
Tokyo 34, 40–43, 70, 138
Toronto 52, 103
Toulouse 126, 139
Tunis 125
Turin 139
Turkey 128, 129, 133
Ufa 139
Ukraine 131, 134, 136
United Arab Emirates 131
United Kingdom 6, 8, 78, 86, 92, 120, 129, 135, 138, 140, 141
Uruguay 135
United States 6, 8, 9, 16, 21, 41, 48, 74, 75, 129–131, 133–138
Uzbekistan 138
Valencia (Spain) 26, 104, 134
Valencia (Ven.) 139
Valparaiso 139
Vancouver 126, 139
Venezuela 130, 135, 139
Vienna 8, 105
Vietnam 132
Volgograd 134, 139
Wales 6
Warsaw 115
Washington DC 17, 45, 49, 74, 75, 108, 136
West Midlands 140
West Yorkshire 8
Wuhan 139
Wuppertal 122
Xi'an 139
Yekaterinburg 139
Yerevan 139
Yokohama 126, 139
Zurich 125

Subject Index

45-degree angles 22, 25, 26, 30, 32, 34, 38, 39, 41, 44, 46, 48, 50, 56, 68, 70, 78, 79, 81–84, 90, 92, 94, 99, 100, 112, 119, 120, 121, 124, 125, 140
Archaeology 60, 79, 97
Architects, see Designers
Architecture 22, 25, 27, 28, 36, 50, 60, 62, 72, 81, 83, 90, 97, 104, 113
  Tiling 50, 58, 86, 94
Art 8, 9, 11, 63, 72, 97, 99, 101, 113, 140
  Airbrush 72, 87
  Abstraction 9, 29, 30, 48, 72
  Art Nouveau 36, 38
  Cubism 88
  Functionalism 48, 50, 64, 86, 144
  Kandinsky 9
  Mondrian 9
  Patterson, Simon 22
  Sculpture 63, 113
  Tate Gallery, the 22
Ashworth, Michael 5,
Beck, Harry, see Map-makers
Cartography 4, 5, 7–9, 20, 22, 23, 28, 46, 62, 77, 117
  3-D 8, 28, 30, 31, 72, 73, 123, 134
  Beading 30, 31
  Black-background 22, 48, 52, 62, 95, 103, 130, 133
  Color coding of lines 29, 60, 62
  Geometry 4, 13, 22, 29, 30, 34, 88
  Icon 13, 20, 26, 30, 32, 75, 144
  Legibility 18, 21, 36, 41, 52, 58, 62, 68, 74, 104, 112, 119, 123
  Orientation 16, 39, 46, 62, 126
  Scale 8, 9, 22, 99
  Symbols 5, 8, 24, 58, 70, 81, 104, 112
    Black circle/white center 9, 12, 13, 16, 20, 25, 28, 30, 40, 41, 46, 56, 64, 66, 74, 79, 83, 85, 86, 88, 92, 94, 100, 103–105, 108, 111, 112, 114, 118–120, 122, 125, 130, 137
    Diamond 26, 81, 92
    Open circle 16, 18, 20, 30, 41, 63, 66, 99, 102, 108, 114, 115,
    Pictogram 58, 60, 131
    Roundel 22, 23, 38, 40, 124
    Taefuk 68
    Ticks 4, 22, 46, 54, 62, 64, 78, 79, 81, 83, 85, 86, 89, 92, 98, 100, 104, 118, 121, 124
    White-line connectors 20, 30, 46, 76, 98, 100, 137
  Topography 4, 8, 9, 12, 13, 16, 21, 22, 25, 28, 30, 34, 36, 37, 41, 46, 48, 50, 56, 58, 68, 74, 76, 78, 79, 86, 93, 94, 99, 104, 115
Cars, see Transit modes
Commuting, see Passengers
Congressional Subway, the 75
Design 4, 5, 7, 9, 13–15, 21, 27, 30, 32, 38, 45, 46, 48, 50, 52, 56, 58, 60, 74, 76, 83, 86, 95, 103, 112, 119, 124
  Advertising/design agencies
    Artemy Lebdev Studio 30
    BDC Conseil 39
    Bright International 42
    Bureau Mijksenaar 78, 98
    Diseño Shakespear 45
    Kick Design Inc. 35
    Kommet 72, 73
    LS London 1, 23
    QuickMap 23
    Steer Davis Gleave 97
    Top Result Metro 80
  Branding 13, 18, 32, 41, 45, 50, 52, 62, 80, 91, 92
  Corporate Identity 14, 22, 23, 27, 32, 39, 52, 65, 75, 81, 86
  Engraving 8,
  Front covers 10, 11, 23, 38, 46, 48, 64, 78, 113, 124
  Graphic design 4, 5, 7, 11, 15, 21, 27, 33, 48, 86
  Letterpress 8
  Logos and emblems 14, 23, 27, 28, 38, 39, 41, 50, 58, 60, 62, 66, 72, 75, 79, 81, 96, 100, 106, 107, 109, 115, 137–139, 144
  Marketing and publicity 5, 7, 17, 22–25, 46, 49, 52, 58, 84, 86, 103, 104, 118, 120
  Poster 1, 2, 8, 9, 14, 20, 23, 29, 35, 36, 54, 73, 113
  Printing 5, 8, 14, 16, 20, 22, 25, 46, 28, 41, 52
  Signs, see Signage
  Tickets 4, 5, 20, 28, 41, 56, 70, 91
Designers and Architects
  D'Adamo, Raleigh 32, 33
  Beck, Harry, 1, 4, 9, 13, 21–23, 25, 30, 34, 37–39, 78, 100, 123
  Brennan, Joseph 34, 35
  Calvert, Margaret 92
  Dow, George 9, 140
  Foale, Alan 1, 2, 22
  Foster, Sir Norman 81
  Frutiger, Adrian 39

Index

***(Subject Index** cont.)*
Garbutt, Paul E. 2, 22
Gill, McDonald 21
Goldstein, Stanley 32, 33
Guimard, Hector 36, 38, 60
Hagstrom, Andrew 32, 33
Hertz, Mike 34
Jabbour, Eddie 34
Johnston, Edward 21, 22
Jourda and Perraud 90
Keil, Maria 58
Kinneir, Jock 92
Lagoutte, F. 37, 38
Lewis, Ken 38, 39
Mijksenaar, Paul 22
Pick, Frank 11, 23
Porchez, Jean-François 39
Redon, Gerorges 38
Rouxel, Patrice 38
Salomon, George 32-34
Shakespear, Diseño and Lorenzo 45, 52, 53
Spandonide, C. 38, 39
Spiekermann, Eric 14
Stingemore, F. H. 9, 20–22, 30
Tauranac, John 34
Vignelli, Massimo 32, 34
Weese, Harry 74
Diagrams and schematics 4, 5, 8, 9, 13, 14, 18, 20–23, 25, 29, 31, 32, 35, 37–41, 46, 48, 52, 54, 56, 58, 60, 62, 64, 66, 68, 70, 72, 74, 78, 79, 81, 83, 86, 91, 94, 96–98, 100–105, 115, 123, 140
Development of 9, 15, 23, 34
Electronic/backlit 56, 113
First use 9, 12–14, 17, 18, 21, 22, 28, 29, 33, 38, 46, 95
Geometry on 13, 29, 30, 34, 41
In-car diagram 9, 13, 18, 20, 21, 28, 37, 52, 56, 58, 62, 75, 81, 86, 111, 120, 140
Environment 4, 5, 80, 85, 89, 136
Congestion 5, 48, 79, 85, 128
Ecology 5
Pollution 4, 56, 79, 89
Renewable energy 129
Urban regeneration 50, 84
Engineering 6, 23, 36, 56, 81, 107
Bienvenue, Fulgence 36
Bridges 49, 56, 133, 137
Construction 6, 7, 17, 25, 29, 33, 48–50, 52, 59, 60, 63, 64, 69–74, 78, 79, 81, 83–85, 91, 97–100, 102, 103, 107, 108, 114, 118, 121, 126, 128–131, 133, 135, 136, 137, 139
Cut-and-cover 6, 7, 83, 103
Earthquakes 60, 68, 133, 136
Electrification 9, 13, 16, 48, 79, 86, 120
Evans, Oliver 6
Fire, Flood 73, 128, 133
Geology 83, 97, 101, 112, 133, 136
Leonida, Dimitrie 83
Pearson, Charles 6
Power supply 6, 7, 66, 67, 69, 86, 91, 93, 120, 129, 134, 136
Roy, William 8
Signaling 9, 16
Stephenson, George 6
Tunneling shield 7
Yerkes, Charles Tyson 16
Fantasy maps 4, 68, 69, 140, 141
Fonts, see Typefaces
Industrial Revolution 6, 8
Internet, the 1–5, 28, 31, 35, 79, 85, 88, 131, 140, 142
Language and culture 40, 41, 44, 62, 66, 69, 82, 79, 102,
Bi- /multilingual 24, 41, 56, 82
English/Anglo 16, 28, 34, 40–42, 48, 62, 66, 67, 70, 86
Dutch 82
French/Francophile 62, 63, 82
Gallic 25
Goethe Institute 140
Japanese 41–43, 66
Kanji 41
Korean 70
Latin 41, 43
Spanish 26, 52
Light rail, see Transit modes
Mapmakers, see also Designers
Airey, John 9
BD Conseil 39
Bureau Mijksenaar 22, 23, 98
FWT 140
LS London 1, 22
Michael Hertz Associates 34
Ordnance Survey 6–8
QuickMap 22, 23
Railways Clearing House 9
Unofficial versions 20, 22, 36, 38, 41
Media, film & TV
Dark City 140
Dr. Who 140
Film location 51
Koyaanisqatsi 69
Newspapers 28
Petit Journal, The 36
Simpsons, The 69
Who Framed Roger Rabbit? 89
Monorail, see Transit modes
Museums 64, 125
Budapest Transport Museum 142
Glasgow Transport Museum 142
London's Transport Museum 1, 5, 23, 142
Louvre, the 37
New York Transit Museum 5, 35, 142
Stockholm's Transport Museum 101, 142
Tate Gallery, the 22
Passenger usage
Airport 17, 49, 55, 57, 59, 69, 74, 79, 84, 88, 92, 97, 104, 107, 114, 118, 119, 121, 130, 134, 135, 137
Commuting 7, 40, 49, 68, 70, 78, 84, 97, 107, 109, 112, 117–121, 126, 128, 135, 140
Events and attractions 28, 38, 60, 78
Exhibitions and expos 64, 66, 67, 72, 113
Integrated transport 15, 86, 88, 90, 92, 99, 112, 121, 118, 133
Leisure time use 8
Monuments 29, 74, 97
Olympics, the 60, 62, 64, 79, 80, 139
Souvenirs 22, 41
Tourism 20, 24, 25, 28, 29, 34, 36, 38, 40, 41, 50, 69, 70, 78, 82, 95, 97, 113, 125
World Cup, the 132
Zoos 59, 64, 119, 137
Pré-Métro, see Transit modes
Rat, Hans 5
Railways
Heavy 7, 40, 94, 118, 132, 140
AMTRAK 108
British Rail 92
Cercanias 7, 25, 27
Danish Railways 140
Inter-City 140
Japan Rail 40
LNER 9, 23
Stadtbahn 7, 12, 13, 103,
Commuter
CityRail 124
EastRail 7
Euskotrain 81
DART 7, 118
HEV 50
Metro-North Railroad 6
RER 7, 26, 39
Sceaux Line 26
S-Bahn 7, 9, 12–14, 54, 55, 64, 112, 119, 122, 125, 128, 135
S-Tog 84
Vorortbahn 12, 13
Early/pioneering,
Baltimore & Ohio 6
Bermondsey–Deptford 6
Blackwall 6
Bolton & Leigh 6
City & South London 7, 50
Gran Metropolitano de Catalunya 47
Greek and Roman 6
Lewiston, NY 6
Lancashire & Yorkshire 9
Liverpool & Manchester 6, 8
Liverpool Overhead 120
Long Island 7
Los Angeles Pacific 9
Manhattan 7
Metropolitan West Side 16
Mumbles 6
New York & Harlem 6, 8
New York Municipal 32, 33
Pen-y-darren 6
Sarria Line 46
South Carolina Canal 8
Stockton & Darlington 6
Surrey Iron 6
Tower Subway, the 6
UERL 20
Union Pacific 8
Woolaton Wagonway 6
Gauge 47, 56, 86, 104
Old alignments 7, 92, 112, 114, 129, 131, 135, 137, 138
Rack 90, 134
Railwaymania 6
S-Bahn, see Railways
Sanyo Electric Railway 133
Schwandl, Robert 2, 5
Signage 15, 18, 22, 23, 26, 32, 34, 38, 41, 50, 58, 62, 67, 75, 92, 98, 109, 115, 118
Information graphics 5, 22, 75
Totems 45
Sociopolitical issues 12, 94, 113, 133
Colonies 8, 56, 100
Costs and funding 6, 7, 59, 68, 74, 83, 84, 86, 87, 94, 108, 109, 113, 125, 128, 133
Eastern Bloc 83, 87, 95, 110
Communism 29, 80, 83
Government 8, 22, 40, 50
Great Depression, the 16, 32
Maps echoing trends 12, 52, 70
Marshall Plan 58
Planned economy 50, 83
Profits/financing 16, 32, 52
Referenda 68, 125
Revolution 136
Socialist 83, 136
Soviet 50, 51, 72, 83, 87, 110, 113, 128, 134
Spanish Civil War 47
Wars 4, 8, 12, 14, 28, 38, 47, 59, 83, 95
Stations
Architecture 17, 27–30, 36, 38, 50, 58–60, 62–65, 72, 75, 79, 81, 83, 86, 87, 90, 95, 101, 104, 110, 113, 125, 128, 136, 139
Deepest, the 136
Ghost 13, 14
Naming of on maps 21, 22, 38, 45, 54, 66, 82, 98
Numbering of 41, 43, 45, 66, 70, 91, 100
Platform 17, 22, 38, 64, 77, 82, 86, 91, 94, 105, 115, 129, 130
Platform-edge doors/lights 73, 75, 90, 100
Spacing of 9, 13, 25, 30, 41, 56, 58, 62, 74, 78, 84, 87, 91–95, 102, 104, 108, 112, 124, 125, 129
Streetcars, see Transit modes
Tickets, see Print
Trams, see Transit modes
Transit modes
Air/airport 17, 55, 57, 59, 69, 74, 79, 84, 88, 92, 97, 104, 107, 114, 118, 119, 121, 130, 134, 135, 137
Bicycle 80
Bus 5, 17, 18, 49, 69, 89, 91, 92, 114, 140
Cable car 68
Car (automobile) 16, 17, 48, 68, 79, 80, 83, 85, 89, 129
Elevated (Els or L) 6, 7, 16, 32, 48, 94
Ferries 48
Freeway/road/street 4–9, 12, 16–18, 20, 25, 26, 28, 33, 34, 36, 38, 39, 45, 46, 48, 49, 52, 58, 60, 62, 68, 75, 78–81, 83, 85, 86, 89, 92, 94, 99, 107, 108, 110, 112, 115, 118, 130, 131, 133, 136, 138,
Funicular 90, 132–134
Hackney carriage 6
Horse-drawn 5, 6, 48
Light rail or LRT 5–7, 49, 55, 68, 69, 86, 88, 89, 93, 103, 108, 109, 113, 117–119, 122, 128–138, 142, 135
Light or mini-metro 25, 125, 128, 129, 136
Monorail 88, 114, 119, 122, 128, 133, 134, 136, 138
People-mover 119, 131
Pré-Métro 7, 86, 96, 128, 130, 131, 135,
Rolling stock 30
Rubber-tired 7, 62, 90, 102, 114, 134
Streetcar/tram/trolley 5–7, 9, 16, 46, 48, 49, 64, 66–68, 78, 79, 82, 89, 90, 92–95, 97, 104, 111, 118–125, 128–136, 139
Schwebebahn 95
Tram-Train 7
Trolleybus 5, 69, 90
VAL 102, 111, 129, 137, 139
Waterways 22, 25, 34, 50, 70, 76, 78, 81, 84, 86, 87, 98, 113, 139
Transit operators
BART 68, 69, 137
BKV 51
BMCL 109
BMT 32, 33
Brooklyn Rapid Transit 32
BTS 109
BVG 14, 15
Chicago rapid Transit 16
CMP 36–38
CTA 18, 19
CTM 25, 26
Dopravni Podnik 95
Ferrocarril Barcelona 26, 47
Valencia FGV 104
GVB 78
Hudson & Manhattan 113
HVV 55
IND 32, 33
IRT 32, 33
ISAP 79
KTM 88
LUAS 118
London Transport 9, 12, 14, 124
MARC 108
MBTA 45, 49
Merseyrail 120
Metro de Madrid 27
Metropolitana de Lisboa 59
Metrorec 114
Metrorex 83
Metrovias 53
MTA 32, 34, 35, 89
MTR 56, 57
MVG 66
MVV 64
Nord-Sud 36
PATCO 94
PATH 113
RATP 38–39, 62
SEPTA 94
SL 101
SMRT 70, 71, 100
SMS 71
SPT 86
STAR 88
STCUM 63
TOEI 40–43
Transport for London 1, 22
TRTA 40–43
TTC 103
WMATA 75
Transit system names
Ankaray 128
Aksaray 133
BART 68, 137
Bursaray 129
Carmelit Subway 132
Clockwork Orange, the 86
Docklands Light Railway 6
Eidan Subway 40
Földalatti 7, 50
Hochbahn 54
London Underground 1, 4, 5, 9, 11, 13, 20, 22, 23, 29, 78, 140
MAX, the 136
Metro, the 25, 28, 36–39, 46, 50, 74, 75, 78, 85, 92, 95, 99, 104, 108, 111
Metropolitan Railway 5–9, 20, 26
Metrolink 120, 137
MetroMover 121
MetroNorte 27
MetroRail 121
Metrostar 134
MetroSur 25, 26
Metro Transit 135
Muni Metro 7, 68
NET, the 135
O-Train 136
RandstadRail 132
Rapid, the 118
Rocket, the 103
Schwebebahn 122
Skytrain 109, 139
Skyway 119
Subte 45
Subway, the 17, 32, 86, 103, 144
T-bane 93
T, the 49, 122
TheRide 131
Tube, the 4, 16, 20, 21, 22, 34, 138
Tunnelbanenätet 101
Tyne and Wear Metro 92
U-Bahn 5, 7, 12–14, 54, 64, 102, 118
Üstra, the 119
Wiener Linien 105
Type 39, 58
Breaking over lines 13, 54, 50, 54, 58, 64, 72, 74, 82, 91–95, 102, 104, 108, 112, 119, 124, 125, 129
Fonts
Akzidenz-Grotesk 2, 32, 34
Frutiger 39
Handwritten 8, 21, 49
Helvetica 32, 39
Johnston, Edward 20, 21, 22, 142
Libre Sans Serif 81
Parisine 39
Rail Alphabet 92
Rotis 81
Transport 92
Univers 39
Sans-serif 21, 22, 62, 86
UITP 2, 5